Selected Titles in This Series

D0141081

An Introduction to Game-Theoretic Modelling

STUDENT MATHEMATICAL LIBRARY
Volume 11

An Introduction to Game-Theoretic Modelling

Second Edition

Michael Mesterton-Gibbons

AMERICAN MATHEMATICAL SOCIETY

Editorial Board

David Bressoud   Carl Pomerance
Robert Devaney, Chair   Hung-Hsi Wu

2000 *Mathematics Subject Classification.* Primary 91–01, 91A10, 91A12, 91A22, 91A40, 91A80; Secondary 92–02, 92D50.

ABSTRACT. An introduction to both cooperative and noncooperative games from the perspective of a mathematical modeller. It deals in a unified manner with the central concepts of both classical and evolutionary game theory, and includes numerous applications of population games to the study of animal behavior.

Library of Congress Cataloging-in-Publication Data

Mesterton-Gibbons, Michael.
    An introduction to game-theoretic modelling / Michael Mesterton-Gibbons.— 2nd ed.
        p. cm. — (Student mathematical library ; v. 11)
    Includes bibliographical references and index.
    ISBN 0-8218-1929-1 (alk. paper)
    1. Game theory.   2. Mathematical models.   I. Title.   II. Series.

QA269.M464   2000
519.3—dc21   00-059399

Copying and reprinting. Individual readers of this publication, and nonprofit libraries acting for them, are permitted to make fair use of the material, such as to copy a chapter for use in teaching or research. Permission is granted to quote brief passages from this publication in reviews, provided the customary acknowledgment of the source is given.

Republication, systematic copying, or multiple reproduction of any material in this publication is permitted only under license from the American Mathematical Society. Requests for such permission should be addressed to the Assistant to the Publisher, American Mathematical Society, P. O. Box 6248, Providence, Rhode Island 02940-6248. Requests can also be made by e-mail to reprint-permission@ams.org.

© 2001 held by the American Mathematical Society. All rights reserved.
The American Mathematical Society retains all rights
except those granted to the United States Government.
Printed in the United States of America.

∞ The paper used in this book is acid-free and falls within the guidelines
established to ensure permanence and durability.
Visit the AMS home page at URL: http://www.ams.org/

10 9 8 7 6 5 4 3 2 1   06 05 04 03 02 01

To my three girls:

Karen, **Sarah** and **Alex**

for their strategy of cooperation.

Contents

Preface

This is the second edition of a short introduction to game theory
and applications from the perspective of a mathematical modeller,
and includes several original pedagogical games. It covers a range
of concepts that have proven useful, or are likely to prove useful, for
quantitative modelling in the life, social and management sciences. Its
approach is heuristic, but systematic, and it deals in a unified manner
with the central ideas of both classical and evolutionary game theory.
In many ways, it is a sequel to my earlier work, *A Concrete Approach
to Mathematical Modelling* [**144**],[1] in which games were not discussed.
The mathematical prerequisites are correspondingly modest: calculus, a rudimentary knowledge of matrix algebra and probability, a
passing acquaintance with differential equations and that intangible
quantity, a degree of mathematical maturity. Naturally, the greater
one's maturity, the more contemptuous one can be of formal prerequisites, and the more one is able to proceed ad hoc when the need
arises.

Reactions to the first edition of this book have been positive and
pleasing. Although I haven't fixed what isn't broken, I have paid
attention to feedback: I have added a brand new chapter on population games, I have included some new examples and exercises, I
have corrected all known errors (and tried hard not to introduce new

---

[1] Bold numbers in square brackets are references listed on pp. 347-361.

ones), and I have brought the references up to date. By design, the references are selective: it would conflict with the goals of this series to cite all potentially relevant published literature. Nevertheless, I have tried to make the vast majority of it easily traceable through judicious references to more recent work.[2]

There exist several excellent texts on game theory, but their excellence is largely for the mathematical purist. Practices that are de rigueur to the purist are often merely distracting to the modeller—for example, lingering over the elegant theory of zero-sum games (nonzero-sum conflicts are much more common in practice), or proving the existence of a Nash equilibrium in bimatrix games (for which the problem in practice is usually to distinguish among a superabundance of such equilibria); or, more fundamentally, beginning with the most general possible formulation of a game and only later proceeding to specific examples (the essence of modelling is rather the opposite). Such practices are therefore honored in the breach. Instead—and as described more fully in the agenda that follows—the emphasis is on concrete examples, and the direction of pedagogy throughout the book is from specific to general. However bright or well motivated, students often have limited appetites for rigor and generality, yet have much still to gain from the mathematical experience of capturing ideas and giving them substance—the experience, that is, of modelling. I hope that such students will find that this book helps not only to make game theory accessible, but also to convey both its power and scope in a variety of applications.

---

[2] And for the very latest, there exists the internet; see [**136**, pp. 147-149] for links.

Acknowledgements

I am very grateful to Eldridge Adams, Phil Crowley, Sasha Dall, David Glick and especially Lindi Wahl for reading parts of the manuscript and supplying many valuable comments and suggestions; to Karen Mesterton-Gibbons for moral support and for carefully proofing the manuscript; to my former game theory students, in particular John Murkerson and Gabe Bouch, for helping to identify several inaccuracies in the first edition; to the American Mathematical Society for technical support; and to the National Science Foundation for supporting my research on game-theoretic modelling.

Agenda

This is a book about mathematical modelling of strategic behavior, which arises whenever the outcome of an individual's actions depends on actions to be taken by other individuals. The individuals may be either human or non-human beings, the actions either premeditated or instinctive. Thus models of strategic behavior are applicable in both the social and the natural sciences.

Examples of humans interacting strategically include store managers fixing prices—the number of customers who buy at one store's price will depend on the price at other stores in the neighborhood; and drivers negotiating a 4-way junction—whether it is advantageous for a driver to assume right of way depends on whether the other drivers concede right of way. Examples of non-humans interacting strategically include spiders disputing a territory—the risk of injury to one animal from being an aggressor will depend on whether the other is prepared to fight; and insects foraging for oviposition sites— the number of one insect's eggs that mature into adults at a given site (where food for growth is limited) will depend on the number of eggs laid there by other insects. These and other strategic interactions will be modelled in detail, beginning in Chapter 1.

To fix ideas, however, it will be helpful first to consider an example that—if somewhat fanciful—will serve to delineate the important distinction between strategic and non-strategic decision making. Let

Table 0.1. Student achievement (number of satisfactory solutions) in mathematics as a function of effort.

	Very hard $(E = 5)$	Quite hard $(E = 3)$	Hardly at all $(E = 1)$
Student 1	10	8	6
Student 2	8	7	5
Student 3	7	5	3
Student 4	7	4	3
Student 5	5	4	2
Student 6	4	2	1

us therefore suppose that the enrollment for some mathematics course is a mere six humans, and that grades for this course are based exclusively on answers to ten questions. Answers are judged to be either satisfactory or unsatisfactory, and the number of satisfactory solutions determines the final letter grade for the course—A, B, C, D or F. In the usual way, A corresponds to 4 units of merit, and B, C, D and F correspond to 3, 2, 1 and 0 units of merit, respectively. The students vary in intellectual ability, and all are capable of working either very hard, or only quite hard, or hardly at all; but there is nothing random about student achievement as a function of effort, which is precisely defined as in Table 0.1. Thus, for example, Student 5 will produce five satisfactory solutions if she works very hard, but only four if she works quite hard; whereas Student 4 will produce seven satisfactory solutions if he works very hard, but only three if he works hardly at all. The students have complete control over how much effort they apply, and so we refer to effort as a decision variable. Furthermore, for the sake of definiteness, we assume that working very hard corresponds to 5 units of effort, quite hard to 3, and hardly at all to 1. Thus, if we denote effort by E and merit by M, then working very hard corresponds to $E = 5$; obtaining the letter grade A corresponds to $M = 4$; and similarly for the other values of E and M.

Let us now suppose that academic standards are absolute, i.e., the number of satisfactory solutions required for each letter grade is prescribed in advance. Then no strategic behavior is possible. This doesn't eliminate scope for decision making—quite the contrary. If, for example, 9 or 10 satisfactory solutions were required for an A, 7 or 8 for a B, 5 or 6 for a C and 3 or 4 for a D, then Student 3 would earn 3 units of merit for $E = 5$, 2 for $E = 3$, and only 1 for $E = 1$; and if she wished to maximize merit per unit of effort, or M/E, then she would still have to solve a simple optimization problem, namely, to determine the maximum of $3 \div 5 = 0.6$, $2 \div 3 = 0.67$ and $1 \div 1 = 1$. The answer, of course, is 1, corresponding to $E = 1$: to maximize M/E, Student 3 should hardly work at all. Nevertheless, such a decision would not be strategic, because its outcome would depend solely on the individual concerned; it would not depend in any way on the behavior of other students.

The story is very different, however, if academic standards are relative, i.e., if letter grades depend on collective student achievement. To illustrate, let s denote number of satisfactory solutions; let b denote the number of satisfactory solutions obtained by the best student, and w the number obtained by the worst; and let grades be assigned according to the following scheme, which awards A to any student in the top fifth of the range, B to any student in the next fifth of the range, and so on:

$$
\begin{aligned}
\text{A}: \quad & \tfrac{1}{5}(4b + w) && \leq s \leq && b \\
\text{B}: \quad & \tfrac{1}{5}(3b + 2w) && \leq s < && \tfrac{1}{5}(4b + w) \\
\text{C}: \quad & \tfrac{1}{5}(2b + 3w) && \leq s < && \tfrac{1}{5}(3b + 2w) \\
\text{D}: \quad & \tfrac{1}{5}(b + 4w) && \leq s < && \tfrac{1}{5}(2b + 3w) \\
\text{F}: \quad & w && \leq s < && \tfrac{1}{5}(b + 4w).
\end{aligned}
$$

Thus if all students chose $E = 5$, then Student 1 would get A, Student 2 would get B, Students 3 and 4 would get C, and Students 5 and 6 would both fail; whereas if all students chose $E = 1$, then Students 1 and 2 would get A, Students 3 and 4 would get C, Student 5 would get D, and only Student 6 would fail.

Students who wish to maximize M or M/E must now anticipate (and perhaps seek to influence) how hard the others will work, and choose E accordingly. In other words, students who wish to attain their goals must behave strategically. For example, Student 1 is now guaranteed an A if she works at least quite hard; and $E = 3$ yields a higher value of M/E than $E = 5$. But $4/3 = 1.33$ is not the highest value of M/E that Student 1 can obtain. If she knew that Student 2 would work at most quite hard, and if she also knew that either

Student 6 would choose $E \leq 3$ or Student 5 would choose $E = 1$, then the grading scheme would award her an A no matter how hard she worked—in particular, if she chose $E = 1$. But Students 5 and 6 can avoid failing only if they obtain at least 4 satisfactory solutions and either Student 3 or Student 4 chooses $E = 1$; in which case, $E = 1$ and its six satisfactory solutions could earn Student 1 only B (although it would still maximize M/E). And so on. We have made our point: changing from absolute to relative standards brings ample scope for strategic behavior to an interaction among individuals that otherwise has none.

If an interaction among individuals gives rise to strategic behavior, and if the interaction can be described mathematically, then we will refer to this description as a *game*, and to each individual in the game as a *player*. In other words, game means mathematical model of conflict or bargaining. Correspondingly, *game theory* is an assemblage of ideas and theorems that attempts to provide a rational basis for resolving conflicts, with or without cooperation. You won't often find game so defined in a dictionary. But new meanings take time to diffuse into dictionaries; and besides, the study of decision-laden interactions among individuals should perhaps be called something other than game theory. On the other hand, there are at least as many similarities between games in the mathematical modeller's sense, and games in the everyday sense, as there are between—say—the gadget I use to control my computer and a mouse. In any event, game is the word and game theory the phrase we shall use.

As an acknowledged field of study in its own right, game theory began with the publication in 1944 of a treatise on games and economic behavior by John Von Neumann and Oskar Morgenstern.[3] Nevertheless, some game-theoretic concepts have been traced to earlier work by Cournot [41], Edgeworth [56], Böhm-Bawerk [20], Borel [21] and Zeuthen [243] in the context of economics, and to Fisher [63] in the context of evolutionary biology. Given the way things are in the world, it is probably only a matter of time before somebody finds that the first book on game theory was written by Aristotle.[4]

Because—as we saw in the opening paragraphs—conflict can arise

---

[3] Nowadays, however, it is customary to consult the third edition [231].

[4] For an entertaining tour of game theory's history, I recommend Mehlmann [136].

among so many different kinds of player, and in such a wide variety of circumstances, we should never expect game-theoretic ideas that are fruitful in one context to be relevant in another (although we should always entertain the possibility that they might be). Nevertheless, many once promising solution concepts—i.e., concepts of what is the best strategy or compromise—have ultimately failed to be satisfactory even in the circumstances for which they were designed, and so we do not discuss them.[5] We concentrate instead on introducing and applying ideas that still hold promise, in particular, Nash equilibrium (Chapter 1), evolutionarily stable strategy (Chapters 2 and 6), Pareto-optimality (Chapter 3), core, nucleolus, Shapley value (Chapter 4) and cooperation via reciprocity (Chapter 5).

By tradition, games are classified as either cooperative or noncooperative, although this dichotomy is universally acknowledged to be imperfect: almost every conflict has an element of cooperation, and almost all cooperation has an element of conflict. In this regard we abide, more or less, by tradition: Chapters 1, 2 and 6 are about noncooperative solution concepts, whereas Chapters 3 and 4 are about cooperative ones. But the distinction is blurred in Chapter 5, where we study cooperation within the context of a noncooperative game. Games are also classified as having either strategic form or characteristic function form.[6] For present purposes, it will suffice to say that a game is in characteristic function form if the conflict is analogous to sharing a pie among players who would each like all of it, and who collectively can obtain all of it, but who as individuals cannot obtain any of it; and that otherwise a game is in strategic form. Games are studied in characteristic function form in Chapter 4, where we discuss, for example, how to split the costs of a car pool fairly. Games are studied in strategic form in Chapters 1-3, 5 and 6, where we study, among other things, the behavior of motorists at a 4-way junction; price setting by store managers; territorial conflict among crustaceans, insects or spiders; and food sharing among ravens.

To interpret games in the wider context of optimization theory,

---

[5] In particular, there is no discussion of the stable set, Von Neumann and Morgenstern's solution concept for characteristic function games.

[6] Game theorists further distinguish between strategic games in extensive form and strategic games in normal form, but we have no use for this distinction. See, for example, [**146**, pp. 1017-1021] or [**173**, pp. 1-5].

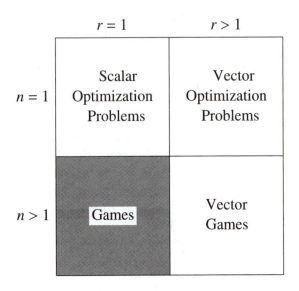

Figure 0.1. Games in the context of optimization theory;
r is the number of rewards per decision maker and n is the
number of decision makers.

it will be helpful now to return for a while to our six mathematics
students. Let us suppose that Student 2 will work quite hard ($E = 3$),
whereas Students 3 to 6 will work very hard ($E = 5$); and that Student 1 already knows this. If Student 1 is rewarded *either* by high
achievement *or* by high achievement per unit effort, then she has a
single reward—either M or M/E—and a single decision variable, E,
with which to maximize it. If, on the other hand, Student 1 is rewarded *both* by high achievement and by high achievement per unit
effort, then she has two rewards—M *and* M/E—but still only a single
decision variable, E, with which to maximize them. In the first case,
with a single reward, we say that Student 1 faces a *scalar* optimization problem (whose solution is clearly $E = 5$ or $E = 3$ if rewarded
by M, but $E = 1$ if rewarded by M/E). In the second case, with two
rewards, we say that Student 1 faces a *vector* optimization problem
(whose solution is far from clear, because the value of E that maximizes M fails also to maximize M/E, and vice versa). More generally,

an optimization problem requires a single decision maker to select a single decision variable (over which this decision maker has complete control) to optimize r rewards; and if $r = 1$ then the problem is a scalar optimization problem, whereas if $r \geq 2$ then the problem is a vector optimization problem.[7]

By contrast, the strategic games that we are about to study require each of n decision makers, called players, to a select a single decision variable—called a strategy—to optimize a single reward. But each player's reward depends on the other players' strategies; i.e., it depends on decision variables that are completely controlled by the other players. That players lack control over all decision variables affecting their rewards is what makes a game a game, and what distinguishes it from an optimization problem. In this book, we do not discuss optimization problems in their own right, nor do we allow players to have more than one reward—in other words, we do not consider "vector games." Thus our agenda for games falls in the shaded region of Figure 0.1.[8]

Now, a strategic interaction can be very complicated, and a game in our sense does not exist unless the interaction can be described mathematically. But this step is often exceedingly difficult, especially if players are many; and especially if we insist—and as modellers we should—that players' rewards be explicitly defined. Therefore, we confine our agenda to strategic interactions that lend themselves readily to a concrete mathematical description—and hence, for the most part, to games with few players. Indeed six is many in this regard: Students 1 to 6, having served us so well, must now depart the scene.

Throughout the book we introduce concepts by means of specific models, later indicating how to generalize them; however, we avoid rigorous statements and proofs of theorems, referring instead to the standard texts. Our approach to games is thus largely the opposite of the classical approach, but has the clear advantage in an introductory text that it fosters substantial progress. We can downplay the

---

[7]As we shall see in due course, a decision variable can itself be a vector. What distinguishes a scalar from a vector optimization problem (or a game from a vector game), however, is the number of rewards per decision maker.

[8]For further examples of optimization problems see, e.g., Chapters 3, 7 and 12 of [**144**]. For vector games, see [**244**]. Figure 0.1 and all other figures in this book were drawn with *Mathematica*® [**239**].

issue of what—in most general terms—constitutes a decision maker's reward, because the reward is self-evident in the particular examples we choose.[9] We can demonstrate the usefulness and richness of games while avoiding unnecessary abstractions; even 2-player games in strategic form have enormous potential, which has scarcely begun to be realized. Moreover, we can rely on intuitions about everyday conflicts to strengthen our grasp of game theory's key ideas, and we can be flexible and creative in applying those ideas.

Our agenda is thus defined. We carry out this agenda in Chapters 1-6, giving summaries and suggestions for further reading in commentaries at the ends of the chapters; and in Chapter 7 we critique our accomplishments. There follow solutions—or at least strong hints— for most of the exercises, and so it is assumed of the reader that he or she is sufficiently mature not to consult a solution until a problem has at least been seriously attempted. Which reminds me: We assume throughout that a protagonist of indeterminate sex is female in odd-numbered chapters and male in even-numbered chapters. This convention is simply the reverse of that which I adopted in *A Concrete Approach to Mathematical Modelling* [**144**], and it renders unnecessary the continual use of "his or her" and "he or she" in place of epicene pronouns.

---

[9]Books that discuss this issue thoroughly include [**122**], [**206**], [**207**] and [**231**]. The last of these books defines the classical approach to the theory of games; the first is an excellent later text covering the same ground and more, but with much less technical mathematical detail; and the other two constitute one of the most comprehensive works on game theory ever published.

Chapter 1

Noncooperative Games

Motoring behavior provides our first example of a game, which will pave the way for several solution concepts, both in this chapter and later in Chapters 2 and 3. Here the example will introduce what is probably game theory's most enduring concept, John Nash's concept of noncooperative equilibrium.

1.1. Crossroads: a motorist's dilemma

Consider a pair of motorists who are driving in opposite directions along a 2-lane road when they arrive simultaneously at a 4-way junction, where each would like to cross the path of the other. For the sake of definiteness, let us suppose, as in Figure 1.1, that the first motorist, say Nan, is travelling north but would like to go west; whereas

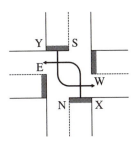

Figure 1.1. The scene of the action: a crossroad

the second motorist, say San, is travelling south but would like to go east. Nan and San cannot proceed simultaneously; one must proceed before the other. Then who should it be? How should the motorists behave? Here's potential for conflict that's fit for a game. We will call the game Crossroads.

To keep matters simple, let's suppose that a motorist has but two choices: she can either wait for the other motorist to turn first, in which case we shall say that she selects pure strategy W; or she can proceed to turn and hope that the other motorist will refrain from doing so, in which case we shall say that she selects pure strategy G.[1] We use the word strategy because we wish to think of Nan as the first "player" or decision maker, and of San as the second player, in a 2-player game; and we call G and W pure strategies to distinguish them from mixed strategies, which we shall introduce in §1.3. If Nan selects pure strategy X, and if San selects pure strategy Y, then we shall say that the players have jointly selected the pure strategy combination XY. Thus our game has precisely four pure strategy combinations, namely, GG, GW, WG and WW.

In the case where each player decides to defer to the other (WW), let ϵ denote the time they spend dithering and frantically waving to each other, before one of them eventually moves. Likewise, in the case where each decides not to defer (GG), let δ denote the time they spend intimidating each other in the middle of the junction, until one of them eventually backs down. It seems reasonable to suppose that the time they waste if both are selfish (GG) exceeds that which they waste if both are altruistic (WW), even if not by much; and we shall therefore assume throughout the text that

$$(1.1) \qquad\qquad 0 < \epsilon < \delta < \infty,$$

even if ϵ/δ is close to 1. Let τ_1 denote the time it takes Nan to negotiate the turn without interruption, i.e., the time that elapses (if San lets her go) between her front bumper crossing NX in Figure 1.1 and her back bumper crossing EY; and let τ_2 denote the corresponding time for San. We are now in a position to analyze the confrontation from Nan's point of view.

---

[1] Recall from the agenda that, in place of epicene pronouns, female pronouns will be used in Chapter 1, male pronouns in Chapter 2, and so on.

Suppose, first, that pure strategy combination GW is selected: Nan decides to go, San decides to wait. Then Nan's delay is zero. Now suppose that WG is selected: San decides to go, Nan decides to wait. Then Nan's delay is τ_2, the time it takes San to cross the junction. Suppose, next, that WW is selected: both decide to wait. There follows a bout of rapid gesticulation, after which it is still the case that either Nan or San is first to proceed; they can't just sit there all day. Quite how it is decided who—given WW—should go first is, in a sense, a game within the game of Crossroads, but we shall not attempt to model it explicitly; rather, we shall simply assume that the two motorists are then equally likely to be first to turn. Accordingly, let F denote the motorist who (given WW) turns first. Then F is a random variable, whose sample space is {Nan, San}, and

$$(1.2) \qquad \text{Prob}(F = \text{Nan}) = \tfrac{1}{2}, \quad \text{Prob}(F = \text{San}) = \tfrac{1}{2}.$$

We note in passing that F could easily be converted to an integer-valued random variable by labelling Nan as 1 and San as 2, but it is more convenient not to do so.

If Nan turns first ($F = \text{Nan}$), then she suffers a delay of only ϵ; whereas if San turns first ($F = \text{San}$), then Nan—from whose viewpoint we are analyzing the confrontation—suffers a delay of $\epsilon + \tau_2$. Thus the expected value of Nan's delay (given WW) is

$$(1.3) \qquad \epsilon \cdot \text{Prob}(F = \text{Nan}) + (\epsilon + \tau_2) \cdot \text{Prob}(F = \text{San}) = \epsilon + \tfrac{1}{2}\tau_2.$$

Suppose, finally, that GG is selected: both decide to go. Then there follows a minor skirmish, of duration δ, which one of the players must eventually win. Let random variable V, with sample space {Nan, San}, denote the victor (given GG); and suppose that if Player k is the victor, then the time she takes to negotiate the junction (given GG) is simply δ greater than she would have taken anyway, i.e., $\delta + \tau_k$. If neither player is especially aggressive, then it seems reasonable to suppose that each is as likely as the other to find her path cleared; in which case, (1.2) holds with V in place of F. If, given GG, Nan is the victor ($V = \text{Nan}$), then her delay is only δ; whereas if San is the victor ($V = \text{San}$), then Nan's delay is $\delta + \tau_2$. So the expected value of her delay, given GG, is $\delta \cdot \text{Prob}(V = \text{Nan}) + (\delta + \tau_2) \cdot \text{Prob}(V = \text{San}) = \delta + \tfrac{1}{2}\tau_2$, by analogy with (1.3).

Table 1.1. Payoff
matrix for Nan

	San	
Nan	G	W
G	$-\delta-\frac{1}{2}\tau_2$	0
W	$-\tau_2$	$-\epsilon-\frac{1}{2}\tau_2$

Table 1.2. Payoff
matrix for San

	San	
Nan	G	W
G	$-\delta-\frac{1}{2}\tau_1$	$-\tau_1$
W	0	$-\epsilon-\frac{1}{2}\tau_1$

We are tacitly assuming that δ and ϵ are both independent of τ_1 and τ_2. You may be tempted to criticize this assumption—but tread daintily if you do so. In the real world, it is more than likely that δ and ϵ would depend upon various aspects of the personalities of the drivers in conflict. But our model ignores them. It differentiates between Nan and San solely by virtue of their transit times, τ_1 and τ_2; and any dependence of δ and ϵ on these may well be weak.

Everyone likes delays to be as short as possible, but in noncooperative game theory it is traditional to like payoffs as large as possible. So, because making a delay small is the same thing as making its negative large, we agree that the payoff to Nan is the negative of her delay. Thus the payoffs to Nan associated with pure strategy combinations GG, GW, WG and WW are, respectively, $-\delta - \frac{1}{2}\tau_2, 0, -\tau_2$ and $-\epsilon - \frac{1}{2}\tau_2$. It is customary to store these payoffs in a matrix, as in Table 1.1, where the rows correspond to strategies of Player 1 and the columns correspond to strategies of Player 2.

Now, in any particular confrontation, the actual payoff to Nan from pure strategy combination GG or WW is a random variable. If the game is played repeatedly, however, then Nan's average payoff from GG or WW over an extended period should be well approximated by the random variable's expected value; and this is how we justify using expected values as payoffs. Furthermore, for the game to be played repeatedly, it is not necessary that the person Nan confronts, when she finds herself at a 4-way junction in the circumstances described above, be the same individual every time. Rather, San is a generic name for all individuals whose behavior is the same for the purposes of our model (though for other purposes it might be very different); if you like, San is any individual who exhibits San-like simultaneous-left-turning-at-a-4-way-junction behavior. Likewise, for the game to be played repeatedly, the 4-way junction at which Nan confronts a San need not be the same junction every time—a similar junction will suffice.

Nevertheless, some qualifying remarks are in order. Sans that are identical for the purposes of our model must, in theory, all take time τ_2 to negotiate a junction unimpeded—or, which is more to the point, Nan must think that they will all take time τ_2. In practice, however, Nan has limited ability to size up the driver who confronts her momentarily across the junction. Perhaps the best she can do is to place her opponent in one of a finite number of classes. She may, for example, classify her opponents as fast, intermediate or slow; in which case, as Nan flits from junction to junction, not one game but three games are played repeatedly, a game for slow Sans, a game for intermediate Sans and a game for fast Sans. (More generally, some finite number of games would be played repeatedly.) On the other hand, we should not imagine that it is totally unrealistic to suppose that Nan's opponent is the same San every time—perhaps they meet at the same time, and at the same place, as they travel to work in the morning in opposite directions.

To obtain the matrix in Table 1.1, we analyzed the game from Nan's point of view. A similar analysis, from San's viewpoint, yields the payoff matrix in Table 1.2. Indeed it is hardly necessary to repeat the analysis, because the only difference between Nan and San that is incorporated into our model of their conflict—or, as game theorists prefer to say, the only asymmetry between the players—is that Nan's transit time may be different from San's ($\tau_1 \neq \tau_2$). Thus San's payoff matrix is the transpose of Nan's with suffix 2 replaced by suffix 1. Transposition is necessary because rows correspond to strategies of Player 1, and columns to those of Player 2, in both tables.

In terms of game theory, the payoff matrices in Tables 1.1 and 1.2 define a 2-player game in which each player has two pure strategies, G and W. If we assume that Nan and San act out of rational self-interest, *and that they cannot communicate prior to the game*, then the game becomes a noncooperative one. (As we shall discover in Chapter 3, what really distinguishes a noncooperative game from a cooperative game is the inability to make commitments; but the players cannot possibly make binding agreements if they cannot even communicate prior to the game.) Not being especially selfish is not necessarily a violation of rational self-interest on the part of Nan or San; and waving at one another does not constitute prior communication. It is therefore legitimate to regard Tables 1.1 and 1.2 as the payoff matrices for a noncooperative, 2-player game.

More generally, a noncooperative, 2-player game in which Player 1 has m_1 pure strategies and Player 2 has m_2 pure strategies is defined by a pair of $m_1 \times m_2$ matrices, A, B, in which a_{ij} denotes the payoff to Player 1, and b_{ij} the payoff to Player 2, from the strategy combination (i, j). If $m_1 = m_2 = m$ and

$$(1.4) \qquad\qquad b_{ij} = a_{ji}, \qquad 0 \le i, j \le m$$

i.e., if B is the transpose of A, then the game is *symmetric*. If

$$(1.5) \qquad a_{ij} + b_{ij} = c, \qquad 0 \le i \le m_1, \ 0 \le j \le m_2$$

where c is a constant, then the game is *constant-sum*; and if, in addition, $c = 0$ then the game is *zero-sum*. Thus Crossroads is symmetric (with $m = 2$) if, and only if, $\tau_1 = \tau_2$. Even if $\tau_1 = \tau_2$, however, the game is not constant-sum because $\delta > \epsilon > 0$, by (1.1).

Crossroads is an example of what game theorists call a *bimatrix* game in *strategic* form—or simply matrix game in strategic form, especially if (1.4) holds. The additional terminology may appear gratuitous now, but will be useful later when we study games that are either not bimatrix or not in strategic form. In §1.3, we shall attempt to "solve" Crossroads by saying which strategies the players should adopt. Meanwhile, we digress to describe a second example of a noncooperative, 2-player game, which arises in evolutionary biology.

1.2. The Hawk-Dove game

Suppose that two animals, I and II, are in conflict over a territory. They have two pure strategies, which—to follow tradition—we will label "Hawk" and "Dove" (but the animals belong to the same species). To play Hawk, or H, one must "escalate"—i.e., act fierce; and if that doesn't scare away the opponent, then one must fight until injury determines a victor. To play Dove, or D, one must first "display"—i.e., merely look fierce, and hope that the opponent is scared away; but if she starts to act fierce, then one must retreat and search for real estate elsewhere. As in Crossroads, there are four pure strategy combinations, namely, HH, HD, DH and DD.

Let ρ be the reproductive value of the territory to I, and let C be the reproductive cost of being injured in a fight. By reproductive value we mean the incremental number of offspring—as opposed to the absolute number—that the territory would yield to the animal (or rather, since the increment is a random variable, its expected value). Thus, if an animal who averages five little ones per breeding season (from the kind of territory that nobody scraps over) can raise this number to eight by acquiring the territory that is in dispute, then $\rho = 8 - 5 = 3$. Similarly, by C we mean the amount by which I's expected offspring would be reduced by virtue of injury. Thus reproductive cost and value are both in terms of "expected future reproductive success" or EFRS (or sometimes "fitness") for short.

Let us analyze this territorial conflict from I's point of view. Suppose, first, that strategy combination DH is selected by the players: I plays Dove, II plays Hawk. Then I retreats as soon as II acts fierce, and the payoff in terms of EFRS is zero. Now suppose that HD is selected: I plays Hawk, II plays Dove. Then II retreats as soon as I acts fierce, and the payoff to I is ρ. Suppose, next, that DD is selected: both I and II play Dove. Then there follows a staring match, during which both animals look fierce but refrain from acting fierce. Let F denote the animal who (given DD) first gets tired of staring and retreats; then F is a random variable, whose sample space is {I, II}. If neither animal is especially nervous, then it seems reasonable to suppose that each is as likely as the other to retreat first, so that $\text{Prob}(F = \text{I}) = \frac{1}{2} = \text{Prob}(F = \text{II})$. If I retreats first ($F = \text{I}$), then her payoff is zero; whereas if II retreats first ($F = \text{II}$), then the payoff to I is ρ. Thus the expected value of I's payoff (given DD) is

$$(1.6) \qquad 0 \cdot \text{Prob}(F = \text{I}) + \rho \cdot \text{Prob}(F = \text{II}) = \tfrac{1}{2}\rho.$$

Suppose, finally, that HH is selected: both animals play Hawk. Then there follows a major skirmish, which one of the animals must eventually win. Let random variable W, with sample space {I, II}, denote the winner (given HH); and suppose that neither I nor II is an especially good fighter. Then it seems reasonable to suppose that each is as likely as the other to win; in which case, $\text{Prob}(W = \text{I}) = \frac{1}{2} =$

Table **1.3.** Payoff
matrix for I

	II	
	H	D
I H	$\frac{1}{2}(\rho-C)$	ρ
D	0	$\frac{1}{2}\rho$

Table **1.4.** Payoff
matrix for II

	II	
	H	D
I H	$\frac{1}{2}(\rho-C)$	0
D	ρ	$\frac{1}{2}\rho$

Prob(W = II). If I wins, then her payoff is ρ; but if II wins, then the payoff to I is $-C$. Thus the expected value of I's payoff (given HH) is $\rho \cdot \text{Prob}(W = \text{I}) + (-C) \cdot \text{Prob}(W = \text{II}) = \frac{1}{2}(\rho - C)$. It follows that I's payoff matrix is as shown in Table 1.3. Moreover, if we assume that the territory's reproductive value and the cost of injury are the same for II as they are for I—i.e., if no asymmetry between the animals appears in our model of their conflict—then it follows immediately that II's payoff matrix is the transpose of I's, or Table 1.4.

We will revisit our animals later, when we shall refer to their game as the Hawk-Dove game.[2] Meanwhile, however, we return to Nan and San.

1.3. Rational reaction sets and Nash equilibria

To determine which strategy Nan should adopt in Crossroads, let's begin by supposing that San is so slow that

$$(1.7) \qquad\qquad \tau_2 > 2\delta > 2\epsilon.$$

Then whether San chooses W or G is quite irrelevant because

$$(1.8) \qquad\qquad -\delta - \tau_2/2 > -\tau_2,$$

and so it follows immediately from Table 1.1 that Nan's best strategy is to hit the gas: every element in the first row of her payoff matrix is greater than the corresponding element in the second row of her payoff matrix. We say that strategy G *dominates* strategy W for Nan, and that G is a *dominant* strategy for Nan. More generally, if A is Player 1's payoff matrix (defined at the end of §1.1), pure strategy i is said to *dominate* pure strategy k for Player 1 if $a_{ij} \geq a_{kj}$

---

[2]Maynard Smith, who introduced the Hawk-Dove game (see §6.9), associates payoff $\rho/2$ with strategy combination DD by supposing in place of (1.6) that the resource is equally shared by contestants who both play Dove [**132**].

for *all* $j = 1, \ldots, m_2$ and $a_{ij} > a_{kj}$ for *some* j; and if i dominates k for all $k \neq i$, then i is called a *dominant strategy*. Dominance is *strong* if the above inequalities are all strictly satisfied, and otherwise (i.e., if even one inequality is not strictly satisfied) *weak*. Thus, in particular, (1.8) implies that G is strongly dominant for Player 1. Similarly, if B is Player 2's payoff matrix, pure strategy i dominates pure strategy k for Player 2 if $b_{ji} \geq b_{jk}$ for all $j = 1, \ldots, m_1$ with $b_{ji} > b_{jk}$ for some j; and i is a dominant strategy if i dominates k for all $k \neq i$. Again, dominance is weak unless all inequalities are strictly satisfied. Note that we speak of *a* strongly dominant strategy, rather than *the* strongly dominant strategy, because both players may have one. Clearly, if one strategy (weakly or strongly) dominates another, then the other is (weakly or strongly) dominated by the one.

In practice, if (1.7) is used to define a slow San, then we could interpret our model as yielding the following advice: "If you think the driver across the road is a slowpoke, then put down your foot and go." Furthermore, if (1.7) is used to define a slow San, then it might well be appropriate to define an intermediate San by

(1.9) $2\delta > \tau_2 > 2\epsilon$

and a fast San by

(1.10) $2\delta > 2\epsilon > \tau_2.$

Then Nan would be slow for $\tau_1 > 2\delta > 2\epsilon$, intermediate for $2\delta > \tau_1 > 2\epsilon$ and fast for $2\delta > 2\epsilon > \tau_1$. Note that G is a dominant strategy for San if Nan is slow, because then $-\delta - \tau_1/2 > -\tau_1$.[3]

Suppose, now, that either (1.9) or (1.10) is satisfied. Then (1.8) is false; and what's best for Nan is no longer independent of what San chooses, because the second element in row 1 of Nan's payoff matrix is still greater than the second element in row 2, whereas the first element in row 1 is now smaller than the first in row 2. No pure strategy is obviously better for Nan. Then what should Nan choose?

If sometimes G is better and sometimes W is better (depending on what San chooses), then shouldn't Nan's choice be in some sense a mixture of strategies G and W? One way to mix strategies would

---

[3] We shall ignore as fanciful the possibility that, say, τ_2 and 2δ might actually be equal. In general, only *in*equalities between parameters that aggregate behavior can ever be meaningful.

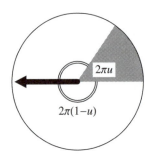

Figure 1.2. Nan's spinning arrow and disk

be to play G with probability u, and hence W with probability $1 - u$. Accordingly, let N denote Nan's choice of pure strategy. Then N is a random variable, with sample space $\{G, W\}$; and $\text{Prob}(N = G) = u$, $\text{Prob}(N = W) = 1 - u$. If Nan plays G with probability u, then we will say that Nan selects *mixed strategy* u, where $u \in [0, 1]$.[4]

Similarly, if San plays G with probability v and hence W with probability $1 - v$, then we shall say that San selects mixed strategy v, where $v \in [0, 1]$; if San's pure strategy is the random variable S with sample space $\{G, W\}$, then $\text{Prob}(S = G) = v = 1 - \text{Prob}(S = W)$. If Nan selects mixed strategy u and San selects mixed strategy v, then we shall refer to the row vector (u, v) as their mixed strategy combination. Thus GG, GW, WG and WW are the same as $(1, 1)$, $(1, 0)$, $(0, 1)$ and $(0, 0)$, respectively. All strategy combinations can be represented as points in the unit square of the Cartesian coordinate plane, i.e., the set $\{(u, v) \mid 0 \le u, v \le 1\}$.[5]

But how could Nan and San arrange all this? Let's suppose that the spinning arrow depicted in Figure 1.2 is mounted on Nan's dashboard. When confronted by San, Nan gives the arrow a quick twirl. If it comes to rest in the shaded sector of the disk, then she plays G; if it comes to rest in the unshaded sector, then she plays W. Thus selecting strategy u means having a disk with a shaded sectoral angle $2\pi u$; and changing one's strategy means changing the disk. What about

---

[4]Here "$\in$" means "belonging to," and $[a, b]$ denotes the set of all numbers between a and b, inclusive.

[5]The vertical bar means "such that." In general, $\{x \mid P\}$ denotes the set of all x such that P is satisfied. For example, $[a, b] = \{x \mid a \le x \le b\}$.

the time required to spin the arrow—does it matter? Not if San also has a spinning arrow mounted on her dashboard and takes about as long to twirl it (of course, San's shaded sector would subtend angle $2\pi v$ at the center). And if you think it's a bit far-fetched that motorists would drive around with spinning arrows on their dashboards, then you can think of Nan's spinning arrow as merely the analogue of a mental process through which she decides whether to go or wait at random—but in such a way that she goes, on average, fraction u of the time. Similarly for San. Note, incidentally, the important point that strategies are selected prior to interaction: the players arrive at the junction with their disks already shaded.

Let F_1 denote the payoff to Nan. Then F_1 is a random variable with sample space

$$(1.11) \qquad \left\{ -\delta - \tfrac{1}{2}\tau_2, 0, -\tau_2, -\epsilon - \tfrac{1}{2}\tau_2 \right\}.$$

If Nan and San choose their strategies independently, then

$$(1.12) \qquad \begin{aligned} \mathrm{Prob}\big(F_1 = -\delta - \tfrac{1}{2}\tau_2\big) &= \mathrm{Prob}(N = G \text{ and } S = G) \\ &= \mathrm{Prob}(N = G) \cdot \mathrm{Prob}(S = G) \\ &= uv. \end{aligned}$$

Similarly, $\mathrm{Prob}(F_1 = 0) = u(1 - v)$, $\mathrm{Prob}(F_1 = -\tau_2) = (1 - u)v$ and $\mathrm{Prob}\big(F_1 = -\epsilon - \tfrac{1}{2}\tau_2\big) = (1 - u)(1 - v)$. Let E denote expected value, and let $f_1(u, v)$ denote the expected value of Nan's payoff from mixed strategy combination (u, v). We will refer to the expected value of a payoff as a *reward*. Thus Nan's reward from strategy combination (u, v) becomes $f_1(u, v) = \mathrm{E}[F_1] = -(\delta + \tfrac{1}{2}\tau_2) \cdot \mathrm{Prob}(F_1 = -\delta - \tfrac{1}{2}\tau_2) + 0 \cdot \mathrm{Prob}(F_1 = 0) - \tau_2 \cdot \mathrm{Prob}(F_1 = -\tau_2) - (\epsilon + \tfrac{1}{2}\tau_2) \cdot \mathrm{Prob}(F_1 = -\epsilon - \tfrac{1}{2}\tau_2)$ or

$$(1.13) \qquad f_1(u, v) = \big(\epsilon + \tfrac{1}{2}\tau_2 - \{\delta + \epsilon\}v\big) u + \big(\epsilon - \tfrac{1}{2}\tau_2\big) v - \epsilon - \tfrac{1}{2}\tau_2$$

after simplification. San's reward from the same combination is

$$(1.14) \qquad f_2(u, v) = \big(\epsilon + \tfrac{1}{2}\tau_1 - \{\delta + \epsilon\}u\big) v + \big(\epsilon - \tfrac{1}{2}\tau_1\big) u - \epsilon - \tfrac{1}{2}\tau_1$$

after a similar calculation—or simply exchanging Nan's strategy and San's transit time in (1.13) for San's strategy and Nan's transit time.

More generally, let Player 1 have m_1 pure strategies with payoff matrix A, and let Player 2 have m_2 pure strategies with payoff matrix B; A and B are both $m_1 \times m_2$. Defining $s_1 = m_1 - 1, s_2 = m_2 - 1$, let

Player 1 select pure strategy i with probability $u_i, 1 \leq i \leq s_1$; and let Player 2 select pure strategy j with probability $v_j, 1 \leq j \leq s_2$. Then because probabilities must sum to 1 for each player, we have $u_{m_1} = 1 - \sum_{i=1}^{s_1} u_i$ and $v_{m_2} = 1 - \sum_{j=1}^{s_2} v_j$. Thus Player 1 has only s_1 decision variables, and Player 2 has only s_2 decision variables. Accordingly, let the s_1-dimensional row vector $u = (u_1, u_2, \dots, u_{s_1})$ be called Player 1's strategy, let the s_2-dimensional row vector $v = (v_1, v_2, \dots, v_{s_2})$ be called Player 2's strategy, and let the $(s_1 + s_2)$-dimensional row vector (u, v) be the players' joint strategy combination. Then the players' rewards from strategy combination (u, v) are

$$(1.15) \quad f_1(u,v) = (u, u_{m_1}) A (v, v_{m_2})^T, \ f_2(u,v) = (u, u_{m_1}) B (v, v_{m_2})^T$$

where a superscript T denotes transpose.

Both Nan and San would like their reward to be as large as possible. Unfortunately, Nan does not know what San will do (and San does not know what Nan will do), because this is a noncooperative game. Therefore, Nan should reason as follows: "I do not know which v San will pick—but for every v, I will pick u to make $f_1(u, v)$ as large as possible." In this way, Nan obtains a set of points in the unit square. Each of these points corresponds to a strategy combination (u, v) that is rational for Nan, in the sense that for each v (over which Nan has no control) a corresponding u is a strategy that makes Nan's reward as large as possible. We will refer to this set of strategy combinations as Nan's *rational reaction set*, and denote it by R_1. In mathematical terms, we have

$$(1.16) \quad R_1 = \left\{ (u,v) \, \middle| \, 0 \leq u, v \leq 1, \ f_1(u,v) = \max_{0 \leq \overline{u} \leq 1} f_1(\overline{u}, v) \right\}.$$

For each (u, v) in R_1, if Player 2 selects v then a best reply for Player 1 is to select u (*a* best reply rather than *the* best reply because there may be more than one). Note that R_1 is obtained in practice by holding v constant and maximizing f_1 as a function of a single variable (whose maximum will depend on v).[6]

Likewise, San should reason as follows: "I do not know which u Nan will pick—but for every u, I will pick v to make $f_2(u, v)$ as large as possible." In this way, San obtains a set of points in the unit square.

---

[6]It would perhaps be more accurate to describe R_1 as Player 1's *individually* rational reaction set; nevertheless, we abide by tradition. The distinction between individual and group rationality will emerge in Chapter 3.

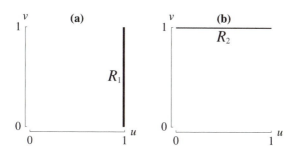

Figure 1.3. Rational reaction sets when $\tau_1 > 2\delta, \tau_2 > 2\delta$.

Each of these points corresponds to a strategy combination that is rational for San, in the sense that for each u (over which San has no control) a corresponding v is a strategy that makes San's reward as large as possible. We will refer to this set of strategy combinations as San's rational reaction set, and denote it by R_2. That is,

$$(1.17) \qquad R_2 = \left\{ (u,v) \,\middle|\, 0 \le u, v \le 1, \; f_2(u,v) = \max_{0 \le \overline{v} \le 1} f_2(u,\overline{v}) \right\}.$$

For each (u,v) in R_2, if Player 1 selects u then a best reply for Player 2 is to select v. In practice, R_2 is obtained by holding u constant and maximizing f_2 as a function of a single variable (with the maximum depending on u). Note the important point that each player can determine her rational reaction set without any knowledge of the other player's reward.

Suppose, for example, that $\delta < \frac{1}{2}\min(\tau_1, \tau_2)$, i.e., both drivers are slow. Then, because $0 \le v \le 1$ implies that the coefficient of $\overline{u}$ in

$$(1.18) \quad f_1(\overline{u},v) = \left(\epsilon\{1-v\} + \tfrac{1}{2}\tau_2 - \delta v \right)\overline{u} + \left(\epsilon - \tfrac{1}{2}\tau_2 \right)v - \epsilon - \tfrac{1}{2}\tau_2$$

is always positive, $f_1(\overline{u},v)$ is maximized for $0 \le \overline{u} \le 1$ by choosing $\overline{u} = 1$. Therefore, Nan's rational reaction set is

$$(1.19) \qquad R_1 = \left\{ (u,v) \mid u = 1, 0 \le v \le 1 \right\},$$

the edge of the unit square that runs between $(1,0)$ and $(1,1)$; see Figure 1.3(a), where R_1 is represented by a thick solid line. Similarly, because $0 \le u \le 1$ implies that the coefficient of $\overline{v}$ in

$$(1.20) \quad f_2(u,\overline{v}) = \left(\epsilon\{1-u\} + \tfrac{1}{2}\tau_1 - \delta u \right)\overline{v} + \left(\epsilon - \tfrac{1}{2}\tau_1 \right)u - \epsilon - \tfrac{1}{2}\tau_1$$

is always positive, $f_2(u, \overline{v})$ is maximized for $0 \le \overline{v} \le 1$ by choosing $\overline{v} = 1$. Therefore, San's rational reaction set is

$$(1.21) \qquad R_2 = \big\{(u, v) \mid 0 \le u \le 1, v = 1\big\},$$

the edge of the unit square that runs between $(0, 1)$ and $(1, 1)$; see Figure 1.3(b), where R_2 is represented by a thin solid line. Of course, all that Figure 1.3 tells us is that the best strategy against a slow driver is G, which we knew long before we began to talk about mixed strategies. But something new will shortly emerge.

Notice that the rational reaction sets R_1 and R_2 have a non-empty intersection $R_1 \cap R_2 = \{(1, 1)\}$. The strategy combination $(1, 1)$ which lies in both sets has the following property: if either player selects strategy 1, then the other cannot obtain a greater reward by selecting a strategy other than 1. In other words, no player can increase her reward by a *unilateral* departure from the strategy combination $(1, 1)$. By virtue of having this property, $(1, 1)$ is said to be a Nash-equilibrium strategy combination. More generally, (u^*, v^*) is a Nash-equilibrium strategy combination, or simply *Nash equilibrium*, of a noncooperative, 2-player game when, if one player sticks rigidly to her Nash-equilibrium strategy (u^* in the case of Player 1, v^* in the case of Player 2), then the other player cannot increase her reward by selecting a strategy other than her Nash-equilibrium strategy. Alternatively, (u^*, v^*) is a Nash equilibrium if u^* is a best reply to v^* AND v^* is a best reply to u^*. A still more general definition of Nash equilibrium, for games among arbitrary numbers of players, will be given in §1.6.

Now, if we were interested solely in finding the best pair of strategies for two slow drivers ($\tau_1, \tau_2 > 2\delta$), then introducing the concept of Nash equilibrium would be like using a sledgehammer to burst a soap bubble. It is obvious from Figure 1.3 that $(1, 1)$ is the only pair of strategies that two rational players would select, for (u, v) will be selected only if u lies in R_1 and v in R_2. Things get a little bit more complicated, however, if either driver is either fast or intermediate.

To determine R_1 and R_2 in these circumstances, it will be convenient first to define parameters θ_1 and θ_2 by

$$(1.22) \qquad (\delta + \epsilon)\theta_k = \epsilon + \tfrac{1}{2}\tau_k, \quad k = 1, 2.$$

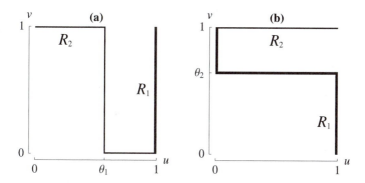

Figure 1.4. R_1 and R_2 for **(a)** $\tau_1 < 2\delta < \tau_2$, **(b)** $\tau_2 < 2\delta < \tau_1$.

Then, from (1.13)-(1.14),

(1.23a) $\qquad f_1(\overline{u}, v) = (\delta + \epsilon)(\theta_2 - v)\overline{u} + \left(\epsilon - \frac{1}{2}\tau_2\right)v - \epsilon - \frac{1}{2}\tau_2$

(1.23b) $\qquad f_2(u, \overline{v}) = (\delta + \epsilon)(\theta_1 - u)\overline{v} + \left(\epsilon - \frac{1}{2}\tau_1\right)u - \epsilon - \frac{1}{2}\tau_1.$

If $\frac{1}{2}\tau_1 < \delta < \frac{1}{2}\tau_2$ (slow San, fast or intermediate Nan) then $\theta_1 < 1 < \theta_2$, and so the $\overline{u}$ that maximizes $f_1(\overline{u}, v)$ for $0 \leq \overline{u} \leq 1$ is still $\overline{u} = 1$; whereas the $\overline{v}$ that maximizes $f_1(u, \overline{v})$ is[7]

(1.24) $\qquad \overline{v} = \begin{cases} 1 & \text{if } 0 \leq u < \theta_1 \\ \text{any } \overline{v} \in [0, 1] & \text{if } u = \theta_1 \\ 0 & \text{if } \theta_1 < u \leq 1. \end{cases}$

Thus R_1 is the same as before, whereas R_2 consists of three straight-line segments as shown in Figure 1.4(a). We see that, if San has no knowledge of Nan's reward function f_1, then any $v \in [0, 1]$ could be rational for San because, for all she knows, Nan could select the strategy $u = \theta_1$, to which any $v \in [0, 1]$ is a best reply. If, on the other hand, San knows Nan's reward function, then the only rational choice for San is $v = 0$, because only $v = 0$ is a best reply to $u = 1$.

---

[7]Note that $\overline{v}$ defined by (1.24) is strictly not a function of u, because $\overline{v}$ takes more than one value where $u = \theta_1$. In general, a function is a rule that assigns a unique element of a set V (called the range), to each element of a set U (called the domain); and a rule that assigns a subset of V to each element of U is called a multi-valued function, or *correspondence*. Thus (1.24) defines a correspondence with $U = V = [0, 1]$.

Of course, $u = 1$ is also a best reply to $v = 0$, because it's the best reply to anything. Thus $(1, 0)$, the only point in the intersection of R_1 and R_2, is a Nash equilibrium. In terms of pure strategies, the Nash equilibrium is GW: G is a best reply to W (regardless), and W is a slow driver's best reply to an intermediate or fast driver's G.

Similarly, if $\frac{1}{2}\tau_2 < \delta < \frac{1}{2}\tau_1$ (slow Nan, fast or intermediate San) then, either from symmetry or from an expression similar to (1.24), R_1 and R_2 are as shown in Figure 1.4(b). If Nan has no knowledge of San's reward function f_2, then any $u \in [0, 1]$ could be rational for Nan because, for all she knows, San could select the strategy $v = \theta_2$, to which any $u \in [0, 1]$ is a best reply. If, on the other hand, Nan knows San's reward function, then the only rational choice for Nan is $u = 0$, because only $u = 0$ is a best reply to $v = 1$. Because $v = 1$ is also a best reply to $u = 0$, $(0, 1)$, the only point in the intersection of R_1 and R_2, is a Nash equilibrium. In terms of pure strategies, this time the Nash equilibrium is WG, but the interpretation is otherwise the same: G is a best reply to W, and W is a slow driver's best reply to an intermediate or fast driver's G. We see that the concept of Nash equilibrium depends crucially on each player knowing the other player's reward (whereas the concept of rational reaction set does not). If such is the case, then it is customary to say that the players have *complete information*.

There is no denying that the concept of Nash equilibrium has desirable properties. Provided each player has knowledge of the other's reward, there's a certain sense in which the solution is self-enforcing, even though there is no explicit cooperation between the players. Consider, however, the case in which neither driver is slow, so that $\max(\tau_1, \tau_2) < 2\delta$, or $\max(\theta_1, \theta_2) < 1$. The $\overline{u}$ that maximizes $f_1(\overline{u}, v)$ for $0 \leq \overline{u} \leq 1$ is now readily shown to be

$$(1.25) \qquad \overline{u} = \begin{cases} 1 & \text{if } 0 \leq v < \theta_2 \\ \text{any } \overline{u} \in [0, 1] & \text{if } v = \theta_2 \\ 0 & \text{if } \theta_2 < v \leq 1 \end{cases}$$

whereas the $\overline{v}$ that maximizes $f_2(u, \overline{v})$ is still (1.24). Thus R_1 and R_2 are as shown in Figure 1.5. We observe at once that $R_1 \cap R_2 = \{(1, 0), (\theta_1, \theta_2), (0, 1)\}$: there are three Nash equilibria. Then which do we regard as the solution?

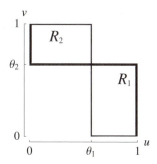

Figure 1.5. Rational reaction sets when $\max(\tau_1, \tau_2) < 2\delta$.

Table 1.5. Nash-equilibrium rewards in Crossroads

(u, v)	$f_1(u, v)$	$f_2(u, v)$
$(1, 0)$	0	$-\tau_1$
(θ_1, θ_2)	$-\left(\delta + \frac{1}{2}\tau_2\right)\theta_2$	$-\left(\delta + \frac{1}{2}\tau_1\right)\theta_1$
$(0, 1)$	$-\tau_2$	0

The rewards associated with the three Nash equilibria are given in Table 1.5. You can readily show that

$$(1.26) \qquad -\left(\delta + \tfrac{1}{2}\tau_k\right)\theta_k - \left(-\tau_k\right) = \tfrac{(2\delta - \tau_k)(\tau_k - 2\epsilon)}{4(\delta + \epsilon)}, \qquad k = 1, 2.$$

Thus $(1, 0)$ is always the best Nash equilibrium for Nan; and (θ_1, θ_2) is second or third best, according to whether $2\delta > \tau_2 > 2\epsilon$ (intermediate San) or $2\delta > 2\epsilon > \tau_2$ (fast San). Again, $(0, 1)$ is always the best Nash equilibrium for San; and (θ_1, θ_2) is second or third best according to whether $2\delta > \tau_1 > 2\epsilon$ (intermediate Nan) or $2\delta > 2\epsilon > \tau_1$ (fast Nan). Even though θ_1 is a best reply to θ_2 and θ_2 is a best reply to θ_1, there is no reason to expect the players to select these strategies, because for each player there is another strategy combination with the best-reply property that yields a higher reward. But if Nan selects her best Nash-equilibrium strategy $u = 1$, and if San selects her best Nash-equilibrium strategy $v = 1$, then the resulting strategy combination $(1, 1)$ belongs to neither player's rational reaction set! Then which—if any—of the Nash equilibria should we regard as the solution of the game? We will return to this matter in Chapter 2.

Table 1.6. Nan's payoff matrix in Four Ways

	G	W	C
G	$-\delta - \frac{1}{2}\tau$	0	0
W	$-\tau$	$-\epsilon - \frac{1}{2}\tau$	$-\tau$
C	$-\tau$	0	$-\delta - \frac{1}{2}\tau$

Table 1.7. San's payoff matrix in Four Ways

	G	W	C
G	$-\delta - \frac{1}{2}\tau$	$-\tau$	$-\tau$
W	0	$-\epsilon - \frac{1}{2}\tau$	0
C	0	$-\tau$	$-\delta - \frac{1}{2}\tau$

1.4. Four Ways: a motorist's trilemma

Nan and San's dilemma becomes even more intriguing if we allow a
third strategy, denoted by C, in which each player's action is contingent upon that of the other. A player who adopts C will select G if
the other player selects W, but she will select W if the other player
selects G. Let us suppose that, if Nan is a C-strategist, then the first
thing she does when she arrives at the junction is to wave San on; but
if San replies by waving Nan on, then immediately Nan puts down
her foot and drives away. If, on the other hand, San replies by hitting
the gas, then Nan waits until San has traversed the junction. But
what happens if San is also a C-strategist? As soon as they reach
the junction, Nan and San both wave at one another. Nan interprets
San's wave to mean that San wants to wait, so Nan drives forward;
San interprets Nan's wave to mean that Nan wants to wait, so San
also drives forward; and the result is the same as if both had selected
strategy G. Thus if a G-strategist can be described as selfish and
a W-strategist as an altruist, then a C-strategist could perhaps be
described as an impatient altruist.

   For the sake of simplicity, let us assume that the game is symmetric, i.e., $\tau_1 = \tau_2$, and denote the common value of these two parameters by τ. Then Nan and San's payoff matrices A and B, respectively,
are as shown in Tables 1.6 and 1.7. As always, the rows correspond

to strategies of Player 1 (Nan), and the columns correspond to strategies of Player 2 (San); thus the entry in row i and column j is the payoff, to the player whose payoffs are stored in the matrix, if Player 1 selects strategy i and Player 2 selects strategy j. Because the game is symmetric, B is just the transpose of A. To distinguish this game from Crossroads, we will refer to it as Four Ways.

If the drivers are so slow that $\tau > 2\delta$ or $\sigma > 1$, where

$$(1.27) \qquad \sigma = \tau/2\delta,$$

then their best strategy is to hit the gas, because G dominates C and strictly dominates W for Nan, from Table 1.6; and similarly for San, from Table 1.7. Thus G is a (weakly) dominant strategy for both players: neither has an incentive to depart from it, which makes strategy combination GG a Nash equilibrium. Furthermore, GG is the only Nash equilibrium when $\sigma > 1$ (Exercise 1.3), and so we do not hesitate to regard it as the solution of the game: when there is only one Nash equilibrium, there is no indeterminacy to resolve.[8]

The game becomes interesting, however, when $\tau < 2\delta$ or $\sigma < 1$, which we assume for the rest of this section. As in Crossroads, no pure strategy is now dominant. We therefore consider mixed strategies. If Nan selects pure strategy G with probability u_1 and pure strategy W with probability u_2, then we shall say that Nan selects strategy u, where $u = (u_1, u_2)$ is a 2-dimensional row vector. Thus Nan selects pure strategy C with probability $1 - u_1 - u_2$, where

$$(1.28a) \qquad 0 \le u_1 \le 1, \qquad 0 \le u_2 \le 1, \qquad 0 \le u_1 + u_2 \le 1.$$

So Nan's strategies correspond to points of a closed triangle in 2-dimensional space. Similarly, if San selects G with probability v_1 and W with probability v_2, then we shall say that San selects strategy v, where $v = (v_1, v_2)$ is also a 2-dimensional vector; and because San selects C with probability $1 - v_1 - v_2$, we have

$$(1.28b) \qquad 0 \le v_1 \le 1, \qquad 0 \le v_2 \le 1, \qquad 0 \le v_1 + v_2 \le 1.$$

Subsequently, we shall use Δ to denote the closed triangle in 2-dimensional space defined EITHER as the set of all points that satisfy

---

[8] Even if there were more than one Nash equilibrium, there would be no indeterminacy if all combinations of Nash-equilibrium strategies yielded the same payoffs. This equivalence holds in general only for zero-sum games; see, for example, Owen [**173**] or Wang [**233**]. For an example of a zero-sum game, see Exercise 1.33.

(1.28a) OR as the set of all points that satisfy(1.28b); the sets are identical, because this triangle exists independently of whether we use u or v to label a point in it. If Nan selects $u \in \Delta$ and San selects $v \in \Delta$, then we shall say that they jointly select strategy combination (u, v), where $(u, v) = (u_1, u_2, v_1, v_2)$ is a 4-dimensional vector.

The sample space of N, Nan's choice of pure strategy, is now $\{G, W, C\}$ instead of $\{G, W\}$; $\text{Prob}(N = G) = u_1$, $\text{Prob}(N = W) = u_2$ and $\text{Prob}(N = C) = 1 - u_1 - u_2$. San's choice of pure strategy, S, has the same sample space, but with $\text{Prob}(S = G) = v_1$, $\text{Prob}(S = W) = v_2$ and $\text{Prob}(S = C) = 1 - v_1 - v_2$. The payoff to Nan, F_1, now has sample space $\left\{-\delta - \frac{1}{2}\tau, 0, -\tau, -\epsilon - \frac{1}{2}\tau\right\}$; and if strategies are still chosen independently, then $\text{Prob}(F_1 = -\delta - \tau/2) = \text{Prob}(N = G, S = G$ or $N = C, S = C) = \text{Prob}(N = G, S = G) + \text{Prob}(N = C, S = C) = \text{Prob}(N = G) \cdot \text{Prob}(S = G) + \text{Prob}(N = C) \cdot \text{Prob}(S = C) = u_1 v_1 + (1 - u_1 - u_2)(1 - v_1 - v_2)$. Similarly, $\text{Prob}(F_1 = 0) = u_1 v_2 + u_1(1 - v_1 - v_2) + (1 - u_1 - u_2)v_2$, $\text{Prob}(F_1 = -\tau) = u_2 v_1 + u_2(1 - v_1 - v_2) + (1 - u_1 - u_2)v_1$ and $\text{Prob}(F_1 = -\epsilon - \tau/2) = u_2 v_2$. Thus Nan's reward from the mixed strategy combination (u, v) is $f_1(u, v) = \text{E}[F_1] = -\left(\delta + \frac{1}{2}\tau\right) \cdot \text{Prob}\left(F_1 = -\delta - \frac{1}{2}\tau\right) + 0 \cdot \text{Prob}\left(F_1 = 0\right) - \tau \cdot \text{Prob}\left(F_1 = -\tau\right) - \left(\epsilon + \frac{1}{2}\tau\right) \cdot \text{Prob}\left(F_1 = -\epsilon - \frac{1}{2}\tau\right)$ or, after simplification,

$$
\begin{aligned}
(1.29) \qquad f_1(u, v) = &- \left(2\delta v_1 + \left\{\delta + \tfrac{1}{2}\tau\right\}\{v_2 - 1\}\right)u_1 \\
&- \left(\left\{\delta - \tfrac{1}{2}\tau\right\}\{v_1 - 1\} + \{\delta + \epsilon\}v_2\right)u_2 \\
&+ \left(\delta - \tfrac{1}{2}\tau\right)v_1 + \left(\delta + \tfrac{1}{2}\tau\right)(v_2 - 1).
\end{aligned}
$$

Similarly, San's reward from the strategy combination (u, v) is

$$
\begin{aligned}
(1.30) \qquad f_2(u, v) = &- \left(2\delta u_1 + \left\{\delta + \tfrac{1}{2}\tau\right\}\{u_2 - 1\}\right)v_1 \\
&- \left(\left\{\delta - \tfrac{1}{2}\tau\right\}\{u_1 - 1\} + \{\delta + \epsilon\}u_2\right)v_2 \\
&+ \left(\delta - \tfrac{1}{2}\tau\right)u_1 + \left(\delta + \tfrac{1}{2}\tau\right)(u_2 - 1).
\end{aligned}
$$

Note that, by virtue of symmetry,

$$
(1.31) \qquad\qquad f_2(u, v) = f_1(v, u)
$$

for all u and v satisfying (1.28). Note also that (1.29) and (1.30) are special cases of (1.15).

Although u and v are now vectors, as opposed to scalars, everything we have said about rational reaction sets and Nash equilibria with respect to Crossroads remains true for Four Ways, provided only that we replace $0 \leq u \leq 1$ by $u \in \Delta$ and $0 \leq v \leq 1$ by $v \in \Delta$ (and therefore also $0 \leq \overline{u} \leq 1$ by $\overline{u} \in \Delta$ and $0 \leq \overline{v} \leq 1$ by $\overline{v} \in \Delta$). Thus the players' rational reaction sets in Four Ways are defined by

$$(1.32a) \qquad R_1 = \left\{ (u,v) \mid u,v \in \Delta, \; f_1(u,v) = \max_{\overline{u}} f_1(\overline{u},v) \right\}$$

$$(1.32b) \qquad R_2 = \left\{ (u,v) \mid u,v \in \Delta, \; f_2(u,v) = \max_{\overline{v}} f_2(u,\overline{v}) \right\},$$

but the set of all Nash equilibria is still $R_1 \cap R_2$. On the other hand, because the rational reaction sets now lie in a 4-dimensional space, as opposed to a 2-dimensional space, we cannot locate the Nash equilibria by drawing diagrams equivalent to Figures 1.3-1.5. Instead, we proceed as follows. We first define dimensionless parameters

$$(1.33) \quad \gamma = \frac{\epsilon}{\delta}, \; \alpha = \frac{(\sigma+\gamma)(\sigma+1)}{1+2\gamma+\sigma^2}, \; \beta = \frac{(1-\sigma)^2}{1+2\gamma+\sigma^2}, \; \omega = \frac{2\sigma}{1+\sigma}$$

and

$$(1.34) \qquad\qquad \theta = \frac{2\epsilon+\tau}{2\epsilon+2\delta} = \frac{\sigma+\gamma}{1+\gamma}$$

where σ is defined by (1.27). In view of (1.1), α, β, γ, σ, θ and ω all lie between 0 and 1. If the coefficients of u_1 and u_2 in (1.29) are both negative, then clearly $f_1(u,v)$ is maximized by selecting $u_1 = 0$ and $u_2 = 0$, or $u = (0,0)$; moreover, $(0,0)$ is the only maximizing strategy for Player 1. If these coefficients are merely nonpositive, then there will be more than one maximizing strategy; nevertheless, $u = (0,0)$ will continue to be one of them. But the coefficient of u_1 in (1.29) is nonpositive when the point (v_1, v_2) lies on or above the line in 2-dimensional space that joins the point $(\sigma/\omega, 0)$ to the point $(0,1)$; whereas the coefficient of u_2 in (1.29) is nonpositive when the point (v_1, v_2) lies on or above the line that joins the point $(1,0)$ to the point $(0, 1-\theta)$. Thus the coefficients of u_1 and u_2 in (1.29) are both nonpositive when the point (v_1, v_2) lies in that part of Δ which corresponds to (the interior or boundary of) the triangle marked C in Figure 1.6. Let us denote by $v^C = (v_1^C, v_2^C)$ any strategy for San that corresponds to a point in C. Then what we have shown is that all 4-dimensional vectors of the form $\left(0, 0, v_1^C, v_2^C\right)$ must lie in R_1.

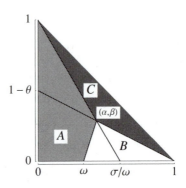

Figure 1.6. Subsets A, B and C of Δ defined by (1.28).

Extending our notation in an obvious way, let us denote by $v^A = (v_1^A, v_2^A)$ any strategy for San that corresponds to a point in A, by $v^{AC} = (v_1^{AC}, v_2^{AC})$ any strategy for San that corresponds to a point lying in both A and C, and so on. Then, by considering the various cases in which the coefficient of u_1 or the coefficient of u_2 or both in (1.29) are nonpositive, nonnegative or zero, it is readily shown that all strategy combinations in Table 1.8 must lie in Nan's rational reaction set, R_1; see Exercise 1.5. Furthermore, if we repeat the analysis for f_2 and San (as opposed to f_1 and Nan), and if we denote by $u^A = (u_1^A, u_2^A)$ any strategy for Nan that corresponds to a point in A, by $u^{AC} = (u_1^{AC}, u_2^{AC})$ any strategy for Nan that corresponds to a point in both A and C, and so on, then we readily find that all strategy combinations in Table 1.9 must lie in San's rational reaction set, R_2. Indeed, in view of symmetry condition (1.31), it is hardly necessary to repeat the analysis.

A strategy combination is a Nash equilibrium if, and only if, it appears both in Table 1.8 and in Table 1.9. Therefore, to find all Nash equilibria, we must match strategy combinations from Table 1.8 with strategy combinations from Table 1.9 in every possible way. For example, consider the first row of Table 1.8. It does not match the first, fourth or sixth row of Table 1.9 because $(1, 0)$ does not lie in A. It does not match the last row of Table 1.9, even for $(v_1, v_2) \in A$, because $\alpha < 1$ (or because $\beta > 0$). Because $(1, 0)$ lies in B and

Table 1.8. R_1 for Four Ways.

u_1	u_2	v_1	v_2	CONSTRAINTS
1	0	v_1^A	v_2^A	
0	1	v_1^B	v_2^B	
0	0	v_1^C	v_2^C	
u_1	0	v_1^{AC}	v_2^{AC}	$0 \le u_1 \le 1$
0	u_2	v_1^{BC}	v_2^{BC}	$0 \le u_2 \le 1$
u_1	u_2	v_1^{AB}	v_2^{AB}	$u \in \Delta,\ u_1 + u_2 = 1$
u_1	u_2	α	β	$u \in \Delta$

Table 1.9. R_2 for Four Ways.

u_1	u_2	v_1	v_2	CONSTRAINTS
u_1^A	u_2^A	1	0	
u_1^B	u_2^B	0	1	
u_1^C	u_2^C	0	0	
u_1^{AC}	u_2^{AC}	v_1	0	$0 \le v_1 \le 1$
u_1^{BC}	u_2^{BC}	0	v_2	$0 \le v_2 \le 1$
u_1^{AB}	u_2^{AB}	v_1	v_2	$v \in \Delta,\ v_1 + v_2 = 1$
α	β	v_1	v_2	$v \in \Delta$

Table 1.10. Nash equilibria for Four Ways.

u_1	u_2	v_1	v_2	CONSTRAINTS
1	0	0	1	
0	1	1	0	
1	0	0	0	
0	0	1	0	
1	0	0	v_2	$0 \le v_2 < 1$
0	u_2	1	0	$0 \le u_2 < 1$
0	1	v_1	0	$\omega \le v_1 < 1$
u_1	0	0	1	$\omega \le u_1 < 1$
α	β	α	β	

$(0, 1)$ lies in A, however, we can match the first row of Table 1.8 with
the second row of Table 1.9, and so $(1, 0, 0, 1)$ is a Nash equilibrium.
Likewise, because $(1, 0)$ lies in C and $(0, 0)$ in A, we can match the

first row of Table 1.8 with the third row of Table 1.9, so that $(1,0,0,0)$
is a Nash equilibrium. Finally, we can match the first row of Table
1.8 with the fifth row of Table 1.9 to deduce that $(1,0,0,v_2)$ is a
Nash-equilibrium strategy combination when $0 \leq v_2 < 1$, because
then $(0,v_2)$ lies in A. The Nash equilibria we have found in this way
are recorded in rows 1, 3 and 5 of Table 1.10.

Repeating the analysis for the remaining six rows of Table 1.8, we
obtain (Exercise 1.6) an exhaustive list of Nash-equilibrium strategy
combinations. They are recorded in Table 1.10. The first four rows of
this table correspond to equilibria in pure strategies: rows 1 and 2 to
equilibria in which one player selects G and the other W, rows 3 and
4 to equilibria in which one player selects G and the other C. The
remaining five rows correspond to equilibria in mixed strategies. We
see that, although rows 1-4 and 9 of the table correspond to isolated
equilibria, there are infinitely many equilibria of the other types. If
you thought that having three equilibria to choose from in Crossroads
was bad enough, then I wonder what are you thinking now. Which, if
any, of all these infinitely many equilibria do we regard as the solution
of Four Ways?

Good question! Perhaps you would like to mull it over, at least
until Chapter 2. Meanwhile, do Exercise 1.29.

1.5. Store Wars: a continuous game of prices

Although it is always reasonable to suppose that decision makers have
only a finite number of pure strategies, when the number is large
it is often convenient to imagine instead that the strategies form a
continuum. Suppose, for example, that the price of some item could
reasonably lie anywhere between five and ten dollars. Then if a cent is
the smallest unit of currency, and if selecting a strategy corresponds
to setting the price of the item, then the decision maker has a finite
total of 501 pure strategies. Because this number is large, however,
it may be preferable to suppose that the price in dollars can take any
value between 5 and 10 (and round to two decimal places). Then
rewards are calculated directly, i.e., without the intermediate step of
calculating payoff matrices; and the game is said to be *continuous*, to
distinguish it from matrix games like Crossroads, Four Ways and the
Hawk-Dove game. The definition of Nash equilibrium is not in the

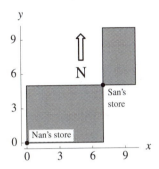

Figure 1.7. Battleground for Store Wars

least affected; but whereas matrix games are guaranteed to have at least one Nash equilibrium, continuous games may have none at all. These ideas are illustrated by the following example.[9]

A subdivision of area 50 square miles consists of two rectangles of land, as shaded in Figure 1.7; the smaller rectangle measures 15 square miles, the larger rectangle 35. If we take the southwest corner of the subdivision to be the origin of a Cartesian coordinate system Oxy, with x increasing to the east and y to the north, then the subdivision contains all points (x, y) such that either $0 \le x \le 7, 0 \le y \le 5$ or $7 \le x \le 10, 5 \le y \le 10$. All roads through the subdivision run either from east to west or from north to south. There are two stores, one at $(0,0)$, the other at $(7,5)$. Each sells a product for which the daily demand is uniformly distributed over the 50 square miles, in the sense that customers are equally likely to live anywhere in the subdivision; the product might, for example, be bags of ice. If buyers select a store solely by weighing the price of the product against the cost of getting there (bags of ice at the first store are identical to those at the second, etc.), and if each store wishes to maximize revenue from the product in question, then how should prices be set? Because the best price for each store depends on the other store's price, their decisions are interdependent; and if they do

---

[9]For a proof that matrix games have at least one Nash equilibrium, see, for example, [**233**]. The example Store Wars was suggested by the Hotelling model described in Phlips [**181**, pp. 42-45]. Phlips assumes that prospective customers are uniformly distributed along a line; whereas Store Wars assumes—in effect—that they are nonuniformly distributed along a line.

not communicate with one another before setting prices, then we have all the necessary ingredients for a noncooperative game. We will call this game Store Wars.

Let Player 1 be Nan, who is manager of the store at $(0,0)$; and let Player 2 be San, who is manager of the store at $(7,5)$.[10] Let p_1 be Nan's price for the product, let p_2 be San's price, and let c be the cost per mile of travel to the store, assumed the same for all customers. Thus the round-trip cost of travel from Nan's store to San's store would be $24c$—no matter how you went, because all roads through the subdivision run from east to west or from north to south. Clearly, if Nan's price were to exceed this round-trip travel cost plus San's price for the item in question, then Nan could never expect anyone to buy from her. Accordingly, we can safely assume that

$$(1.35a) \qquad\qquad p_1 \leq p_2 + 24c.$$

Similarly, because nobody in the larger rectangle can be expected to buy from San if her price exceeds Nan's by the round-trip travel cost between the stores, and assuming that San would like to attract at least some customers from the larger rectangle, we have

$$(1.35b) \qquad\qquad p_2 \leq p_1 + 24c.$$

Furthermore, there are upper and lower limits to the price that a store can charge for a product, and it will be convenient to write these as

$$(1.35c) \qquad\qquad p_1 \leq 4c\alpha, \quad p_2 \leq 4c\alpha$$

$$(1.35d) \qquad\qquad p_1 \geq 4c\beta, \quad p_2 \geq 4c\beta$$

where α and β are dimensionless parameters.[11] Except in Exercise 1.13, however, we shall assume throughout that $\beta = 0$; provided that β is sufficiently small, this assumption will not affect the principal results of our analysis.[12]

---

[10] If Nan were to live near San's store and San were to live near Nan's store, then we could easily explain why they keep meeting each other in Crossroads!

[11] $4c$ is the cost of driving round a square-mile block. Because c is a cost per unit length, we must multiply it by a distance (here 4) to make the right-hand side of each inequality a quantity with the dimensions of price.

[12] In terms of the economist's inverse demand curve, with quantity measured along the horizontal axis and price along the vertical axis, $4\alpha c$ is the price at which the demand curve meets the vertical axis, whereas $4\beta c$ is simply the cost price of the item. Strictly, however, we ignore questions of supply and demand; or, if you prefer, we assume that demand is infinitely elastic at $4\alpha c$ but infinitely inelastic at greater or lower prices.

Now, let (X, Y) be the residential coordinates of the next customer for the product in question. Because all roads run either north and south or east and west, her distance from Nan's store is $|X| + |Y|$ and her distance from San's store is $|7 - X| + |5 - Y|$. Thus, assuming that she selects a store *solely* by weighing the price of the product against the cost of travel from her residence (she doesn't, for example, buy the product on her way home from work), this customer will buy from Nan if

$$(1.36a) \qquad p_1 + 2c(|X| + |Y|) < p_2 + 2c(|7 - X| + |5 - Y|);$$

whereas she will buy from San if

$$(1.36b) \qquad p_1 + 2c(|X| + |Y|) > p_2 + 2c(|7 - X| + |5 - Y|).$$

But $X \geq 0, Y \geq 0$; thus $|X| + |Y|$ is the same as $X + Y$. Furthermore, the shape of the subdivision precludes either $X > 7, Y < 5$ or $X < 7, Y > 5$; therefore $|7 - X| + |5 - Y|$ is the same thing as $|12 - X - Y|$. But if we had $X + Y > 12$ in (1.36a), then it would now reduce to $p_1 + 24c < p_2$, which violates (1.35b). Accordingly, we can assume that $X + Y \leq 12$ in (1.36a) and rewrite it as $p_1 + 2c(X + Y) < p_2 + 2c(12 - X - Y)$. Hence the next customer will buy from Nan if

$$(1.37a) \qquad X + Y < \frac{p_2 - p_1}{4c} + 6.$$

Similarly, if $X + Y \leq 12$, then the next customer will buy from San if

$$(1.37b) \qquad X + Y > \frac{p_2 - p_1}{4c} + 6.$$

If, on the other hand, $X + Y > 12$, then (1.36b) reduces to $p_1 + 24c > p_2$, and the customer will certainly buy from San. Thus the next customer will buy from San either if $X + Y > 12$ or if $X + Y \leq 12$ and (1.37b) is satisfied. But $X + Y > 12$ implies (1.37b) because the right-hand side of (1.37b) is less than or equal to 12 (by virtue of (1.35b)). Thus, in any event, the next customer will buy from San if (1.37b) is satisfied. Of course, San's monopoly over the smaller rectangle was built into the model when we assumed (1.35b).

Because the next customer could live anywhere in the subdivision, X and Y are (continuous) random variables; hence so is $X + Y$. Let G denote its cumulative distribution function, i.e., define

$$(1.38) \qquad G(s) = \text{Prob}(X + Y \leq s), \quad 0 \leq s \leq 20;$$

and let F_1 denote Nan's payoff from the next customer. Then F_1 is also a random variable, which in view of (1.37) is defined by

$$(1.39) \qquad F_1 = \begin{cases} p_1 & \text{if } X + Y < \frac{p_2 - p_1}{4c} + 6 \\ 0 & \text{if } X + Y > \frac{p_2 - p_1}{4c} + 6. \end{cases}$$

Because F_1 is a random variable, it cannot itself be maximized; but instead we can maximize its expected value, which we shall denote by f_1, and define to be Nan's reward.

It will be convenient to make prices dimensionless, by scaling them with respect to $4c$. Let us therefore define u and v by

$$(1.40) \qquad u = \frac{p_1}{4c} \qquad v = \frac{p_2}{4c}$$

where u is Nan's strategy and v is San's. Then, from (1.38)-(1.40),

$$(1.41) \quad \begin{aligned} f_1(u, v) = \mathrm{E}[F_1] &= p_1 \cdot \mathrm{Prob}(X + Y < v - u + 6) + \\ &\quad 0 \cdot \mathrm{Prob}(X + Y > v - u + 6) \\ &= 4cuG(v - u + 6); \end{aligned}$$

of course, $\mathrm{Prob}(X + Y = v - u + 6) = 0$, because $X + Y$ is a continuous random variable.[13] Similarly, San's payoff is the random variable

$$(1.42) \qquad F_2 = \begin{cases} 0 & \text{if } X + Y < v - u + 6 \\ 4cv & \text{if } X + Y > v - u + 6 \end{cases}$$

and her reward is

$$(1.43) \quad \begin{aligned} f_2(u, v) = \mathrm{E}[F_2] &= 4cv \cdot \mathrm{Prob}(X + Y > v - u + 6) \\ &= 4cv \left\{ 1 - G(v - u + 6) \right\}. \end{aligned}$$

Note that, in view of (1.40), (1.35a) requires $u \leq v + 6$, whereas (1.35b) requires $v \leq u + 6$. Thus, in view of (1.35c), the set of all feasible strategy combinations, which we shall denote by D, is

$$(1.44) \qquad D = \left\{ (u, v) \mid 0 \leq u, v \leq \alpha, \, |u - v| \leq 6 \right\}.$$

We shall call D the *decision set* (because this phrase is less cumbersome than strategy combination set); and it will be convenient to

---

[13]See, for example, [**144**, pp. 523-524].

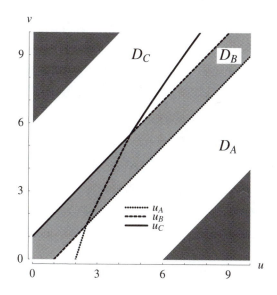

Figure 1.8. The decision set D when $\alpha = 10$. The lighter shaded region is D_B; the darker region lies outside D.

define three subsets of D by

(1.45a) $\qquad D_A = \big\{ (u, v) \mid u \leq \alpha,\, v \geq 0,\, 1 \leq u - v \leq 6 \big\}$

(1.45b) $\qquad D_B = \big\{ (u, v) \mid 0 \leq u, v \leq \alpha,\, |u - v| \leq 1 \big\}$

(1.45c) $\qquad D_C = \big\{ (u, v) \mid u \geq 0,\, v \leq \alpha,\, 1 \leq v - u \leq 6 \big\},$

so that $D = D_A \cup D_B \cup D_C$. For $\alpha = 10$, D is depicted in Figure 1.8. The lighter shaded region is D_B; the darker region lies outside D.

If we assume that customers are uniformly distributed throughout the subdivision, then $G(s)$ is readily calculated with the help of Figure 1.9 (on which the coordinates of some representative points are marked), because $\text{Prob}(X + Y \leq s)$ is just the fraction of the total area of the subdivision that lies below the line $x + y = s$. Suppose, for example, that $0 \leq s \leq 5$. Then the area below the lowest of the thick solid lines in Figure 1.9 is $\frac{1}{2}s^2$; and so the fraction of total area below the line is $\frac{1}{100}s^2$ (because the populated area is 50 square miles). Or suppose that $5 \leq s \leq 7$. Then, similarly, the fraction of total area below the second lowest of the solid lines in Figure 1.9 is $\frac{1}{10}s - \frac{1}{4}$.

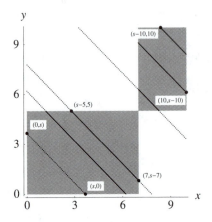

Figure 1.9. Calculation of G defined by (1.38). See text for discussion.

Continuing in this manner, we readily find that

$$(1.46) \qquad G(s) = \begin{cases} \frac{1}{100}s^2 & \text{if } 0 \le s \le 5 \\ \frac{2s-5}{20} & \text{if } 5 \le s \le 7 \\ \frac{7}{10} - \frac{1}{100}(12-s)^2 & \text{if } 7 \le s \le 12; \end{cases}$$

$0 \le s \le 5$ corresponds to D_A, $5 \le s \le 7$ to D_B and $7 \le s \le 12$ to D_C. Because San has a monopoly over the upper rectangle in Figure 1.9, $G(s)$ is not needed for $s \ge 12$ (but see Exercise 1.7).

We can now obtain the rational reaction sets, which are defined by (1.32) with $(u, v) \in D$ in place of $u \in \Delta, v \in \Delta$. That is,

$$(1.47a) \qquad R_1 = \left\{ (u, v) \in D \mid f_1(u, v) = \max_{\overline{u}} f_1(\overline{u}, v) \right\}$$

$$(1.47b) \qquad R_2 = \left\{ (u, v) \in D \mid f_2(u, v) = \max_{\overline{v}} f_2(u, \overline{v}) \right\}.$$

First we find R_1. From (1.41), (1.43) and (1.46),

$$(1.48a) \qquad f_1(u, v) = \frac{cu}{25} \begin{cases} (v - u + 6)^2 & \text{if } (u, v) \in D_A \\ 5(2v - 2u + 7) & \text{if } (u, v) \in D_B \\ 70 - (u - v + 6)^2 & \text{if } (u, v) \in D_C \end{cases}$$

$$(1.48b) \quad f_2(u,v) = \frac{cv}{25} \begin{cases} (u-v+4)(v-u+16) & \text{if } (u,v) \in D_A \\ 5(2u-2v+13) & \text{if } (u,v) \in D_B \\ 30 + (u-v+6)^2 & \text{if } (u,v) \in D_C. \end{cases}$$

From (1.48), if $(u,v) \in D_A$, so that $v+1 \le u \le v+6$, then

$$(1.49) \qquad \partial f_1/\partial u = \tfrac{1}{25}c(v-u+6)(v-3u+6),$$

which is positive for $u < \tfrac{1}{3}(v+6)$ and negative for $\tfrac{1}{3}(v+6) < u < (v+6)$. Thus f_1 has its maximum for $u \le v+6$ where $u = \tfrac{1}{3}(v+6)$. If $\tfrac{1}{3}(v+6) \le v+1$, however, then the maximum for $v+1 \le u \le v+6$ will occur where $u = v+1$. In other words, the maximum of f_1 over the region D_A occurs at $u = u_A(v)$, where

$$(1.50a) \qquad u_A(v) = \begin{cases} \tfrac{1}{3}v + 2 & \text{if } 0 \le v \le \tfrac{3}{2} \\ v+1 & \text{if } \tfrac{3}{2} \le v \le \alpha. \end{cases}$$

The curve $u = u_A$ is represented in Figure 1.8 by dotted lines. If $(u,v) \in D_B$, so that $v-1 \le u \le v+1$, then $\partial f_1/\partial u$ is positive for $u < \tfrac{1}{4}(2v+7)$ and negative for $u > \tfrac{1}{4}(2v+7)$; so that the maximum of f_1 over the region D_B occurs at $u = u_B(v)$, where

$$(1.50b) \qquad u_B(v) = \begin{cases} v+1 & \text{if } 0 \le v \le \tfrac{3}{2} \\ \tfrac{1}{2}v + \tfrac{7}{4} & \text{if } \tfrac{3}{2} \le v \le \tfrac{11}{2} \\ v-1 & \text{if } \tfrac{11}{2} \le v \le \alpha. \end{cases}$$

Similarly (Exercise 1.8), the maximum of f_1 over the region D_C occurs at $u = u_C(v)$, where

$$(1.50c) \quad u_C(v) = \begin{cases} v-1 & \text{if } 1 \le v \le \tfrac{11}{2} \\ \tfrac{1}{3}\big(2v + \sqrt{(v-6)^2 + 210}\big) - 4 & \text{if } \tfrac{11}{2} \le v \le \alpha. \end{cases}$$

The curves $u = u_C$ and $u = u_B$ are shown solid and dashed, respectively, in Figure 1.8; although $u = u_C$ appears to consist of straight line segments, for $v \ge \tfrac{11}{2}$ it has a slight downward curvature.

Now, for any $v \in D$, the maximum of $f_1(u,v)$ as a function of u is found by comparing the maximum over D_A with the maxima over D_B and D_C to identify which of those three numbers is largest:

$$(1.51) \quad \max_{\bar{u}} f_1(\bar{u},v) = \max\big\{f_1(u_A(v),v),\ f_1(u_B(v),v),\ f_1(u_C(v),v)\big\}.$$

Thus, from (1.47a) and Figure 1.8,

$$(1.52) \quad R_1 \; = \; \left\{ (u_A(v), v) \mid 0 \leq v \leq \tfrac{3}{2} \right\}$$
$$\cup \left\{ (u_B(v), v) \mid \tfrac{3}{2} \leq v \leq \tfrac{11}{2} \right\}$$
$$\cup \left\{ (u_C(v), v) \mid \tfrac{11}{2} \leq v \leq \alpha \right\}.$$

To verify (1.52), suppose, for example, that $0 \leq v \leq \tfrac{3}{2}$. Then f_1 is larger along $u = u_A(v)$ than elsewhere in D_A, including the boundary with D_B; but because this boundary is where f_1 is maximized on D_B (for $0 \leq v \leq \tfrac{3}{2}$), f_1 must be larger along $u = u_A(v)$ than elsewhere in both $D_A \cup D_B$, including its boundary with D_C (when $1 \leq v \leq \tfrac{3}{2}$); but because this boundary is where f_1 is maximized on D_C (for $1 \leq v \leq \tfrac{3}{2}$), f_1 must be larger along $u = u_A(v)$ than anywhere else in $D_A \cup D_B \cup D_C = D$ (for $0 \leq v \leq \tfrac{3}{2}$). Similarly for $\tfrac{3}{2} \leq v \leq \alpha$. The result for $\alpha = 10$ is sketched in Figure 1.11(a). Note in particular that Nan's rational reaction to $v = 0$ would be $u = 2$. Thus, even if San were to give away the product ($p_2 = 0$), Nan should still charge $p_1 = 8c$ for it, because she would still attract customers who reside south or west of the line $x + y = 4$.

Although R_1 is *connected*—it's all in one piece—connectedness is not a general property of rational reaction sets. To see this, note that if $\alpha = 10$, and if the maxima of f_2 over subsets D_A, D_B and D_C of D occur where $v = v_A(u)$, $v = v_B(u)$ and $v = v_C(u)$, respectively, then from Exercise 1.9 we have

$$(1.53a) \quad v_A(u) = \begin{cases} u - 1 & \text{if } 1 \leq u \leq \tfrac{17}{2} \\ \tfrac{1}{3}\left(2u + \sqrt{(u-6)^2 + 300}\right) - 4 & \text{if } \tfrac{17}{2} \leq u \leq 10 \end{cases}$$

because $\max(0, u - 6) \leq v \leq u - 1$ for $(u, v) \in D_A$;

$$(1.53b) \quad v_B(u) = \begin{cases} u + 1 & \text{if } 0 \leq u \leq \tfrac{9}{2} \\ \tfrac{1}{2}u + \tfrac{13}{4} & \text{if } \tfrac{9}{2} \leq u \leq \tfrac{17}{2} \\ u - 1 & \text{if } \tfrac{17}{2} \leq u \leq 10 \end{cases}$$

because $\max(0, u - 1) \leq v \leq u + 1$ for $(u, v) \in D_B$; and

$$(1.53c) \quad v_C(u) = \begin{cases} u + 6 & \text{if } 0 \leq u \leq 4 \\ 10 & \text{if } 4 \leq u \leq \tfrac{9}{2} \\ u + 1 & \text{if } \tfrac{9}{2} \leq u \leq 9 \end{cases}$$

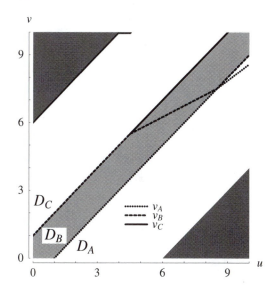

Figure 1.10. A set of points containing R_2 for $\alpha = 10$. The lighter shaded region is D_B; the darker region lies outside D.

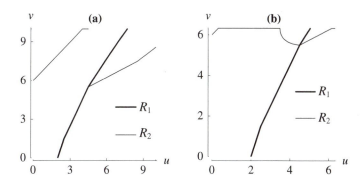

Figure 1.11. R_1 and R_2 for **(a)** $\alpha = 10$ and **(b)** $\alpha = 2\sqrt{10}$.

because $u + 1 \leq v \leq u + 6$ for $(u,v) \in D_C$. Note that $v = v_C(u)$ is not strictly a function but rather a multi-valued function; it is double-valued at $u = \frac{9}{2}$, because $v = \frac{11}{2}$ and $v = 10$ both maximize $f_2(9/2, v)$. The graphs of $v = v_A(u), v = v_B(u)$ and $v = v_C(u)$ are depicted in Figure 1.10. It now follows from

$$(1.54) \quad \max_{\overline{v}} f_2(u, \overline{v}) = \max\{f_2(u, v_A(u)), \ f_2(u, v_B(u)), \ f_2(u, v_C(u))\}$$

and Figure 1.10 that San's rational reaction set is

$$(1.55) \quad R_2 = \{(u, v_C(u)) \mid 0 \le u \le \tfrac{9}{2}\}$$
$$\cup \{(u, v_B(u)) \mid \tfrac{9}{2} \le u \le \tfrac{17}{2}\}$$
$$\cup \{(u, v_A(u)) \mid \tfrac{17}{2} \le u \le 10\}.$$

R_2 is sketched in Figure 1.11(a). Because the maximum of $f_2(9/2, v)$ $= \tfrac{1}{25}c\{30 + \left(v - \tfrac{21}{2}\right)^2\}$ for $\tfrac{11}{2} \le v \le 10$, namely, $\tfrac{121c}{10}$ occurs at both ends of the interval, and because $f_2(9/2, v)$ is less than $\tfrac{121c}{10}$ at every intermediate point, the rational reaction set is disconnected along $u = \tfrac{9}{2}$; it contains both $\left(\tfrac{9}{2}, \tfrac{11}{2}\right)$ and $\left(\tfrac{9}{2}, 10\right)$, but no points that lie in between. Nevertheless, R_1 and R_2 still intersect one another at the (only) Nash equilibrium $(u^*, v^*) = \left(\tfrac{9}{2}, \tfrac{11}{2}\right)$. If this is accepted as the solution of the noncooperative game then, from (1.40), Nan's price is $p_1 = 18c$ and San's price is $p_2 = 22c$.

This result is strongly dependent on the value we chose for α. Indeed $\alpha = 10$ has a critical property: it is the largest value of α for which a Nash equilibrium exists. As α increases beyond 10, the left endpoint of the right-hand segment of R_2 moves away from $D_B \cap D_C$ into the interior of D_B, so that $R_1 \cap R_2 = \varnothing$, the empty set. As α moves below 10, on the other hand, the same endpoint moves into the interior of D_C, and there is a second critical value, namely, $\alpha = 2\sqrt{10}$, at which R_2 becomes connected; for this value of α, R_1 and R_2 are sketched in Figure 1.11(b). These results are best left to the exercises, however; see Exercises 1.10-1.12. Then try Exercise 1.26.

1.6. Store Wars II: a three-player game

We could easily turn Store Wars into a 3-player, noncooperative game by placing a third store, say Zan's, somewhere else in the subdivision, perhaps at the northeast corner; but it would significantly complicate the mathematics. Therefore, we shall devise an example of a 3-player game by supposing instead that the interior of some circular island is uninhabitable (perhaps because of a volcano), so that all prospective

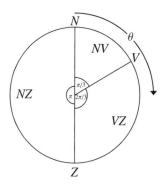

Figure 1.12. Map of battleground for Store Wars II.

customers for a certain product must reside on the island's circumference. To be specific, let us suppose that Nan's store is at the most northerly point of the island, and that Van's store is east of Nan's and one third of the way from Nan's store to the most southerly point of the island, which is also the location of the third store, Zan's; Nan is Player 1, Van Player 2 and Zan Player 3. Let the radius of the island be a miles; and let $a\theta$ denote distance along the circumference, measured clockwise from the most northerly point. Then $0 \le \theta < 2\pi$, and the location of a customer's residence is determined by her θ-coordinate, with Nan's store at $\theta = 0$, Van's at $\theta = \pi/3$ and Zan's at $\theta = \pi$; see Figure 1.12. We will call this game Store Wars II.

We will suppose that customers are uniformly distributed along the circumference. Thus if Θ denotes the θ-coordinate of a randomly chosen customer, then

$$(1.56) \qquad \text{Prob}(0 \le \theta_1 < \Theta < \theta_2 < 2\pi) \;=\; \tfrac{1}{2\pi}(\theta_2 - \theta_1).$$

For example, if NV denotes the event that Θ lies between 0 and $\pi/3$, VZ the event that Θ lies between $\pi/3$ and π, and NZ the event that Θ lies between π and 2π (see Figure 1.12), then from (1.56) we have

$$(1.57) \qquad \text{Prob}(NV) = \tfrac{1}{6}, \quad \text{Prob}(VZ) = \tfrac{1}{3}, \quad \text{Prob}(NZ) = \tfrac{1}{2}.$$

Let p_i denote Player i's price for the product in question, for $i = 1, 2, 3$. Then we shall assume, as in §1.5, that the difference in prices between adjacent stores does not exceed the round-trip cost of travel between them. Thus if travel costs c dollars per mile, then

(1.58) $|p_1 - p_2| \leq \frac{2\pi ac}{3}$, $|p_2 - p_3| \leq \frac{4\pi ac}{3}$, $|p_1 - p_3| \leq 2\pi ac$.

As in §1.5, there are lower and upper bounds on the prices:

(1.59) $8\pi ac\beta \leq p_i \leq 8\pi ac\alpha$, $i = 1, 2, 3$.

But again as in §1.5, we shall assume throughout that $\beta = 0$.

Now, let Θ be the residential coordinate of the next customer (hence $0 \leq \Theta < 2\pi$); and suppose, as in §1.5, that this customer selects a store solely by weighing the price of the product against the cost of travel from her residence. Then, in view of (1.58), she will always buy from one of the two stores between which she lives. For example, the customer will buy from Nan if she resides in the sector denoted by NV in Figure 1.12 and the total cost of buying from Nan is less than the total cost of buying from Van, i.e., if $0 < \Theta < \pi/3$ and $p_1 + 2ac\Theta < p_2 + 2ac(\pi/3 - \Theta)$ or, equivalently, $0 < \Theta < \pi/6 + (p_2 - p_1)/4ac$.[14] The customer will also buy from Nan, however, if she resides in the sector denoted by NZ in Figure 1.12 and the total cost of buying from Nan is less than the total cost of buying from Zan, i.e., if $\pi < \Theta < 2\pi$ and $p_1 + 2ac(2\pi - \Theta) < p_3 + 2ac(\Theta - \pi)$ or, equivalently, $3\pi/2 + (p_1 - p_3)/4ac < \Theta < 2\pi$. As usual, we need not worry about the event that, for example, $p_1 + 2ac\Theta$ equals $p_2 + 2ac(\pi/3 - \Theta)$ precisely, because the event is associated with probability zero. Thus the next customer will buy from Nan if

(1.60) $0 < \Theta < \frac{\pi}{6} + \frac{p_2 - p_1}{4ac}$ or $\frac{3\pi}{2} + \frac{p_1 - p_3}{4ac} < \Theta < 2\pi$.

From (1.56), the probability of this event is

(1.61a) $\frac{1}{2\pi}\left(\frac{\pi}{6} + \frac{p_2 - p_1}{4ac}\right) + \frac{1}{2\pi}\left(2\pi - \left\{\frac{3\pi}{2} + \frac{p_1 - p_3}{4ac}\right\}\right) = \frac{1}{3} + \frac{p_2 - 2p_1 + p_3}{8ac\pi}$.

Similarly, the customer will buy from Van if either $0 < \Theta < \pi/3$ and $p_2 + 2ac(\pi/3 - \Theta) < p_1 + 2ac\Theta$ or $\pi/3 < \Theta < \pi$ and $p_2 + 2ac(\Theta - \pi/3) < p_3 + 2ac(\pi - \Theta)$, i.e., if $\frac{\pi}{6} + \frac{p_2 - p_1}{4ac} < \Theta < \frac{2\pi}{3} + \frac{p_3 - p_2}{4ac}$. From (1.56), the probability of this event is

(1.61b) $\frac{1}{4} + \frac{p_1 - 2p_2 + p_3}{8ac\pi}$.

---

[14]Note that (1.58) ensures $\pi/6 + (p_2 - p_1)/4ac \leq \pi/3$.

A similar calculation (Exercise 1.14) shows that the next customer will buy from Zan if $2\pi/3 + (p_3 - p_2)/4ac < \Theta < 3\pi/2 + (p_1 - p_3)/4ac$, and that the probability of this event is

$$(1.61c) \qquad \frac{5}{12} + \frac{p_1 - 2p_3 + p_2}{8ac\pi}.$$

Of course, the three probabilities in (1.61) must sum to 1.

For $i = 1, 2, 3$, let the random variable F_i denote Player i's payoff from the next customer; its expected value, $f_i = \mathrm{E}[F_i]$, is Player i's reward. By analogy with (1.39), F_1 is p_1 if (1.60) is satisfied and zero otherwise, so that Nan's reward is simply p_1 times (1.61a). Likewise, Van's reward is simply p_2 times (1.61b), and Zan's reward is p_3 times (1.61c). It will be convenient, however, to make prices dimensionless by scaling them with respect to $8\pi ac$. Accordingly, we define strategies u, v and z for Nan, Van and Zan, respectively, by

$$(1.62) \qquad u = p_1/8\pi ac, \quad v = p_2/8\pi ac, \quad z = p_3/8\pi ac.$$

Then the players' rewards are

$$(1.63a) \qquad f_1(u, v, z) = 8\pi acu\left(\tfrac{1}{3} + v - 2u + z\right),$$

$$(1.63b) \qquad f_2(u, v, z) = 8\pi acv\left(\tfrac{1}{4} + u - 2v + z\right),$$

$$(1.63c) \qquad f_3(u, v, z) = 8\pi acz\left(\tfrac{5}{12} + u - 2z + v\right);$$

and from (1.58) and (1.59) with $\beta = 0$, the players' decision set D consists of all (u, v, z) such that

$$(1.64) \qquad |u - v| \le \tfrac{1}{12}, \ |v - z| \le \tfrac{1}{6}, \ |u - z| \le \tfrac{1}{4}, \ 0 \le u, v, z \le \alpha.$$

Extending (1.47) in the obvious way, rational reaction sets R_1, R_2 and R_3 for a 3-player, noncooperative game are defined by

$$(1.65a) \qquad R_1 = \left\{(u, v, z) \in D \mid f_1(u, v, z) = \max_{\overline{u}} f_1(\overline{u}, v, z)\right\}$$

$$(1.65b) \qquad R_2 = \left\{(u, v, z) \in D \mid f_2(u, v, z) = \max_{\overline{v}} f_2(u, \overline{v}, z)\right\}$$

$$(1.65c) \qquad R_3 = \left\{(u, v, z) \in D \mid f_3(u, v, z) = \max_{\overline{z}} f_3(u, v, \overline{z})\right\}.$$

Furthermore, the strategy combination (u^*, v^*, z^*) is a Nash equilibrium if u^* is a best reply to (v^*, z^*), v^* a best reply to (u^*, z^*) and z^* a best reply to (u^*, v^*); i.e., if Player 1 has nothing to gain by selecting $u \ne u^*$ when Players 2 and 3 have already selected (v^*, z^*), Player 2 has nothing to gain by selecting $v \ne v^*$ when Players 1 and 3 have already selected (u^*, z^*) and Player 3 has nothing to gain by selecting $z \neq z^*$ when Players 1 and 2 have already selected (u^*, v^*). Thus (u^*, v^*, z^*) is a Nash equilibrium if $f_1(u^*, v^*, z^*) \geq f_1(u, v^*, z^*)$ for all $(u, v^*, z^*) \in D, f_2(u^*, v^*, z^*) \geq f_2(u^*, v, z^*)$ for all $(u^*, v, z^*) \in D$ and $f_3(u^*, v^*, z^*) \geq f_3(u^*, v^*, z)$ for all $(u^*, v^*, z) \in D$ or

$$(1.66) \qquad\qquad (u^*, v^*, z^*) \in R_1 \cap R_2 \cap R_3.$$

To obtain R_1, we must maximize f_1 as a function of u for all v and z, subject to the constraint that $(u, v, z) \in D$; to obtain R_2, we must maximize f_2 as a function of v for all u and z, subject to the same constraint; and similarly for R_3. Typically, the rational reaction sets of a 3-player game are much more difficult to calculate and visualize than those of a 2-player game; but in the particular case of Store Wars II they are all readily calculated—at least when α is sufficiently large, which we assume henceforward to simplify the analysis. It will also help to simplify matters if we define quantities $\hat{u}, \hat{v}$ and $\hat{z}$ by

$$(1.67a) \qquad\qquad \hat{u} = \tfrac{1}{4}\left(\tfrac{1}{3} + v + z\right)$$

$$(1.67b) \qquad\qquad \hat{v} = \tfrac{1}{4}\left(\tfrac{1}{4} + u + z\right)$$

$$(1.67c) \qquad\qquad \hat{z} = \tfrac{1}{4}\left(\tfrac{5}{12} + u + v\right).$$

First we calculate R_3. From (1.64) in the limit as $\alpha \to \infty$, for all u, v satisfying $|u - v| \leq \tfrac{1}{12}$ we must maximize f_3 as a function of z subject to $u - \tfrac{1}{4} \leq z \leq u + \tfrac{1}{4}, v - \tfrac{1}{6} \leq z \leq v + \tfrac{1}{6}$ and $z \geq 0$ or

$$(1.68a) \qquad \max\left(0, u - \tfrac{1}{4}, v - \tfrac{1}{6}\right) \leq z \leq \min\left(u + \tfrac{1}{4}, v + \tfrac{1}{6}\right).$$

But $|u - v| \leq \tfrac{1}{12}$ implies both $v + \tfrac{1}{6} \leq u + \tfrac{1}{4}$ and $v - \tfrac{1}{6} \geq u - \tfrac{1}{4}$, so that (1.68a) reduces to

$$(1.68b) \qquad \max\left(0, v - \tfrac{1}{6}\right) \leq z \leq v + \tfrac{1}{6}.$$

Moreover, it is straightforward (Exercise 1.15) to show that the maximum of f_3 on the interval $[0, \infty)$ occurs at $z = \hat{z}$, where $\hat{z}$ is defined by (1.67c); and that $|u - v| \leq \tfrac{1}{12}$ implies $\hat{z} \leq \tfrac{1}{2}v + \tfrac{1}{8}$, which in turn implies $\hat{z} < v + \tfrac{1}{6}$. Thus, because f_3 is increasing on $[0, \hat{z}]$ and decreasing on $[\hat{z}, \infty)$, the maximum of f_3 on subinterval (1.68b) must occur at $z = \max\left(0, v - \tfrac{1}{6}\right)$ if $\max\left(0, v - \tfrac{1}{6}\right) \geq \hat{z}$, but at $z = \hat{z}$ if $\max\left(0, v - \tfrac{1}{6}\right) < \hat{z}$. Because zero cannot exceed a positive number, the first of these two inequalities is satisfied if, and only if, $v - \tfrac{1}{6} \geq \hat{z}$

Table 1.11. R_1 for Store Wars II.

u	v	z
$v^A + \frac{1}{12}$	v^A	z^A
$\frac{1}{4}\left(\frac{1}{3} + v^B + z^B\right)$	v^B	z^B
$v^C - \frac{1}{12}$	v^C	z^C

Table 1.12. R_2 for Store Wars II.

u	v	z
u^D	$u^D + \frac{1}{12}$	z^D
u^E	$\frac{1}{4}\left(\frac{1}{4} + u^E + z^E\right)$	z^E
u^F	$z^F - \frac{1}{6}$	z^F
u^G	$u^G - \frac{1}{12}$	z^G

Table 1.13. R_3 for Store Wars II.

u	v	z
u^H	v^H	$\frac{1}{4}\left(\frac{5}{12} + u^H + v^H\right)$
u^J	v^J	$v^J - \frac{1}{6}$

or $36v \geq 12u + 13$ (in addition to $|u - v| \leq \frac{1}{12}$). In other words, the first inequality is satisfied when the point (u, v) belongs to region J of Figure 1.13(c), which extends all the way to infinity in the northeasterly direction; then f_3 is maximized by $z = v - \frac{1}{6}$ (as indicated in Figure 1.13). Correspondingly, f_3 is maximized by $z = \hat{z}$ when $36v < 12u + 13$, or when (u, v) belongs to region H. Thus R_3 contains the strategy combinations in Table 1.13 (where (u^K, v^K) is an arbitrary point in region K of Figure 1.13(c), for $K = H, J$).

Next we calculate R_1. From (1.64) in the limit as $\alpha \to \infty$, for all v, z satisfying $|v - z| \leq \frac{1}{6}$ we must maximize f_1 as a function of u subject to $v - \frac{1}{12} \leq u \leq v + \frac{1}{12}$, $z - \frac{1}{4} \leq u \leq z + \frac{1}{4}$ and $u \geq 0$ or

(1.69a) $\qquad \max\left(0, v - \frac{1}{12}, z - \frac{1}{4}\right) \leq u \leq \min\left(v + \frac{1}{12}, z + \frac{1}{4}\right).$

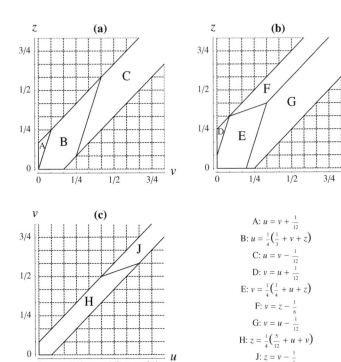

Figure 1.13. Best replies to other players' prices in Store Wars II for **(a)** Player 1, **(b)** Player 2 and **(c)** Player 3. In each case, the triangular regions criss-crossed by dashed grid lines lie outside the decision set. In **(a)**, region B is bounded by parallel line segments with equations $z = 3v$ between $(0,0)$ and $(1/12, 1/4)$ and $9v = 3z + 2$ between $(1/4, 1/12)$ and $(5/12, 7/12)$. In **(b)**, region E is bounded by parallel line segments with equations $12z = 36u + 1$ from $(0, 1/12)$ to $(1/12, 1/3)$ and $36u = 12z + 7$ from $(7/36, 0)$ to $(1/3, 5/12)$, and by a line segment with equation $36z = 12u + 11$ between $(1/12, 1/3)$ and $(1/3, 5/12)$; and region F is separated from region G by a line segment with equation $12z = 12u + 1$ extending from $(1/3, 5/12)$ to infinity. In **(c)**, regions H and J are separated by a line segment with equation $36v = 12u + 13$ from $(5/12, 1/2)$ to $(2/3, 7/12)$.

But $v \leq z + \frac{1}{6}$ implies $v + \frac{1}{12} \leq z + \frac{1}{4}$ and $v \geq z - \frac{1}{6}$ implies $v - \frac{1}{12} \leq z - \frac{1}{4}$, so that (1.69a) reduces to

$$(1.69b) \qquad \max\left(0, v - \tfrac{1}{12}\right) \leq u \leq v + \tfrac{1}{12}.$$

Again, because the maximum of f_1 on $[0, \infty)$ occurs at $u = \hat{u}$ and f_1 is increasing on $[0, \hat{u}]$ but decreasing on $[\hat{u}, \infty)$, the maximum of f_1 for (1.69b) must occur where $u = \max\left(v - \frac{1}{12}, 0\right)$ if $\max\left(v - \frac{1}{12}, 0\right) \geq \hat{u}$, where $u = v + \frac{1}{12}$ if $v + \frac{1}{12} \leq \hat{u}$ and where $u = \hat{u}$ if $\max\left(v - \frac{1}{12}, 0\right) < \hat{u} < v + \frac{1}{12}$. The first of these three inequalities is satisfied where $v - \frac{1}{12} \geq \hat{u}$ or $9v \geq 3z + 2$ (in addition to $|v - z| \leq \frac{1}{6}$), i.e., where (v, z) belongs to region C of Figure 1.13(a); then f_1 is maximized by $u = v - \frac{1}{12}$. Similarly, the second inequality is satisfied when $z \geq 3v$ or (v, z) belongs to region A of Figure 1.13(a), with f_1 maximized by $u = v + \frac{1}{12}$; and the remaining pair of inequalities is satisfied when (v, z) belongs to region B with f_1 maximized by $u = \hat{u}$. Thus R_1 contains the strategy combinations in Table 1.11 (where (v^K, z^K) is an arbitrary point in region K of Figure 1.13(a), for $K = A, B, C$). A similar calculation shows that R_2 contains the strategy combinations in Table 1.12 (where (u^K, z^K) is an arbitrary point in region K of Figure 1.13(b), for $K = D, E, F, G$); see Exercise 1.15.

To satisfy $(u^*, v^*, z^*) \in R_1 \cap R_2 \cap R_3$, we must identify all possible ways in which a row from Table 1.11 can also lie in both Table 1.12 and Table 1.13. There are three ways to choose the row from Table 1.11, four ways to choose the row from Table 1.12 and two ways to choose the row from Table 1.13, yielding 24 choices in all. It turns out, however, that 23 of these choices are impossible; see Exercise 1.16. For example, we cannot match the first rows of Tables 1.11 and 1.12 because $u = v^A + \frac{1}{12}$ implies $u \geq \frac{1}{12}$, which—by inspection of Figure 1.13—allows $u = u^D$ only if $z^D = \frac{1}{3}$; which precludes $z^D = z^A$, because $z^A = \frac{1}{3}$ is impossible. This argument rules out two of the 24 choices, because it is valid for either row from Table 1.13. More obviously, we cannot match the second row of Table 1.12 (which requires $u \leq 1/3$) with the second row of Table 1.13 (which requires $u \geq 5/12$), regardless of the row we choose from Table 1.11; this observation rules out a further three possibilities. Continuing in this manner, we find that the only legitimate possibility is for the second rows of Tables 1.11 and 1.12 to match the first row of Table 1.13. Hence (Exercise

1.16) the unique Nash equilibrium is $(u^*, v^*, z^*) = \left(\frac{1}{6}, \frac{3}{20}, \frac{11}{60}\right)$. Because of its uniqueness, we can safely regard this equilibrium as the solution of Store Wars II; in which case, Nan should charge $\frac{4}{3}\pi ac$ dollars, Van $\frac{6}{5}\pi ac$ dollars and Zan $\frac{22}{15}\pi ac$ dollars for the product in question.

The concepts of rational reaction set and Nash equilibrium generalize in the obvious way to n-player noncooperative games. Let the players correspond to the integers between 1 and n and set $N = \{1, 2, ..., n\}$. Let Player k's strategy be denoted by w^k, for all $k \in N$; thus, for example, in Store Wars II we have $w^1 = u, w^2 = v$ and $w^3 = z$. Possibly w^k is a vector; for example, in Four Ways we have $w^1 = (u_1, u_2)$ and $w^2 = (v_1, v_2)$. Let $w = (w^1, w^2, \ldots, w^n)$ be the players' joint strategy combination. Note that w is a "vector of vectors"—if Player k controls s_k variables, i.e., w^k is an s_k-dimensional vector, then the dimension of w is $s_1 + s_1 + \ldots + s_n$. The n rewards can now be written succinctly as $f_1(w), f_2(w), \ldots, f_n(w)$. Let $w\|\overline{w}^k$ denote the joint strategy combination that is identical to w except for Player k's strategy, which is $\overline{w}^k$, i.e., define

$$(1.70) \qquad w\|\overline{w}^k = (w^1, \ldots, w^{k-1}, \overline{w}^k, w^{k+1}, \ldots, w^n);$$

thus, in particular, $w\|w^k = w$. Let the set of all feasible w—the decision set—be denoted as usual by D. Then, for $k \in N$, Player k's rational reaction set is defined by

$$(1.71) \qquad R_k = \{w \in D \mid f_k(w) = \max_{\overline{w}^k} f_k(w\|\overline{w}^k)\}.$$

If we define $w\backslash w^k$ to be that part of the joint strategy combination which is not under the control of Player k, i.e., if we define $w\backslash w^k = (w^1, \ldots, w^{k-1}, w^{k+1}, \ldots, w^n)$, then $w^* = ((w^*)^1, (w^*)^2, \ldots, (w^*)^n)$ is a Nash equilibrium if, for all $k \in N$, Player k's $(w^*)^k$ is a best reply to the other players' $w^*\backslash(w^*)^k$: no player has an incentive to deviate from her Nash-equilibrium strategy if all other players adhere to theirs. In other words, w^* is a Nash equilibrium if, for any $\overline{w} \in D$,

$$(1.72) \qquad\qquad f_k(w^*) \geq f_k(w^*\|\overline{w}^k)$$

for all $k \in N$; or, which is the same thing, if

$$(1.73) \qquad\qquad w^* \in R_1 \cap R_2 \cap \ldots \cap R_n.$$

Furthermore, if (1.72) is satisfied with strict inequality (for $\overline{w}^k \neq (w^*)^k$) for all $k \in N$, then we say that w^* is a *strong* Nash equilibrium (see, e.g., Exercises 1.24, 1.27 and 1.30); and otherwise w^* is a *weak* Nash equilibrium.

For games with more than three players, it is usually a difficult problem to calculate rational reaction sets and Nash equilibria (and even for 2- and 3-player games the exercise need not be trivial). We will therefore not dwell on such cases. There do exist theorems on necessary conditions for Nash equilibria, however; see, for example, [**230**]. These theorems enable one to eliminate points in D that are not equilibria; and remaining candidates can be tested for the Nash equilibrium property by applying definition (1.72), as in Exercise 1.16.

1.7. Max-min strategies

Our studies of bimatrix games have revealed a difficulty with Nash equilibrium as a solution concept for noncooperative games: a game may have more than one equilibrium. In our studies of continuous games, however, we have seen that a Nash equilibrium *can* be unique; in which case, is it not reasonable to regard it as the game's solution? Not necessarily. A further difficulty with our concept of Nash equilibrium is that we have had to assume complete information: every player must know every other player's reward function. But suppose that each player knows only her own reward function—what is then her best strategy? A possible answer involves the concept of max-min strategy, which we describe in this section.

Let us first define the *minimizing functions, $m_1, m_2, \ldots, m_n$* by

$$(1.74) \qquad m_k\big(w^k\big) = \min_{\overline{w}:\,\overline{w}^k = w^k} f_k\big(\overline{w}\big).$$

For each $k \in N$, $m_k(w^k)$ yields the minimum value of Player k's reward with respect to the variables controlled by the other players, i.e., $w \backslash w^k$. In particular, for a 2-player game we have

$$(1.75) \qquad m_1(u) = \min_v f_1(u, v), \quad m_2(v) = \min_u f_2(u, v).$$

Suppose, for example, that San is an intermediate or slow driver in Crossroads, for which Nan's reward function is

$$(1.76) \quad f_1(u, v) = \big(\epsilon + \tfrac{1}{2}\tau_2 - \{\delta + \epsilon\}v\big)u + \big(\epsilon - \tfrac{1}{2}\tau_2\big)v - \epsilon - \tfrac{1}{2}\tau_2,$$

from (1.13). Then because $\epsilon < \frac{1}{2}\tau_2$, the coefficient of v in f_1, namely, $\epsilon - \tau_2/2 - (\delta + \epsilon)u$, is always negative, and so f_1 is minimized with respect to v by $v = 1$. Thus, from (1.75), Nan's minimizing function is given by

$$(1.77) \qquad m_1(u) \;=\; f_1(u, 1) \;=\; (\tau_2/2 - \delta)u - \tau_2.$$

For $k \in N$, we now define a *max-min* strategy for Player k to be a w^k that maximizes $m_k(w^k)$; we denote this max-min strategy by $\tilde{w}^k$, and we refer to $\tilde{w}$ as a joint max-min strategy combination. In Crossroads with $\epsilon < \tau_2/2$, for example, it is clear from (1.77) that $\tilde{u} = 0$ (always wait) is the unique max-min strategy for Nan if $\tau_2/2 < \delta$, whereas $\tilde{u} = 1$ (always go) is the unique max-min strategy for Nan if $\tau_2/2 > \delta$. Thus (provided $\tau_2/2 \neq \delta$) Nan's max-min strategy when San is intermediate or slow is always a pure strategy: W if San is intermediate, G if San is slow. But a max-min strategy for a matrix game need not be a pure strategy. For example, in Crossroads with $\epsilon > \tau_2/2$ (hence $\delta > \tau_2/2$), it follows from (1.75)-(1.76) that

$$(1.78) \qquad m_1(u) \;=\; \begin{cases} (\epsilon + \frac{1}{2}\tau_2)(u - 1) & \text{if } (\delta + \epsilon)u \le \epsilon - \frac{1}{2}\tau_2 \\ (\frac{1}{2}\tau_2 - \delta)u - \tau_2 & \text{if } (\delta + \epsilon)u > \epsilon - \frac{1}{2}\tau_2, \end{cases}$$

so that Nan's unique max-min strategy when San is fast is the mixed strategy $\tilde{u} = (\epsilon - \tau_2/2)/(\delta + \epsilon)$.[15]

The concept of max-min strategy rests on the idea that, no matter which w^k is chosen, the other players will do their worst by making the reward f_k as small as possible—equal to $m_k(w^k)$, in fact; and it responds by selecting the best of these worst rewards, namely, $m_k(\tilde{w}^k)$. A max-min strategy is a fail-safe strategy. It is absolutely fail-safe when it is a pure strategy (see Exercise 1.17); and it is fail-safe on the average when it is a mixed strategy in a bimatrix game, or when the rewards in a continuous game are expected values. But a max-min strategy is also in general a very pessimistic strategy, because if the other players do not know Player k's reward function, then how could they minimize it—except, perhaps, by chance? Not surprisingly, $\tilde{w}$ rarely belongs to every player's rational reaction set

---

[15]A max-min strategy need not be unique; for example, any strategy would be a max-min strategy for Nan with $\epsilon < \tau_2/2$ and $\delta = \tau_2/2$ because m_1 would then be independent of u, from (1.77). But see Footnote 3.

and frequently belongs to no player's rational reaction set; see Exercise 1.19. Indeed there is no guarantee that $\tilde{w}$ is even feasible, i.e., that $\tilde{w} \in D$; see Exercise 1.23. But there is one important exception: if a 2-player game is zero-sum, ie., if $f_1(u,v) + f_2(u,v) = 0$ for all $(u,v) \in D$, then $\tilde{w} \in R_1 \cap R_2$; see Exercise 1.18. In this very special case, there is no need to argue over the merits of max-min strategies versus Nash-equilibrium strategies because the two coincide. Such happy circumstances are rare, however, in game-theoretic modelling.

1.8. Commentary

In Chapter 1 we have introduced the concepts of pure strategy, payoff matrix (§1.1), mixed strategy, rational reaction set and Nash equilibrium (§§1.3 and 1.6). We have used them to analyze bimatrix games with two (§§1.1-1.3) or three (§1.4) pure strategies, as well as 2-player (§1.5) and 3-player (§1.6) continuous games. We have discovered that neither existence nor uniqueness of Nash equilibrium is assured in general; however, existence is assured for bimatrix games (if we allow mixed strategies). A proof that every bimatrix game has at least one Nash equilibrium strategy combination, based on Nash's [**164**] application of the Brouwer fixed-point theorem to n-player games, appears in Owen [**173**, pp. 127-128]. Existence theorems do not tell us how to compute Nash equilibria, however, and we have seen that this task can be far from trivial—even for 2-player games.

Our treatment of Nash equilibrium, the central concept in noncooperative game theory, has been predicated on complete information; but the concept extends to games of *incomplete* information, in which players do not know their opponents' rewards but are able to quantify their feelings about them. Indeed there exists a large literature on so-called Bayesian Nash equilibrium, which derives from the work of Harsanyi [**89, 90, 91**]; see, e.g., Rasmusen [**187**] and references therein. But this literature is largely against the spirit and beyond the scope of our agenda, and so we have confined our treatment of incomplete information to the concept of max-min strategy (§1.7).

Exercises 1

1. Obtain the reward functions f_1, f_2 for the animals engaged in the Hawk-Dove contest of §1.2. Sketch the rational reaction sets **(a)** when ρ (reproductive value of territory) $> C$ (reproductive cost of injury) and **(b)** when $\rho < C$. Find all Nash equilibria. How is this game related to Crossroads mathematically?[16]

2. Suppose that Crossroads is symmetric with $\tau_1 = \tau = \tau_2$ and that neither driver is especially fast or slow, i.e., $2\delta > \tau > 2\epsilon$. Show directly from the payoff matrices in Tables 1.1 and 1.2 that the pure strategy combinations GG and WW cannot be Nash equilibria. Thus WW cannot be chosen by two rational players, because if either selected W then the other would have an incentive to deviate from W; and similarly for GG.

3. Show that a pair of dominant strategies in a 2-player, noncooperative game is always the unique Nash equilibrium.

4. Show that (1.29) and (1.30) are special cases of (1.15).

5. Verify Tables 1.8 and 1.9.[17]

6. Verify Table 1.10.

7. Verify (1.46), and calculate $G(s)$ for $12 \le s \le 20$. Deduce the probability density function g of the random variable $X+Y$, and verify that g is continuous.

8. Obtain (1.48) and (1.50).

9. Obtain (1.53).

10. Show that Store Wars has no Nash equilibrium if $\alpha = 11$.

11. Find R_2 for Store Wars when $\alpha = 2\sqrt{10}$.

12. According to Figure 1.11(a), when $\alpha = 10$ there is a value of p_1 for which San's rational reaction to an increase in p_1 should be to lower her price. Does this make sense? Interpret.

13. How would the rational reaction sets sketched in Figures 1.11(a) and 1.11(b) differ if β were greater than zero in (1.35d)?

14. Verify (1.61).

15. Verify Tables 1.11-1.13.

---

[16] You can add or subtract the same number from every entry in the payoff matrix without altering the nature of the game.

[17] The line joining $(\omega, 0)$ to (α, β) in Figure 1.6 corresponds to equal coefficients, the point (α, β) to vanishing coefficients.

16. (a) Verify that $(u^*, v^*, z^*) = \left(\frac{1}{6}, \frac{3}{20}, \frac{11}{60}\right)$ is a Nash equilibrium of Store Wars II by showing that it matches a row from each of Tables 1.11-1.13.

(b) Verify that $(u^*, v^*, z^*) = \left(\frac{1}{6}, \frac{3}{20}, \frac{11}{60}\right)$ is a Nash equilibrium of Store Wars II by applying (1.72).

(c) Verify that $(u^*, v^*, z^*) = \left(\frac{1}{6}, \frac{3}{20}, \frac{11}{60}\right)$ is the unique Nash equilibrium of Store Wars II by showing that it is the only strategy combination that lies in each of Tables 1.11-1.13.

17. Show that the rewards to Player 1 and Player 2 from a bimatrix game can never be less than $\max_i\{\min_j a_{ij}\}$ and $\max_j\{\min_i b_{ij}\}$, respectively, where A is Player 1's $m_1 \times m_2$ payoff matrix, B is Player 2's payoff matrix and $0 \leq i \leq m_1, 0 \leq j \leq m_2$.

18. Show that for a 2-player, noncooperative, zero-sum game, a max-min strategy is always a Nash-equilibrium strategy, and vice versa.

19. Use Crossroads to establish that
(a) A 2-player game need not be zero-sum for a joint max-min strategy combination to be a Nash equilibrium.
(b) A joint max-min strategy combination may lie in no player's rational reaction set.
(c) Even if a joint max-min strategy combination does not lie in any player's rational reaction set, it may be equivalent to a Nash equilibrium in terms of associated rewards.

20. Find all max-min strategies for Store Wars II, as described in §1.6.

21. Find all Nash equilibria of Store Wars II in the case where Zan's store is located at $\theta = 3\pi/2$ (as opposed to $\theta = \pi$).

22. Find all max-min strategies for Four Ways.

23. Find all max-min strategies for Store Wars in §1.5.

24. (a) To allow for asymmetry between owner and intruder in a territorial contest, Maynard Smith [**132**, p. 22] adds a third strategy, Bourgeois, to the Hawk-Dove game in §1.2. A Bourgeois plays Hawk if an owner but Dove if an intruder; thus Bourgeois, denoted by B, is a conditional strategy (like C in Four Ways). If, in a contest chosen at random, two contestants are equally likely to be owner or intruder, write

down the payoff matrices for the resulting H-D-B game and
show that BB is a strong Nash equilibrium if $\rho < C$.[18]

(b) How realistically is the asymmetry modelled if the game is
still symmetric?

25. The symmetric, 2-player bimatrix game with payoff matrices $A = \begin{bmatrix} R & S \\ T & P \end{bmatrix}$ and $B = A^T$ satisfying $T > R > P > S$ and $2R > S + T$
is known in the literature of the social and biological sciences as
the "prisoner's dilemma." Sketch the rational reaction sets and
find all Nash equilibria.

26. Store Wars III is a noncooperative game of prices between two
stores, whose potential customers are uniformly distributed along
the line segment $0 \leq x \leq L$. The first store is at $x = a$ and the
second at $x = b$, where $0 \leq a < b \leq L$. Assuming that the price
difference does not exceed the cost of travel between the stores,
i.e, $|p_1 - p_2| \leq 2c(b - a)$, where p_i is Player i's price and c the
unit cost of travel, sketch the players' rational reaction sets and
find all Nash equilibria (if any).

27. (a) Greater sophistication is added to the Hawk-Dove game of
§1.2 by defining a third strategy, R, for Retaliator [**132**, pp.
17-18]. A Retaliator always begins by displaying, but then
escalates and prepares for battle if her opponent escalates;
thus R, like C in Four Ways or Bourgeois in Exercise 1.24, is
a conditional strategy. In confrontations between D and R,
the Retaliator will sometimes intuit that her opponent is a
really a Dove and exploit her by escalating. The probability
that a Retaliator will secure a disputed territory in such a
confrontation is therefore of the form $(1+\lambda)/2$, where $\lambda > 0$
is small (and hence the probability that the Dove will secure
the territory is $\frac{1}{2}(1 - \lambda)$, as opposed to $\frac{1}{2}$ in §1.2). Write
down the payoff matrices for the resulting H-D-R game and
show that RR is a strong Nash equilibrium.

(b) Is R the only Nash-equilibrium strategy?

---

[18]You can show that B is a strong Nash-equilibrium strategy without computing
f_1 or f_2.

28. A noncooperative, bimatrix game with three pure strategies has payoff matrices

$$A = \begin{bmatrix} -\lambda & \rho & -\rho \\ -\rho & -\lambda & \rho \\ \rho & -\rho & -\lambda \end{bmatrix}, \quad B = A^T$$

where $|\lambda|$ is much smaller than ρ. What could this game model? Find all Nash-equilibrium strategies for **(a)** $\lambda > 0$ and **(b)** $\lambda < 0$.

29. In §1.4 we assumed that the payoff associated with the strategy combination $(0,0,0,0)$, or CC, was identical to the payoff associated with the strategy combination $(1,0,1,0)$, or GG; in both cases, the payoff was $-\delta - \tau/2$. In practice, however, if Nan and San both selected pure strategy $C = (0,0)$, then the time they spent negotiating or intimidating each other might not be quite the same as if they had both selected $G = (1,0)$. Accordingly, let us denote the time by μ (instead of δ); we would still expect it to exceed the dithering time associated with strategy combination $(0,1,0,1)$, or WW, and so (1.1) is replaced by $0 < \epsilon < \delta, \mu < \infty$. Because the payoff associated with CC is now $-\mu - \frac{1}{2}\tau$, the payoff matrices are

$$A = \begin{bmatrix} -\delta - \tau/2 & 0 & 0 \\ -\tau & -\epsilon - \tau/2 & -\tau \\ -\tau & 0 & -\mu - \tau/2 \end{bmatrix}$$

and $B = A^T$. Find all Nash-equilibrium strategy combinations for $\epsilon = 2 = \tau$ and **(a)** $\delta = 3$, $\mu = 4$, **(b)** $\delta = 4$, $\mu = 3$.

30. Show that $(1,0)$ and $(0,1)$ are both strong Nash-equilibrium strategy combinations in Crossroads.

31. How does relaxing the assumption $|p_1 - p_2| \leq 2c(b - a)$ in Store Wars III (Exercise 1.26) affect the existence of a unique Nash equilibrium?

32. Here we modify Crossroads (§1.3) to allow a motorist's behavior to depend on her direction of travel, as suggested by [**219**]. Suppose that a town center lies to the north and west of the crossroad in Figure 1.1, whereas the suburbs lie to the south and east; thus a northbound driver is heading (west) into town, but a southbound driver is heading out. Then a strategy is a 2-dimensional vector whose first component is the probability of selecting pure strategy G if heading into town, and whose second component is the probability of selecting pure strategy G if heading out.

Let τ_2 be the junction transit time of *any* southbound driver and τ_1 that of any northbound driver; thus Player 1's transit time is τ_1 when she heads into town but τ_2 when she heads out, and similarly for Player 2. (The only asymmetry between players is directional; the junction may be slower in one direction and faster in the other because, e.g., the road is on a hill.) If no drivers are slow (i.e., $2\delta > \max(\tau_1, \tau_2)$) and the possible directions of travel are equally likely:

   (a) Obtain the reward functions, verifying that the game is symmetric (i.e., satisfies (1.31)).

   (b) Calculate the rational reaction sets, and hence find all Nash-equilibrium strategy combinations.

33. The zero-sum game of Chump is played between two camels, a dromedary (Player 1) and a bactrian (Player 2). Player k must simultaneously flash F_k humps and guess that her opponent will flash G_k. Possible pure strategies (F_k, G_k) satisfy $0 \le F_1, G_2 \le 1$ and $0 \le F_2, G_1 \le 2$. Thus Player 1 has six pure strategies, namely, $(0,1)$, $(0,2)$, $(1,0)$, $(1,1)$, $(1,2)$ and $(0,0)$; and Player 2 likewise has six pure strategies, namely, $(0,1)$, $(1,0)$, $(1,1)$, $(2,0)$, $(2,1)$ and $(0,0)$. If both players are right or wrong, then the game is a draw; but if one is wrong and the other is right, then the first pays $F_1 + F_2$ piasters to the second.

   (a) Write down the payoff matrices A and B for Chump.

   (b) With rewards defined according to (1.15), a strategy for each player becomes a 5-dimensional vector. Verify that, if $u^* = \left(\frac{3}{35}, \frac{18}{35}, \frac{8}{35}, \frac{6}{35}, 0\right)$ and $v^* = \left(\frac{4}{7}, \frac{2}{7}, 0, \frac{3}{35}, \frac{2}{35}\right)$, then (u^*, v^*) is a Nash equilibrium.

Chapter 2

Evolutionary Stability and Other Selection Criteria

We discovered in Chapter 1 that a noncooperative game can have several Nash equilibria. If one of them is to be regarded as the solution of the game, then we must introduce criteria for distinguishing it from all the others. Three such criteria are compared in this chapter.

2.1. Harsanyi and Selten's criterion

A general theory of equilibrium selection has been developed by John Harsanyi and Reinhard Selten [**93**]. It is based on what its authors call the tracing procedure. The theory is elaborate, and to discuss it in depth would divert us too far from our goal; nevertheless, we shall at least suggest the thinking that underlies the theory. Some further details of the tracing procedure appear in Appendix A.

Accordingly, consider once more the game of Crossroads. For the sake of definiteness, let us suppose that

(2.1a)
$$2\delta > \tau_1 > \tau_2 > \delta - \epsilon,$$

or equivalently that

(2.1b)
$$1 > \theta_1 > \theta_2 > \tfrac{1}{2},$$

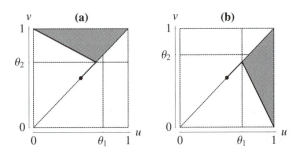

Figure 2.1. Result of players' thought experiment when $p = 1/2 = q$ and **(a)** $1/2 < \theta_2 < \theta_1 < 1$, **(b)** $1/2 < \theta_1 < \theta_2 < 1$.

where θ_1 and θ_2 are defined by (1.22). You will recall that τ_1 is the time it takes our northbound driver, who is called Ned in this chapter, to cross the junction unimpeded; and that τ_2 is the corresponding transit time for our southbound driver, who is called Sed in this chapter. Because $2\delta > \max(\tau_1, \tau_2)$ from (2.1a), neither driver is slow. Let us also suppose that both players know there are three Nash equilibria; and that both are attempting, *without communicating*, to choose a strategy that will make their joint selection an equilibrium.

The two players might achieve their goal by means of the following thought experiment. Let each player guess what is best for the other; specifically, let Ned guess that Sed should select $v = q$, and let Sed guess that Ned should select $u = p$. Then, because Ned is trying to think like a Sed when he selects q for him, and because Sed is trying to think like a Ned when he selects p for him, it is not unreasonable to regard the strategy combination (p, q) as a first *tentative* solution of the game (even if it lies in neither player's rational reaction set). Furthermore, let Ned be aware (for whatever reason) that $u = p$ is the choice that Sed would (tentatively) consider best for him, and let Sed be aware that $v = q$ is the choice that Ned would consider best for him; thus it is common knowledge that (p, q) is the tentative solution. It would be quite unreasonable to regard (p, q) as the firm solution, however, because the u that Sed picks for Ned cannot possibly be as good as the u that Ned would pick for himself; and similarly for Sed. The tentative solution must therefore be revised. How should Ned and Sed revise it?

For the sake of definiteness, let us suppose that $1 > \theta_1 > p > 0$ and $1 > \theta_2 > q > 0$; thus, from Figure 1.5, (p, q) is below, and to the left of, both R_1 and R_2. Because $(1, q) \in R_1$, Ned's best reply to Sed's tentative selection, $v = q$, is $u = 1$. Likewise, because $(p, 1) \in R_2$, Sed's best reply to Ned's tentative selection, $u = p$, is $v = 1$. Does this mean that the players should switch from (p, q) to $(1, 1)$? No, this would be too drastic a correction, because it would thrust each player from a strategy combination that lies below his rational reaction set to a strategy combination that lies above it. On the other hand, it does entice us to believe that Ned should edge toward R_1, and that Sed should edge toward R_2, in the direction of $(1, 1)$. Let us therefore suppose that, without communicating (purely by virtue of a mutual thought experiment), Ned and Sed revise their tentative solution by displacing (u, v) along the straight line that joins (p, q) to $(1, 1)$.

Let Ned and Sed perturb their tentative solution a fraction t of the distance between (p, q) and $(1, 1)$, so that their new tentative solution is at the point $(p + t\{1 - p\}, q + t\{1 - q\})$. A repetition of their thought experiment will now convince them that they should continue to move along the line joining (p, q) to $(1, 1)$ if t is sufficiently small; more precisely, because $1 > \theta_1 > p > 0$ and $1 > \theta_2 > q > 0$, if $t < \min(t_1, t_2)$ where $t_1 = \frac{\theta_2 - q}{1 - q}$ and $t_2 = \frac{\theta_1 - p}{1 - p}$.

Let Ned and Sed perform further repetitions of their fictitious experiment, increasing the value of t continuously. Provided that $t < \min(t_1, t_2)$ is satisfied, they will continue to perturb in the direction of $(1, 1)$. At $t = \min(t_1, t_2)$, however, the tentative solution will reach either R_1 or R_2, according to whether $t_1 < t_2$ or $t_1 > t_2$. Then the player whose rational reaction set has been reached no longer has a single best reply to the other player's (revised) tentative solution; rather, any of his strategies is a best reply. This multiplicity of best replies does not persist, however; t need be increased only infinitesimally beyond $\min(t_1, t_2)$ for the player in question once more to have a single best reply (but 0 instead of 1).

To make things a little clearer, let us suppose that $p = \frac{1}{2} = q$. Thus the tentative initial guess is at the point $\left(\frac{1}{2}, \frac{1}{2}\right)$, and each player believes that the other is equally likely to use either of his two pure strategies. In view of (2.1a), we have $t_1 < t_2$, or $\min(t_1, t_2) = t_1$, so that Ned's rational reaction set is first to be reached, at the point

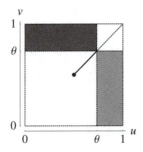

Figure 2.2. Result of experiment for $p = \frac{1}{2} = q, \theta_1 = \theta = \theta_2$.

(θ_2, θ_2); see Figure 2.1(a), where the path traced out by the thought experiment is denoted by a solid black line. At $(\theta_2, \theta_2), v = 1$ is still Sed's unique best reply; but any u such that $0 \leq u \leq 1$ is a best reply for Ned. Thus, if the players continue their train of thought, any $(u, 1)$ could replace $(1, 1)$ as their target point, as indicated by the shaded triangle in Figure 2.1(a). As soon as the tentative solution shifts into this triangle, however, Ned again has a unique best reply, namely, $u = 0$; and Sed's best reply remains $v = 1$. Thus the players would move toward $(0, 1)$. This point lies in both R_1 and R_2, and is therefore a Nash equilibrium. The fictitious experiment has selected $(0, 1)$ as the solution of Crossroads—assuming, of course, that $\theta_1 > \theta_2$.

The briefest of glances at Figure 2.1(b) shows that if the order of the subscripts 1 and 2 in (2.1) were reversed, then the thought process would select $(1, 0)$ as the solution of the game instead. Thus, according to the theory of Harsanyi and Selten, if there is no reason to suppose that either driver should prefer one pure strategy to the other at the outset $(p = \frac{1}{2} = q)$, then the solution of Crossroads is that the slower driver should defer to the faster driver.[1]

At first sight, this solution appears very reasonable; and there is certainly nothing wrong with $p = \frac{1}{2} = q$. Nevertheless, neither player may have reason to suppose that his opponent is either faster or slower; in which case, he should assume that $\tau_1 = \tau_2 = \tau$, say. Then the game becomes symmetric with $\theta_1 = \theta_2 = \theta$, where we recall

---

[1] We have not, however, obtained this solution quite by the method that Harsanyi and Selten advocate; for details of their tracing procedure, see Appendix A.

from (1.34) that $\theta = \frac{2\epsilon + \tau}{2\epsilon + 2\delta}$ (< 1); and a glance at Figure 2.2 now shows that the fictitious experiment would bring the players to the mixed-strategy Nash equilibrium at (θ, θ). Unfortunately, we shall shortly see that this equilibrium may have an undesirable property.

2.2. Kalai and Samet's criterion

Recall that (u^*, v^*) is a Nash equilibrium when, if Player 1 adheres to u^*, then Player 2 cannot increase his reward by selecting a strategy other than v^*; and, if Player 2 adheres to v^*, then Player 1 cannot increase his reward by selecting a strategy other than u^*. Thus (θ, θ) is a Nash equilibrium of the symmetric version of Crossroads because Ned has no better choice than θ if Sed selects θ, and vice versa. Now, let us concede that it is somehow possible to select a mixed strategy; and let us suppose that Sed really wants to select θ, but because of human frailty instead selects $\theta + z$, where $|z|$ is exceedingly small— as small as you please—but not actually zero. In other words, the shaded sector of the disk beneath the spinning arrow in Figure 1.2 is marginally bigger or smaller than he intended it to be. Then $u = \theta$ is no longer a best reply for Ned; rather, $u = 1$ is his best reply if $z < 0$, whereas $u = 0$ is his best reply if $z > 0$. Similar considerations apply, of course, to Sed.

Let the set of all points to which an equilibrium strategy combination, say (u^*, v^*), yields a best reply for both players be known as that equilibrium's *domain of stability*, and denote this domain of stability by $S((u^*, v^*))$. To be quite precise: (u, v) belongs to $S((u^*, v^*))$ if, and only if, for all $(\overline{u}, \overline{v}) \in D$,

$$(2.2) \qquad f_1(u^*, v) \geq f_1(\overline{u}, v) \quad \text{AND} \quad f_2(u, v^*) \geq f_2(u, \overline{v}).$$

From (1.22)-(1.23) with $\theta_1 = \theta_2 = \theta$, f_1 is defined by

$$(2.3\text{a}) \qquad f_1(u, v) = (\delta + \epsilon) \{ (\theta - v)u - \theta(1 + v) \} + 2\epsilon v$$

in the symmetric version of Crossroads; and, by virtue of symmetry, f_2 is defined by

$$(2.3\text{b}) \qquad f_2(u, v) = f_1(v, u) = (\delta + \epsilon) \{ (\theta - u)v - \theta(1 + u) \} + 2\epsilon u.$$

Then the trouble with the equilibrium (θ, θ) is that its domain of stability consists purely of itself. This result follows directly from (2.3), which yields $f_1(\theta, v) - f_1(\overline{u}, v) = (\delta + \epsilon)(\theta - v)(\theta - \overline{u})$ and

$f_2(u, \theta) - f_2(u, \overline{v}) = (\delta + \epsilon)(\theta - u)(\theta - \overline{v})$. Either expression can be made negative for $(\overline{u}, \overline{v}) \in D$ unless $(u, v) = (\theta, \theta)$, and so $S\big((\theta, \theta)\big) = \big\{(\theta, \theta)\big\}$. On the other hand, (2.3) also yields $f_1(1, v) - f_1(\overline{u}, v) = (\delta + \epsilon)(\theta - v)(1 - \overline{u})$ and $f_2(u, 0) - f_2(u, \overline{v}) = (\delta + \epsilon)\overline{v}(u - \theta)$, both of which are nonnegative for all $(\overline{u}, \overline{v}) \in D$ if $\theta \le u \le 1$ and $0 \le v \le \theta$. That is, the domain of stability of the equilibrium $(1, 0)$ is the whole of the lighter shaded rectangle in Figure 2.2 (including its boundary): $S\big((1, 0)\big) = \big\{(u, v) \mid \theta \le u \le 1, 0 \le v \le \theta\big\}$. Similarly, the domain of stability of the equilibrium $(0, 1)$ is the whole of the darker shaded rectangle in Figure 2.2 (again including its boundary), i.e., $S\big((0, 1)\big) = \big\{(u, v) \mid 0 \le u \le \theta, \theta \le v \le 1\big\}$. Thus the equilibria at $(1, 0)$ and $(0, 1)$ do not have the awkward property that if one player makes a small mistake in calculating his Nash-equilibrium strategy, but the other does not, then the latter's strategy is no longer a best reply to the former's.

More formally, following Ehud Kalai and Dov Samet [109], we say that a Nash equilibrium is *persistent* if its domain of stability contains a neighbourhood of the equilibrium—if we can draw a circle of radius z, center the equilibrium, such that the intersection of the circle's interior with the decision set lies completely inside the equilibrium's domain of stability. It does not matter how small the circle is, just so long as its radius is bigger than zero. Thus $(1, 0)$ and $(0, 1)$ are both persistent equilibria; indeed, in either case, the radius z of the circle in question can be as large as $1 - \theta$. But (θ, θ) is not a persistent equilibrium, because no such z exists. Kalai and Samet suggest that we might distinguish among Nash equilibria by eliminating those that are not persistent. This alternative theory of equilibrium selection would therefore suggest that, in the symmetric version of Crossroads, $(1, 0)$ and $(0, 1)$ are both acceptable as equilibria but (θ, θ) is not.

To recapitulate: If we apply the theory of Harsanyi and Selten to the symmetric version of Crossroads, then we select an equilibrium that is not persistent. If, on the other hand, we apply the theory of Kalai and Samet, then we do not know whether $(1, 0)$ or $(0, 1)$ should be the solution of the game. We must therefore decide whether persistence is a desirable property; and if so, then we must somehow break the tie between $(1, 0)$ and $(0, 1)$. Any suggestions?

2.3. Maynard Smith's criterion

While you are still pondering this dilemma, we shall make matters even more intriguing by introducing yet another criterion for equilibrium selection, John Maynard Smith's [**132**]. As in §2.2, we will discuss the criterion for the symmetric version of Crossroads.

Suppose that the following thoughts occur to Ned, who has three Nash-equilibrium strategies, namely, $u^* = 0$, $u^* = \theta$ and $u^* = 1$. From Table 1.5, $u = 1$ is best for Ned if—and only if—Sed selects Nash-equilibrium strategy $v^* = 0$. But why should Sed select $v^* = 0$? No reason at all. On the other hand, if Sed is to select from only three Nash-equilibrium strategies, namely, $v^* = 0$, $v^* = \theta$ and $v^* = 1$, and if Ned doesn't know which strategy Sed will pick, then Ned might as well hope that Sed will choose $v^* = 0$. So let Ned select $u^* = 1$.

Now suppose that Ned's thoughts have also occurred to Sed, who is also rational. Then, clearly, Sed will select strategy $v^* = 1$, which is his best strategy if—and only if—Ned selects $u^* = 0$. But, of course, Ned does not select $u^* = 0$, and Sed does not select $v^* = 0$; rather, they select in this way the strategy combination $(1, 1)$. The associated reward is $f_1(1, 1) = -\frac{1}{2}\tau - \delta = f_2(1, 1)$.

We don't have only one Ned, however, nor only one Sed; rather, we have a huge population of Neds and Seds, all confronting one another across 4-way junctions, all day long, day in, day out, all over the land. All of these Neds and Seds are rational. So why shouldn't the thoughts that have just occurred to our Ned and Sed occur to all of them? No reason at all. So before very long we have a huge population of Neds and Seds, all of whom are playing strategy 1. Indeed there is no longer any reason to distinguish between players by calling one Ned and the other Sed, and so we shall refer to them all as Ed.

Suddenly, one day, it occurs to an Ed that, if everyone but he is playing strategy 1, then there is no longer any uncertainty about the strategy his opponent will choose. Because nobody but this particular Ed has had this brainwave, the next Ed he meets is bound to select $v^* = 1$. Now, the trouble with Nash equilibrium arises because players in a noncooperative game do not know for sure which

Nash-equilibrium strategy their opponent will select. All of a sudden, however, the requisite information is available to an Ed: on the day when he has his brainwave, his reward function reduces to $f_1(u, v) = f_1(u, 1) = (\tau/2 - \delta)u - \tau$. Because $\tau < 2\delta$, $f_1(u, 1)$ is maximized by selecting $u = 0$. Does this mean that the Ed should begin to play $u = 0$? After all, isn't $(0, 1)$ a Nash equilibrium?

It is true that $(0, 1)$ is a Nash equilibrium; but playing $u = 0$ is a rational long-term strategy only if Ed is sure that his next opponent will adhere to $v = 1$. But if one Ed has had this brainwave (and there are so many Eds on the road that, sooner or later, one of them is bound to have the brainwave), do you think he can keep it to himself? Not likely. You know how word gets around. Before very long, all the Eds in the world will have figured out that if everyone else is selecting strategy 1, then they would do better to select strategy 0; 0 is a better reply than 1 to strategy 1 because (from a Ned's point of view) $f_1(0, 1) - f_1(1, 1) = \delta - \tau/2 > 0$ or, which is exactly the same thing (but from a Sed's point of view), $f_2(1, 0) - f_2(1, 1) = \delta - \tau/2 > 0$. But then everyone will be playing strategy 0. Thus the next strategy combination selected will be, not $(0, 1)$, but rather $(0, 0)$; and the associated reward will be $f_1(0, 0) = -\tau/2 - \epsilon = f_2(0, 0)$.

Because $\epsilon < \delta$, we have to admit that the reward associated with $(0, 0)$ is greater than that associated with $(1, 1)$. But the strategy combination $(1, 1)$ did not persist, because an Ed had a brainwave, word got around, and before very long the whole world had evolved to $(0, 0)$. Of course, we should never have expected the world to remain at $(1, 1)$, because $(1, 1)$ is not a Nash equilibrium; indeed it doesn't even lie in an Ed's rational reaction set. Likewise, $(0, 0)$ lies in no Ed's rational reaction set, and so we don't expect the world to remain at $(0, 0)$; it is just as inevitable that some Ed somewhere will try something else as it was when the world stood at $(1, 1)$. This Ed already knows, however, that neither $u^* = 1$ nor $u^* = 0$ is a decent long-term strategy. But $u^* = 0$, $u^* = \theta$ and $u^* = 1$ are his only Nash-equilibrium strategies. Therefore, out of sheer desperation, Ed will select $u^* = \theta$.

Because everyone else is playing $u^* = 0$, our Ed's reward will be $f_1(\theta, 0) = -(1 - \theta)(\epsilon + \frac{1}{2}\tau)$. In the usual way, because $f_1(\theta, 0) - f_1(0, 0) = \theta(\epsilon + \frac{1}{2}\tau) > 0$, it won't be long before word gets around that θ is a better reply to 0 than 0 is, and it won't be much longer before all the Eds in the world are playing it. The world is now at (θ, θ). Thus every Ed in the world who is contemplating strategy u can safely assume that his opponent will select strategy θ; in which case, his reward function reduces to $f_1(u, \theta) = -\left(\delta + \frac{1}{2}\tau\right)\theta = f_2(\theta, u)$, which is independent of u; i.e., $f_1(u, \theta) = f_1(\theta, \theta) = f_2(\theta, \theta) = f_2(\theta, u)$ for all u. No strategy can yield a higher reward against an opponent who selects θ than $u = \theta$ itself yields.

Suppose, however, that an Ed decides, in the usual way, to start playing a different Nash-equilibrium strategy from θ, say $u^* = 1$. Because $f_1(1, \theta) = f_1(\theta, \theta)$, this Ed does no better against an opponent who selects θ than by selecting θ himself. On the other hand, this Ed does no worse; and he may therefore be tempted to continue selecting $u^* = 1$. What happens now? Will 1 become fashionable? Not likely! If 1 begins to catch on, then sooner or later this Ed will meet another Ed who is also using strategy 1, and Ed's reward from this encounter will be $f_1(1, 1)$. If Ned had stuck to using θ then his reward would have been $f_1(\theta, 1)$. But $f_1(\theta, 1) - f_1(1, 1) = (\delta + \epsilon)(1 - \theta)^2$, which is positive. Thus, although a player who switches from θ to 1 will do precisely as well against an opponent who still uses θ, he will fare worse against an opponent who also has switched from θ to 1; therefore, switching from θ to 1 is a bad idea. Similarly, because $f_1(\theta, 0) > f_1(0, 0)$, it is a bad idea to switch from θ to the other Nash-equilibrium strategy, namely, 0. More generally, because

$$(2.4) \qquad f_1(\theta, u) - f_1(u, u) \quad = \quad (\delta + \epsilon)(\theta - u)^2$$

is positive unless $u = \theta$, it would be irrational for an individual to switch from θ to any other strategy. In other words, once the world has arrived at the strategy combination (θ, θ), the world will stay at (θ, θ). It thus appears that the Nash equilibrium (θ, θ) has a measure of long-term stability, which the other two Nash equilibria do not possess. We shall refer to a strategy that is stable in this sense as *uninvadable*. Thus θ is an uninvadable strategy for Crossroads, whereas 0 and 1 are invadable.

The concept of uninvadability yields a further criterion for distinguishing among Nash equilibria: eliminate strategies that are invadable. In the game we have just considered, however, there is complete symmetry between any two players. The model cannot distinguish between them; or, if you prefer, there are no grounds whatsoever for calling one player Ned and the other one Sed. Therefore, any Nash equilibrium (u^*, v^*) that the whole world adopts must show symmetry between strategies as well; i.e., $u^* = v^*$. Thus our latest criterion would eliminate the Nash equilibria $(0, 1)$ and $(1, 0)$ purely on the grounds that symmetry between players requires symmetry between strategies, because a strategy cannot be uninvadable unless first of all it is universally adoptable. But this symmetry argument provides only a necessary condition for uninvadability; a sufficient condition is provided by (2.4) being positive for $u \neq \theta$.

To define the concept of uninvadability more formally, first note that, because of symmetry, i.e., because $f_2(u, v) = f_1(v, u)$, the reward to a player selecting u against a player selecting v is always

$$(2.5) \qquad f(u, v) \; = \; (\delta + \epsilon) \left\{ (\theta - v)u - \theta(1 + v) \right\} + 2\epsilon v,$$

from (2.3); there is no need for a suffix 1 or 2, and so we drop it.[2] Then strategy v^* is uninvadable if, in a large population of players who almost all select it, v^* yields a greater reward than any deviant or "mutant" strategy, say u, that might instead be selected by the diminutive remainder of the population. One discerns an echo of Kant's categorical imperative to behave in such a way that, if everyone did so, then each would benefit; because it is broadly true that the player who selects an uninvadable strategy behaves in such a way that, if virtually everyone did so, then anyone who failed to do so would fail to benefit.

It is only broadly true, however; and it cannot be too strongly emphasized that uninvadability can crucially depend on the assumption that a deviant strategy is uniformly adopted by the diminutive remainder in the previous paragraph—or, which amounts to the same thing, that if there are several deviant strategies, then they are adopted at different times (as in the narrative above). To see this, and at the same time make our definition more precise, let us temporarily assume that there are m deviant strategies, say $v_1, v_2, \dots, v_m$; subsequently, we shall assume that $m = 1$. Let $1 - x$ be the

---

[2]This symmetry between the players' rewards does not, of course, imply $f(u, v) = f(v, u)$. Indeed it is clear from (2.4) that $f(u, v) - f(v, u) = (u - v)\tau \neq 0$ for $u \neq v$

proportion of the population that selects the orthodox strategy v^*, let x_k be the proportion that selects deviant strategy v_k and let V be the strategy selected by a player's next opponent. Then V is a random variable, with sample space $[0,1]$; and, in the absence of any further information, a player's probability distribution over his next opponent is given by

$$(2.6a) \qquad\qquad \text{Prob}(V = v^*) \;=\; 1 - x,$$

$$(2.6b) \qquad \text{Prob}(V = v_k) \;=\; x_k, \quad k = 1, \dots, m, \quad \sum_{k=1}^{m} x_k \;=\; x,$$

where x is a very small positive number (hence so are $x_1, x_2, \dots, x_m$).

If V is a random variable, then so is $f(v^*, V)$. Thus $f(v^*, V)$ is a payoff, not a reward; but we can convert it into a reward, which we shall denote by W, if we calculate its expected value over the distribution of V. So the reward to strategy v^* is $W(v^*) = \text{E}\big[f(v^*, V)\big] = f(v^*, v^*) \cdot \text{Prob}(V = v^*) + \sum_{k=1}^{m} f(v^*, v_k) \cdot \text{Prob}(V = v_k)$ or

$$(2.7a) \qquad W(v^*) \;=\; (1 - x)f(v^*, v^*) + \sum_{k=1}^{m} x_k f(v^*, v_k),$$

by (2.6); whereas the reward to strategy v_j is $W(v_j) = E\big[f(v_j, V)\big] = f(v_j, v^*) \cdot \text{Prob}(V = v^*) + \sum_{k=1}^{m} f(v_j, v_k) \cdot \text{Prob}(V = v_k)$ or

$$(2.7b) \qquad W(v_j) \;=\; (1 - x)f(v_j, v^*) + \sum_{k=1}^{m} x_k f(v_j, v_k).$$

Thus

$$(2.8) \quad W(v^*) - W(v_j) = (1 - x)\left\{ f(v^*, v^*) - f(v_j, v^*) \right\}$$
$$+ \sum_{k=1}^{m} x_k \left\{ f(v^*, v_k) - f(v_j, v_k) \right\}, \quad j = 1, \dots, m.$$

But x is much less than 1, and so are $x_1, x_2, \dots, x_m$. Hence v^* is an uninvadable strategy if

$$(2.9a) \qquad\qquad f(v^*, v^*) \;>\; f(v_j, v^*), \quad j = 1, \dots, m$$

because the first term in (2.8) then dominates the sum of the last m terms. If there exists j such that

$$(2.9b) \qquad\qquad f(v_j, v^*) \;=\; f(v^*, v^*),$$

however, so that the first term in (2.8) is identically zero, then v^* is an uninvadable strategy if

$$(2.9\text{c}) \qquad f(v^*, v_k) > f(v_j, v_k), \quad k = 1, \ldots, m$$

for all such j. On the other hand, v^* is clearly not an uninvadable strategy if

$$(2.10) \qquad f(v^*, v^*) < f(v_j, v^*),$$

for any j. Of particular interest is the case where $m = 1$ (hence $x_1 = x$), so that the deviant strategy is uniformly adopted by a diminutive fraction x of the population. Then we prefer to denote the single deviant strategy by u, rather than v_1, and sufficient conditions (2.9) for an uninvadable strategy require that for all $u \neq v^*$, EITHER

$$(2.11\text{a}) \qquad f(v^*, v^*) > f(u, v^*)$$

OR

$$(2.11\text{b}) \qquad f(v^*, v^*) = f(u, v^*)$$

and

$$(2.11\text{c}) \qquad f(v^*, u) > f(u, u).$$

That is, either u-strategists cannot even enter a population of v^*-strategists; or u-strategists can enter, but they cannot proliferate. An equivalent statement of these conditions is BOTH

$$(2.12\text{a}) \qquad f(v^*, v^*) \geq f(u, v^*) \qquad \text{for all } v$$

AND, for all $u \neq v^*$, either

$$(2.12\text{b}) \qquad f(v^*, v^*) > f(u, v^*)$$

or

$$(2.12\text{c}) \qquad f(v^*, u) > f(u, u),$$

from which it is clear that an uninvadable strategy is of necessity a Nash-equilibrium strategy (Exercise 2.3). We stress that replacing (2.9) by (2.11) or (2.12) does not necessarily limit the number of deviant strategies to one; but it does require that alternative deviant strategies be adopted sequentially. For an illustration of how to analyze simultaneously adopted deviant strategies, see §2.5.

We should also stress that the principal difference between $m = 1$ (uniform deviation) and $m > 1$ (multiple deviation) is the difference between (2.12c) and (2.9c). If (2.11a) is satisfied for all $u \neq v^*$ (including in particular $v_1, ..., v_m$), then (2.9a) is also satisfied; however, (2.12c) does not imply (2.9c), as the following paragraph will illustrate. If, for all $u \neq v^*$, (2.11a)—and hence (2.9a)—is satisfied, then v^* possesses a stronger measure of uninvadability than if only (2.11b) were satisfied, because v^* is then resistant to multiple deviation (provided of course that the total probability of deviation, namely, $x = x_1 + ... + x_m$ is still small). Accordingly, if (2.11a) is satisfied for all $u \neq v^*$, then we say that v^* is *strongly* uninvadable; and otherwise v^* is only *weakly* uninvadable. It is also useful to be able to distinguish between uninvadable strategies and those that merely satisfy (2.12a); and so we will follow Axelrod [6] by saying that if (2.12a) is satisfied, then v^* is *collectively stable*. To say that v^* is collectively stable is really to say neither more nor less than that v^* is a symmetric Nash-equilibrium strategy; nevertheless, the terminology is useful, because it signifies at once that the game is played between an individual and the rest of a population (as opposed to between specific individuals). Likewise, to say that v^* is strongly uninvadable is merely to say that (v^*, v^*) is a strong Nash equilibrium (Exercise 2.3); but the new terminology is again more evocative of the game's inherent symmetry. Obviously, every strongly uninvadable strategy is also uninvadable, and every uninvadable strategy is also collectively stable.

To illustrate the effects of multiple deviation versus uniform deviation on a strategy that is uninvadable, but not strongly uninvadable, let us now apply (2.9)–(2.11) to Crossroads in the case where $\theta < 1$ (for $\theta > 1$, see Exercise 2.1). From (2.5), $\theta < 1$ implies both $f(1,1) < f(0,1)$ and $f(0,0) < f(1,0)$; whence (2.10) implies that neither $v^* = 1$ nor $v^* = 0$ is an uninvadable strategy. We are not surprised: neither $(0,0)$ nor $(1,1)$ is a Nash equilibrium. Thus the only candidate for uninvadable strategy is the remaining Nash-equilibrium strategy, namely, $v^* = \theta$. The two deviant strategies are, say, $v_1 = 0$

and $v_2 = 1$; whence (2.5) implies that (2.9b) is satisfied for both $j = 1$ and $j = 2$. Thus (2.9c) would require both

(2.13a) $$f(\theta, 0) > f(0, 0), \ f(\theta, 1) > f(0, 1)$$

(for $j = 1$) and

(2.13b) $$f(\theta, 0) > f(1, 0), \ f(\theta, 1) > f(1, 1)$$

(for $j = 2$). By (2.4), $f(\theta, 0) > f(0, 0)$ and $f(\theta, 1) > f(1, 1)$; but by (2.5), $f(\theta, 1) < f(0, 1)$ and $f(\theta, 0) < f(1, 0)$; so (2.13) is false. Indeed (2.8) yields $W(\theta) - W(0) = x_1\{f(\theta, 0) - f(0, 0)\} + x_2\{f(\theta, 1) - f(0, 1)\}$ and $W(\theta) - W(1) = x_1\{f(\theta, 0) - f(1, 0)\} + x_2\{f(\theta, 1) - f(1, 1)\}$ or

(2.14a) $W(\theta) - W(0) \ = \ \theta(\delta + \epsilon)(\theta x_1 - \{1 - \theta\}x_2)$

(2.14b) $W(\theta) - W(1) \ = \ (1 - \theta)(\delta + \epsilon)(-\theta x_1 + \{1 - \theta\}x_2)$

on using (2.5), so that

(2.15) $$W(\theta) \ = \ (1 - \theta)W(0) + \theta W(1)$$

must lie between $W(0)$ and $W(1)$ and hence cannot exceed both of them. Yet (2.11) is clearly satisfied by $v^* = \theta$ for any $0 \le u \le 1$.

Is it reasonable to assume that at most one deviant strategy is adopted at any given time? Unfortunately, the answer to this question depends explicitly upon the dynamics of interaction between the players, which our model fails to capture explicitly (in its current state of development). The requisite dynamics is at least quite plausible, however: a lone player selects a deviant strategy, discovers that it rewards him less than the orthodox strategy and reverts to orthodoxy before another player has a chance to deviate. Moreover, the lower the frequency of deviation, the more reasonable the assumption.

The frequency of deviation is widely thought to be sufficiently low when conflict arises in the context of evolutionary biology, because strategies can be identified with inherited behavior and deviations with mutations. Thus the dynamic of gossip and rumor—or whatever it was that made word get around in Crossroads—is replaced by the dynamic of genetic transmission. In repeated plays of Crossroads, the composition of the population (in terms of strategies) changes because successful strategists are imitated by other drivers; whereas, in the course of biological evolution, the composition of a population (again in terms of strategies) changes because successful strategists leave more offspring (who are assumed to inherit genes for the successful strategy). In either case, however, the frequency of a successful strategy increases because "success breeds success"— whether metaphorically, as in the case of Crossroads, or literally, as in the context of evolution—and so the difference between the two dynamics, in terms of their effects on the composition of a population, is largely a matter of time scales. Indeed the concept of uninvadable strategy was first defined in connection with the Hawk-Dove game of §1.2 by Maynard Smith and Price [**129, 134**] who, because of the context, named the concept *evolutionarily stable strategy*, or ESS. It has since been developed extensively by Maynard Smith [**132**] and others.[3] Henceforward, we will find it convenient to refer to uninvadable strategies as evolutionarily stable strategies or ESSes, and to strongly or weakly uninvadable strategies as strong or weak ESSes, regardless of whether the context is biological or sociological.

With the Hawk-Dove game still in mind, let us re-define θ by

$$(2.16) \qquad \theta = \rho/C,$$

where ρ is the reproductive value of the territory in dispute and C the (reproductive) cost of injury. Then, from Exercise 1.1, the reward to an animal who selects u (for his probability of playing Hawk, or H) against an animal who selects v is

$$(2.17) \qquad f(u,v) = \tfrac{1}{2}\left(\rho(1-v) + C(\theta - v)u\right).$$

If $\rho > C$ $(\theta > 1)$, then there is a unique Nash-equilibrium strategy, $v^* = 1$, because H is then a dominant pure strategy: (2.12b) is satisfied with $v^* = 1$ for all $0 \le u < 1$, so that $v^* = 1$ is an ESS. If $\rho < C(\theta < 1)$, on the other hand, then there are three Nash-equilibrium strategies, namely, $v^* = 0$, $v^* = \theta$ and $v^* = 1$; and it is readily deduced from (2.11) or (2.12) that $v^* = \theta$ is an ESS, whereas $v^* = 0$ and $v^* = 1$ are not (Exercise 2.3). Note that θ is small when C is large. Thus a possible explanation for the rarity in nature of protracted fights is that the cost of injury is much too high.

We conclude this section by returning briefly to Four Ways. Despite the profusion of Nash equilibria in Table 1.10, it is clear at once that only one of them is symmetric between strategies, i.e, satisfies

---

[3]See §2.9 for more recent references.

$u = v$. You will recall that $u = (u_1, u_2)$ and $v = (v_1, v_2)$ are vectors in Four Ways, so that symmetry between strategies requires both $u_1 = v_1$ and $u_2 = v_2$. Thus the lone candidate is the last row of Table 1.10, i.e., $(u, v) = (\alpha, \beta, \alpha, \beta)$. Let us define the vector ζ by $\zeta = (\alpha, \beta)$. Then $v^* = \zeta$ is the only candidate for uninvadability.

Now, from (1.31), the reward to a player who selects $u = (u_1, u_2)$ against an opponent who selects $v = (v_1, v_2)$ is

$$
\begin{aligned}
(2.18) \quad f(u, v) \; = \; & - \left(2\delta v_1 + (\delta + \tau/2)(v_2 - 1) \right) u_1 \\
& - \left((\delta - \tau/2)(v_1 - 1) + (\delta + \epsilon) v_2 \right) u_2 \\
& + (\delta - \tau/2) v_1 + (\delta + \tau/2)(v_2 - 1).
\end{aligned}
$$

We readily find (Exercise 2.4) that

$$
(2.19a) \qquad\qquad f(\zeta, \zeta) \; = \; f(u, \zeta)
$$

and that

$$
\begin{aligned}
(2.19b) \quad f(\zeta, u) - f(u, u) \; = \; & \delta \left\{ u_1 + u_2 - \alpha - \beta \right\}^2 \\
& + \delta \left\{ u_1 - \alpha \right\}^2 + \delta \left\{ u_2 - \beta \right\}^2,
\end{aligned}
$$

which is greater than zero for all $u \neq \zeta$. Therefore $v^* = \zeta$ is an uninvadable strategy.

For further analysis of Four Ways, see Exercise 2.5.

2.4. Crossroads as a continuous population game

Four Ways and the symmetric version of Crossroads typify games in which a focal individual, the u-strategist, interacts with individuals drawn at random from a large population of v-strategists. We call such games *population games*. As we saw in §2.3, in analyzing such games, which are inherently symmetric, there is no especially good reason for the focal individual—Ed, say—to be called Player 1, and so we denoted his reward by f (instead of f_1). For the same reason, in this section we denote his rational reaction set by

$$
(2.20) \quad R \; = \; \left\{ (u, v) \mid 0 \leq u, v \leq 1, \quad f(u, v) = \max_{0 \leq \overline{u} \leq 1} f(\overline{u}, v) \right\}
$$

instead of R_1. The v-strategist's rational reaction set

$$
(2.21) \qquad \left\{ (u, v) \mid 0 \leq u, v \leq 1, \quad f(v, u) = \max_{0 \leq \overline{v} \leq 1} f(\overline{v}, u) \right\}
$$

is then easily obtained by reflecting R in $u = v$; however, we will discover that it is largely irrelevant, and so we do not even name it.

Our examples of population games in §2.3 were both bimatrix games. But a population game can also be continuous, without affecting the definition of evolutionarily stable strategy. To illustrate, in this section we generalize the symmetric version of Crossroads studied in §2.3. We will call this new game Crossroads II. Now, in the (symmetric) game of Crossroads, delays due to mutual dithering, mutual impetuosity and waiting for an opponent to traverse the junction are represented by the parameters ϵ, δ and τ, respectively. Players do not distinguish between these three possible sources in their perceptions of the costs of delay. In the game of Crossroads II, however, they perceive delays due to mutual dithering or impetuosity as inherently more wasteful of time—by virtue of being avoidable—than delays due to opponent traversal (which are inevitable, or there would be no conflict). Specifically, drivers perceive traversal delays as very costly only if they perceive themselves to be very late; and the earlier they perceive themselves, the lower they perceive the costs of such delays. Lateness is measured by an index between 0 and 1, with 0 corresponding to the lowest possible perception and 1 to the highest. If X denotes the u-strategist's lateness, then $1 - X$ may be interpreted as his earliness. Similarly, if Y denotes the v-strategist's lateness, then $1 - Y$ may be interpreted as his earliness. Both earliness and lateness are always numbers between 0 and 1. We assume, in fact, that X and Y are continuous random variables, independently distributed between 0 and 1 with probability density function g. Thus the point (X, Y) is distributed over the unit square in Figure 2.3 with joint probability density $g(x)g(y)$ per unit area.

In Crossroads II, players discount traversal delays by a fraction η of their earliness, where $0 \leq \eta < 1$. Thus a traversal delay of τ is perceived as $\tau\{1 - \eta(1 - X)\}$ by Ed, and as $\tau\{1 - \eta(1 - Y)\}$ by any other player. The payoff to Ed is now $-\left(\delta + \frac{1}{2}\tau\{1 - \eta(1 - X)\}\right)$ in the event of GG, $-\tau\{1 - \eta(1 - X)\}$ in the event of WG, and so on. But strategies are no longer probabilities of going; rather, they are critical latenesses above which a player goes, and below which that player waits. In other words, strategy u means go if $X > u$, but otherwise wait; and strategy v means go if $Y > v$, but otherwise

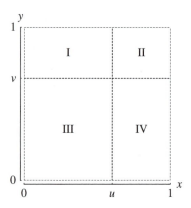

Figure 2.3. The sample space for the joint distribution of latenesses

wait. Thus WG is the event that $X \leq u, Y > v$; GG is the event that $X > u, Y > v$; WW is the event that $X \leq u, Y \leq v$; and GW is the event that $X > u, Y \leq v$. These four events correspond to subregions I, II, III and IV, respectively, of the unit square in the x-y plane.

As in §2.3, Ed's reward f is the expected value of his payoff, which we denote by F. It depends on X and Y, as follows:

$$F(X,Y) = \begin{cases} -\tau\{1 - \eta(1 - X)\} & \text{if } (X,Y) \in \text{I} \\ -\left(\delta + \frac{1}{2}\tau\{1 - \eta(1 - X)\}\right) & \text{if } (X,Y) \in \text{II} \\ -\left(\epsilon + \frac{1}{2}\tau\{1 - \eta(1 - X)\}\right) & \text{if } (X,Y) \in \text{III} \\ 0 & \text{if } (X,Y) \in \text{IV.} \end{cases}$$

So Ed's reward is $f(u,v) = \mathrm{E}[F] = \int_0^1 \int_0^1 F(x,y)\, g(x)\, g(y)\, dx\, dy$, where E denotes expected value (not Ed). Because the text width of these pages only slightly exceeds four inches, however, we will find it a great convenience, especially in Chapter 6, to write

(2.22) $$dA = g(x)\, g(y)\, dx\, dy,$$

so that Ed's reward becomes $f(u,v) = \int_0^1 \int_0^1 F(x,y)\, dA$, which is significantly more compact. Think of (2.22) as merely a notational ruse for avoiding needless clutter: whenever you see dA on the right-hand side of an integral, instantly replace it by $g(x)\, g(y)\, dx\, dy$ in your mind.

Let us now assume, for the sake of simplicity, that lateness is uniformly distributed between 0 and 1, so that g is defined by

$$(2.23) \qquad g(\xi) \;=\; 1, \qquad 0 \le \xi \le 1.$$

Then we can calculate the reward as the sum of four contributions, one for each subregion of the unit square. Integration over subregion I, where $0 \le x \le u$ and $v \le y \le 1$, yields the contribution $f_{\mathrm{I}}(u,v) = \int_{x=0}^{x=u} \int_{y=v}^{y=1} F(x,y)\, dA = -\tau \int_0^u \{1 - \eta(1-x)\}\, dx \int_v^1 dy$, or

$$(2.24) \qquad f_{\mathrm{I}}(u,v) \;=\; -\tau u (1-v)\left\{1 - \eta\left(1 - \tfrac{1}{2}u\right)\right\}$$

after simplification. Similarly, integration over subregion II, where $u \le x \le 1$ and $v \le y \le 1$, yields

$$(2.25) \qquad f_{\mathrm{II}}(u,v) \;=\; -(1-u)(1-v)\left\{\delta + \tfrac{1}{2}\tau\left(1 - \tfrac{\eta}{2}\{1-u\}\right)\right\};$$

integration over subregion III, where $0 \le x \le u$ and $0 \le y \le v$, yields

$$(2.26) \qquad f_{\mathrm{III}}(u,v) \;=\; -uv\left\{\epsilon + \tfrac{1}{2}\tau\left(1 - \tfrac{\eta}{2}\{1-u\}\right)\right\};$$

and integration over subregion IV yields $f_{\mathrm{IV}}(u,v) = 0$. Adding and simplifying, we obtain $f(u,v) = f_{\mathrm{I}}(u,v) + f_{\mathrm{II}}(u,v) + f_{\mathrm{III}}(u,v) + f_{\mathrm{IV}}(u,v) = \left\{\epsilon + \tfrac{1}{2}\tau - (\delta + \epsilon)(1-v)\right\}(1-u) + \left(\epsilon - \tfrac{1}{2}\tau\right)(1-v) - \epsilon - \tfrac{1}{2}\tau - \tfrac{1}{4}\eta\tau\left(u^2 - 2u - 1 + v\right)$ or

$$(2.27) \quad f(u,v) \;=\; (\delta + \epsilon)\left\{(1 - v - \theta)u - \tfrac{1}{2}\lambda\left(u^2 - 2u - 1 + v\right)\right\} \\ - \left\{(1+\theta)\delta - (1-\theta)\epsilon\right\}(1-v),$$

where θ is defined by (1.34) and

$$(2.28) \qquad \lambda \;=\; \tfrac{\eta\tau}{2(\epsilon+\delta)}.$$

The expression for f agrees with (2.5) when $\eta = 0$.[4]

In the case where drivers are either fast or intermediate according to definitions (1.9)-(1.10), i.e., where $\delta > \tfrac{1}{2}\tau$ and so $\lambda < \theta < 1$, it is straightforward to show that $f(u,v)$ is maximized for $0 \le u \le 1$ by

$$(2.29) \qquad u = \begin{cases} 1 & \text{if } 0 \le v \le 1 - \theta \\ \frac{1-\theta+\lambda-v}{\lambda} & \text{if } 1 - \theta < v < 1 - \theta + \lambda \\ 0 & \text{if } 1 - \theta + \lambda \le v \le 1. \end{cases}$$

---

[4]Note that u in Crossroads has a different meaning from u in Crossroads II. In the first case, u is the probability of G for Ed; in the second case, u is a critical lateness above which Ed goes, so that the probability of G for Ed is $1 - u$. Similarly for v.

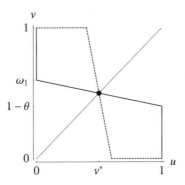

Figure 2.4. Typical rational reaction sets for Crossroads II when drivers are not slow. The quantities ω_1 and v^* are defined by $\omega_1 = 1 - \theta + \lambda$ and (2.30). The figure is drawn for $\delta = 3, \epsilon = 2, \tau = 2$ and $\eta = 1$ (but would have the same topology for any $\delta > \tau/2$).

So the u-strategist's rational reaction set R, defined by (2.20), consists of three straight line segments in the unit square of the u-v plane, as indicated by the solid curve in Figure 2.4. The dashed curve is the v-strategist's rational reaction set, defined by (2.21), and obtained by reflecting R in the line $u = v$ (which is shown dotted). These sets intersect at $(0, 1)$, at $(1, 0)$ and at (v^*, v^*), where

$$(2.30) \qquad v^* = 1 - \frac{\theta}{1 + \lambda}.$$

Thus, as in Crossroads, there are three Nash equilibria, only one of which is symmetric. For any positive η, however, R is the graph of a function (of v), so that—unlike in Crossroads—every strategy has a unique best reply. In particular, because v^* is uniquely the best reply to itself, v^* is a strong ESS, as we can verify by observing that $f(v^*, v^*) - f(u, v^*) = \frac{1}{4}\eta\tau(u - v^*)^2$ is strictly positive for all $u \neq v^*$. So we can regard the weak ESS in Crossroads as rather atypical, because it arises only in the limit as $\eta \to 0$, and it seems unlikely (at least to me) that real-world drivers do not discount traversal delays at all. Of course, in the limit as $\eta \to 0$, the game is indistinguishable from Crossroads itself.

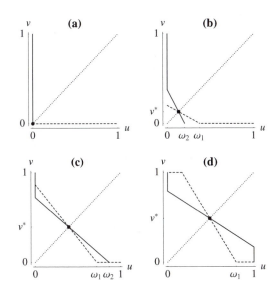

Figure 2.5. Typical rational reaction sets for Crossroads II
when **(a)** $\delta < (1 - \eta)\tau/2$ or $\lambda < \theta - 1$ (implying $\theta > 1$);
(b) $(1 - \eta)\tau/2 < \delta < \eta\tau/2 - \epsilon$ or $\lambda > \max\{1, \theta - 1\}$, a
case that arises only if $1/2 < \eta < 1, \epsilon < (2\eta - 1)\tau/2$; **(c)**
$\max\{(1 - \eta)\tau/2, \eta\tau/2 - \epsilon\} < \delta < \tau/2$ or $\theta - 1 < \lambda < 1$;
and **(d)** $\delta > \tau/2$ or $\lambda < \theta < 1$. The quantities ω_1, ω_2 and
v^* are defined by $\omega_1 = 1 - \theta + \lambda, \omega_2 = \omega_1/\lambda$ and (2.30). In
the first three cases drivers are slow. (The figure is drawn
for $\epsilon = 0, \tau = 5, \eta = 0.75$ with **(a)** $\delta = 0.5$, **(b)** $\delta = 1$, **(c)**
$\delta = 2.25$ and **(d)** $\delta = 3$.)

Figure 2.4 also explains why the v strategist's rational reaction
set is irrelevant in a population game: because we are interested only
in symmetric Nash equilibria, it suffices to know where R intersects
the line $u = v$. Henceforward, therefore, we analyze population games
solely in terms of R.

One good reason for drivers to discount traversal delays is that ev-
erybody benefits. In a population at the ESS, the reward to each driver is $f(v^*, v^*)$, and so the average delay is $-f(v^*, v^*)$. It is straightforward to show that this quantity decreases with respect to η (Exercise 2.10). So a population of drivers who discount traversal delays

experiences shorter average delays than a population of drivers who do not discount; and the more they discount, the less they wait.

The above results are readily generalized to allow for slow drivers, or $\delta < \tau/2$: in place of (2.30), the ESS becomes

$$(2.31) \qquad v^* = \begin{cases} 0 & \text{if } \delta \leq \frac{1}{2}(1 - \eta)\tau \\ 1 - \frac{\theta}{1+\lambda} & \text{if } \delta > \frac{1}{2}(1 - \eta)\tau, \end{cases}$$

as illustrated by Figure 2.5. The game is also readily generalized to allow for an arbitrary nonuniform distribution of lateness, i.e., for any g such that $g(\xi) \geq 0$ and $\int_0^1 g(\xi)\, d\xi = 1$ in place of (2.23); for details, see Exercise 2.10.

2.5. An example of population dynamics

We have already seen in §2.3 that the strategy $v^* = \theta$ is not uninvadable for fast driving in Crossroads if the deviant strategies $v_1 = 0$ and $v_2 = 1$ occur simultaneously. What happens in these circumstances depends explicitly upon the dynamics of interaction among the population of drivers. Here we present a simple model of those dynamics.[5]

Let the proportions of the population of drivers playing deviant strategies 0 and 1 at time t (≥ 0) be $x_1(t)$ and $x_1(t)$, respectively; and let x_1 and x_2 be differentiable functions. Then the proportion of drivers playing strategy θ is $1 - x(t)$, where $x(t) = x_1(t) + x_2(t)$; and so the average reward of the population at time t is

$$(2.32) \qquad \overline{W} = (1 - x)W(\theta) + x_1 W(0) + x_2 W(1),$$

where $W(0)$, $W(\theta)$ and $W(1)$ are defined by (2.7), and we avoid needless clutter by using notation that suppresses their dependence on t. For all $t \geq 0$, the point $(x_1(t), x_2(t))$ must belong to the triangle $\Delta = \{(x_1, x_2) \mid x_1, x_2 \geq 0, x_1 + x_2 \leq 1\}$; see Figure 2.6. We assume that $x_1(0)$ and $x_2(0)$ are both positive; otherwise, there is no danger of strategy θ being invaded.

It seems reasonable to assume that the fraction of drivers using a deviant strategy would increase if the reward from that strategy were greater than average. Thus $dx_1/dt > 0$ if $W(0) > \overline{W}$, and $dx_2/dt >$

---

[5]It requires some familiarity with phase-plane analysis; see, e.g., [**144**, pp. 46-52].

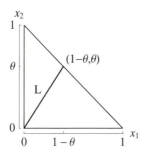

Figure 2.6. Triangle Δ that contains $(x_1(t), x_2(t))$ for all $t \geq 0$.

0 if $W(1) > \overline{W}$. Dynamics consistent with these assumptions are defined by the pair of differential equations

$$(2.33) \qquad \frac{1}{x_1}\frac{dx_1}{dt} = \kappa\{W(0) - \overline{W}\}, \qquad \frac{1}{x_2}\frac{dx_2}{dt} = \kappa\{W(1) - \overline{W}\},$$

where $\kappa(> 0)$ is a constant of proportionality. From (2.15) and (2.32) we have

$$(2.34a) \qquad \begin{aligned} W(0) - \overline{W} &= \{x_2 + \theta(1 - x)\}\{W(0) - W(1)\}, \\ W(1) - \overline{W} &= \{1 - x_2 - \theta(1 - x)\}\{W(1) - W(0)\}, \end{aligned}$$

whereas from (2.14) we have

$$(2.34b) \qquad W(0) - W(1) = (\delta + \epsilon)\{(1 - \theta)x_2 - \theta x_1\}.$$

On substituting (2.34) into (2.33), we obtain a pair of nonlinear ordinary differential equations for the motion of the point $(x_1(t), x_2(t))$:

$$(2.35) \qquad \begin{aligned} \frac{dx_1}{dt} &= kx_1\{\theta(1 - x_1) + (1 - \theta)x_2\}\{(1 - \theta)x_2 - \theta x_1\} \\ \frac{dx_2}{dt} &= kx_2\{\theta x_1 + (1 - \theta)(1 - x_2)\}\{\theta x_1 - (1 - \theta)x_2\}, \end{aligned}$$

where $k = (\delta + \epsilon)\kappa$. From these equations we can deduce the values of $x_1(\infty)$ and $x_2(\infty)$. Then the long-term fraction of drivers adopting strategy θ is $1 - x_1(\infty) - x_2(\infty)$.

It is clear from inspection of (2.35) that, above the line segment in Figure 2.6 defined by

$$(2.36) \qquad L = \{(x_1, x_2) \in \Delta \mid \theta x_1 - (1 - \theta)x_2 = 0\},$$

$dx_1/dt > 0$ and $dx_2/dt < 0$; whereas below L, $dx_1/dt < 0$ and $dx_2/dt > 0$. Therefore, any trajectory that begins in the interior of Δ must end on L as $t \to \infty$ (and every point on L is an equilibrium of dynamical system (2.35)). To check that the trajectory cannot leave Δ across the boundary where $x_1 + x_2 = 1$, we need only observe that along that boundary we have

$$\frac{dx_2}{dx_1} = \frac{dx_2/dt}{dx_1/dt} = -\frac{x_2\{\theta x_1 + (1-\theta)(1-x_2)\}}{x_1\{\theta(1-x_1) + (1-\theta)x_2\}} = -1.$$

Thus trajectories that begin where $x_1 + x_2 = 1$ must end precisely at $(1 - \theta, \theta)$ as $t \to \infty$; and trajectories that begin in the interior must remain in the interior, because trajectories cannot cross.

The long-term fraction of drivers using strategy θ, namely, $x(\infty)$, is indeterminate; it depends upon $x_1(0)$ and $x_2(0)$. Nevertheless, drivers not using θ must use strategies 0 and 1 in the ratio $x_1(\infty)/x_2(\infty)$ $= \frac{1-\theta}{\theta}$. Without the ability to recognize individual drivers, the sub-population in which fraction θ always plays pure strategy G and fraction $1 - \theta$ always plays pure strategy W is indistinguishable from the sub-population in which every driver plays mixed strategy θ (G with probability θ, W with probability $1 - \theta$). Thus, although in theory strategy θ is susceptible to simultaneous invasions by deviant strategies 0 and 1, whether in practice θ is invadable is, at the very least, a moot point. Note, however, that this conclusion is strongly dependent upon equations (2.35), which yield no more than a phenomenological description of the dynamics of interaction.

Now try Exercise 2.11.

2.6. Discrete population games. Multiple ESSes

The population games we analyzed in §§2.3-2.4 have unique evolutionarily stable strategies. But a game may have more than one ESS. Then which ESS will the population adopt? To answer this question, we again require explicit dynamics.

Let us first observe, however, that we are not obliged to allow mixed strategies in matrix population games, and for certain purposes it is preferable to dispense with them. For games so restricted to pure strategies—i.e., for *discrete population games*—it is convenient to frame a new definition of ESS in terms of the payoff matrix A.

Accordingly, let a_{ij} be the payoff to strategy i against strategy j in a symmetric game restricted to m pure strategies. Then, by analogy with (2.12), strategy k is an ESS if BOTH

$$(2.37a) \qquad a_{kk} \geq a_{jk} \qquad \text{for all} \quad j = 1, \ldots, m$$

AND, for all $j \neq k$, either

$$(2.37b) \qquad a_{kk} > a_{jk}$$

or

$$(2.37c) \qquad a_{kj} > a_{jj}.$$

If (2.37b) is satisfied for all $j \neq k$, then strategy k is a strong ESS (and otherwise a weak ESS). If (2.37a) is satisfied, but for some $j \neq k$ neither (2.37b) nor (2.37c) is satisfied, then strategy k is merely collectively stable.

Having made this definition, let us now set $m = 2$ and consider the special case of a discrete population game with payoffs satisfying

$$(2.38) \qquad a_{22} > a_{12} > a_{11} > a_{21}.$$

An example of this game appears, e.g., in §5.3.[6] Because $k = 1$ and $k = 2$ both satisfy (2.37b), both strategies are strong ESSes. Then which will emerge as the winning strategy in a large population of players, some of whom adopt strategy 1, the remainder of whom adopt strategy 2? The answer depends on initial conditions.

By analogy with the previous section, let us suppose that the specific growth rate of the fraction of population adopting strategy k is proportional to the difference between the average payoff to strategy k, denoted by W_k, and the average reward to the entire population, denoted by $\overline{W}$. Let $x_1 = x_1(t)$ and $x_2 = x_2(t)$ be the proportions adopting strategies 1 and 2, respectively, at time t. If the integer-valued random variable $J(t)$ denotes the strategy adopted by a player's opponent at time t, and if the population interacts at random, then with negligible error we have

$$(2.39) \qquad \text{Prob}(J = j) = x_j, \quad j = 1, 2;$$

---

[6]For this particular game, the two pure strategies are the only evolutionarily stable strategies—pure or mixed (Exercise 2.12). Thus, with regard to population dynamics, no generality whatsoever is lost by restricting this game to pure strategies.

whence the expected value of the payoff to strategy k is

$$(2.40) \qquad W_k = \sum_{j=1}^{2} a_{kj} \cdot \text{Prob}(J = j) = a_{k1}x_1 + a_{k2}x_2,$$

for $k = 1, 2$. Similarly, the average reward to the population is

$$(2.41) \qquad \overline{W} = x_1 W_1 + x_2 W_2 = a_{11}x_1^2 + (a_{12} + a_{21})x_1 x_2 + a_{22}x_2^2;$$

and so

$$(2.42) \qquad \frac{1}{x_1}\frac{dx_1}{dt} = \kappa\{W_1 - \overline{W}\}, \quad \frac{1}{x_2}\frac{dx_2}{dt} = \kappa\{W_2 - \overline{W}\}$$

where κ is a constant of proportionality. On using $x_1 + x_2 = 1$, either of these equations reduces to

$$(2.43) \qquad \frac{dx_2}{dt} = \kappa x_2 (1 - x_2)(W_2 - W_1)$$
$$= \kappa x_2 (1 - x_2)(a_{22} - a_{12} + a_{11} - a_{21})(x_2 - \gamma)$$

where

$$(2.44) \qquad \gamma = \frac{a_{11} - a_{21}}{a_{22} - a_{12} + a_{11} - a_{21}};$$

see Exercise 2.13. Thus dx_2/dt is positive or negative according to whether $x_2 > \gamma$ or $x_2 < \gamma$. If $x_2(0) < \gamma$, then (2.43) implies that $x_2(t) \to 0$ as $t \to \infty$, so that strategy 1 wins over the population; whereas if $x_2(0) > \gamma$, then (2.43) implies that $x_2(t) \to 1$ as $t \to \infty$, so that strategy 2 wins instead.

More generally, in a game restricted to m pure strategies, let $x_k = x_k(t)$ be the proportion adopting strategy k at time t, let

$$(2.45) \qquad W_k = \sum_{j=1}^{m} a_{kj} x_j$$

be the reward to strategy k, and let

$$(2.46) \qquad \overline{W} = \sum_{k=1}^{m} x_k W_k$$

be the average reward to the entire population. Then the long-term dynamics can be described by the differential equations

$$(2.47) \qquad \frac{1}{x_k}\frac{dx_k}{dt} = \kappa\{W_k - \overline{W}\}, \quad k = 1, \dots, m,$$

which were introduced by Taylor and Jonker [222]. Note that if

$$(2.48) \qquad x_1(0) + x_2(0) + \dots + x_m(0) = 1$$

—which must, of course, hold—then (2.47) implies

(2.49) $$x_1(t) + x_2(t) + \ldots + x_m(t) = 1, \quad 0 \le t < \infty$$

(Exercise 2.14).

In biological evolution, where success breeds success in the sense that individuals with genes for a successful strategy leave behind more offspring (with the same genes) than players without those genes, the generation in which success is bred is physically distinct from the generation in which, as it were, success originated; and it may be more appropriate to describe game dynamics, not by differential equations, but rather by recurrence equations—especially if successive generations do not overlap (as with many insect populations). Furthermore, even if successful strategies are passed on by imitation (as opposed to genetic transmission), it may be legitimate to think of the population as composed of discrete generations of players—even though the individuals in the population are the same from one "generation" to the next. In repeated plays of Crossroads, for example, success breeds success in the sense that today's successful strategy will be copied tomorrow by drivers who used a different strategy today; moreover, those who used it successfully today will continue to use it tomorrow. Thus the successful strategy will appear with greater frequency in tomorrow 's population, and we can think of today's and tomorrow's populations as discrete generations of drivers—if you like, less savvy drivers and more savvy drivers—even though the two populations are composed of the same individuals.

Accordingly, let us now adapt the dynamics from differential to recurrence equations. For $n = 0, 1, 2, \ldots$, let $\overline{W}(n)$ denote the average expected payoff to the population in the n-th generation; and, for $k = 1, 2, \ldots, m$, let $W_k(n)$ and $x_k(n)$ denote the values of x_k and W_k in the n-th generation. Then, by analogy with (2.45) and (2.46),

(2.50) $$W_k(n) = \sum_{j=1}^{m} a_{kj} x_j(n), \quad k = 1, \ldots, m,$$

(2.51) $$\overline{W}(n) = \sum_{k=1}^{m} x_k(n) W_k(n).$$

We assume that strategy k appears with greater or lesser frequency in generation $n+1$ according to whether $W_k(n) > \overline{W}(n)$ or $W_k(n) <$

$\overline{W}(n)$, in such a way that the proportions playing strategy k at itera-tions $n+1$ and n are in the ratio $W_k(n)/\overline{W}(n)$. So, for $k = 1, \ldots, m$,

$$(2.52a) \qquad x_k(n+1) = \frac{x_k(n)W_k(n)}{\overline{W}(n)}, \quad 0 \leq n < \infty$$

or, equivalently (Exercise 2.14),

$$(2.52b) \qquad \frac{x_k(n+1) - x_k(n)}{x_k(n)} = \sum_{\substack{j=1 \\ j \neq k}}^{m} \frac{\{W_k(n) - W_j(n)\}x_j(n)}{\overline{W}(n)}.$$

We will use (2.52) to describe dynamics both in Exercise 2.22 and in Chapter 5.

In the context of discrete population games, we shall find it con-venient to say that strategy i *infiltrates* strategy k if $x_i(0)$ is a very small positive number but $x_k(0)$ is close to 1. Clearly, up to $m - 1$ strategies can infiltrate strategy k, because the sum of $m - 1$ very small positive numbers is still a very small positive number; and so we will distinguish two cases by saying that strategy k is subject to *pure* infiltration by strategy i if strategy i is the only infiltrator (i.e., if $x_j(0) = 0$ for $j \neq k$ and $j \neq i$), but that strategy k is subject to *mixed* infiltration if there is more than one infiltrator. Now, re-gardless of whether we use (2.47) or (2.52) to describe the subsequent dynamics, the vector $x(\infty) = (x_1(\infty), x_2(\infty), \ldots, x_m(\infty))$ yields the ultimate composition of the population; in other words, the popula-tion evolves to $x(\infty)$. If $x_i(\infty) > x_i(0)$, so that $x_k(\infty) < 1$, then we shall say that strategy i *invades* strategy k; if also $x_i(\infty) = 1$ (hence $x_k(\infty) = 0$), then we shall say that strategy i *eliminates* strategy k. (If, on the other hand, $x_k(\infty) = 1$, then we shall say that strategy k eliminates the infiltrators.) But note that strategy i can invade strat-egy k without eliminating it. We will find this terminology especially useful in Chapter 5.

Now, by analogy with §2.3, if strategy k is an ESS then it is stable against pure infiltration; i.e., if strategy i is the only infiltrator, then strategy k will always eliminate it (see Exercise 2.21). Furthermore, if strategy k is a strong ESS then it is stable against mixed infiltration; i.e., strategy k will eliminate every infiltrator (again, see Exercise 2.21). But if strategy k is not a strong ESS, or if strategy k is only collectively stable, then—as we shall illustrate both in Exercise 2.22

and in §5.4—the outcome of mixed infiltration will depend on specific details of the dynamics, whether (2.47) or (2.52). An important possibility is that $x(\infty) = \xi$, where at least two components of the vector ξ are nonzero. In that case, if $x(0) \approx \xi$ implies $x(\infty) = \xi$, then we shall refer to ξ as an *evolutionarily stable state*, which comprises a mixture of strategies. If, however, $x(0) \approx \xi$ implies only that $x(\infty)$ is close to ξ, then ξ is merely *metastable*; and, as we shall illustrate below and in §5.4, one or more of the strategies in such a metastable mixture may ultimately be eliminated through persistent infiltration by other strategies.

Because evolutionarily stable strategy and evolutionarily stable state are somewhat cumbersome phrases, there is a natural tendency to abbreviate them; and so it is rather unfortunate that they both have the same initials. A frequent workaround is to use "monomorphic ESS" or monomorphism for evolutionarily stable strategy, and "polymorphic ESS" or polymorphism for evolutionarily stable state: thus ESS has two possible meanings, but the morphic qualifier eliminates any ambiguity. Whenever we use ESS all by itself, however, we shall always mean evolutionarily stable strategy.

Before proceeding to the following section, we digress to remark that equations (2.52a) are a special case of first order, nonlinear recurrence equations of the form

$$(2.53a) \qquad x_j(n+1) = G_j(x_1(n), \ldots, x_m(n)), \quad j = 1, \ldots, m,$$

where $G_1, \ldots, G_m$ are functions of $x_1, \ldots, x_m$; or, more succinctly,

$$(2.53b) \qquad x(n+1) = G(x(n)),$$

where the vector x and the vector-valued function G are defined by

$$(2.54) \qquad x(n) = (x_1(n), \ldots, x_m(n)), \quad G = (G_1, \ldots, G_m).$$

In the particular examples that appear in this book, the dynamics of (2.53) will turn out to be rather simple: as n increases, the vector $x(n)$ will progress toward an equilibrium vector $x(\infty) = x^*$, where x^* satisfies $x^* = G(x^*)$. Furthermore, for given $x(0)$, it will be easy to generate the sequence $x(1), x(2), x(3), \ldots$ by computer, recursively from (2.53). Nevertheless, one should still be aware that the dynamics of equations of type (2.53) are potentially very complicated, and a considerable variety of periodic and even chaotic behavior is possible; see, e.g., Devaney [**50, 51**], Devaney and Keen [**52**] or Wiggins [**236**].

2.7. Asymmetry of role: Owners and Intruders

We have yet to consider the possibility that an individual's role during a conflict may vary, and that strategies may therefore be role-dependent. For example, in contests over a lady's heart among knights of The Middle Ages, a much recommended strategy was to be as gentle as a lamb in the role of courtier, but as fierce as a lion in the role of warrior. Insofar as rewards must take account of all possible changes of role throughout a conflict, game-theoretic modelling is more difficult when strategies are role-dependent. Nevertheless, if the strategy set is the same for the entire population, then it may still be possible to describe the conflict as a symmetric game, and to apply the concept of evolutionary stability. We illustrate this possibility by modelling a territorial conflict among animals, who may encounter one another in the role of either owner or intruder.

The existence of two roles—here owner and intruder—introduces an obvious asymmetry among players at any particular encounter. But the conflict consists of all encounters, and a strategy must be selected before the start of the conflict. Thus if all players initially face the same probabilities of being owner or intruder at any particular encounter, for any given strategy combination, then the game is still truly symmetric.

Before proceeding, we note that a role-dependent strategy is just one example of a strategy in which the action taken on some move depends upon the value then realized by some random variable. The nature of this dependence is fully known at the start of the game—otherwise the rule for action couldn't be a strategy—but the precise action taken is conditional upon the value realized by the random variable. We will refer to such strategies as conditional strategies.[7] Thus a role-dependent strategy is a conditional strategy. But the random variable in question need not be a player's role: it could instead, for example, be a physiological state variable, as we shall

---

[7]A concern about conditional strategies has been the extent to which they are heritable. For work that addresses this issue, see [**80, 94**].

briefly discuss at the end of §2.9; or the action taken by an opponent on a previous encounter, as we shall see in Chapter 5.

Having negotiated these preliminaries, let us now imagine that a violent hurricane has wracked the homes of some population of animals, of whom there are, say, $N + 1$. But the hurricane has also created some new habitat, in which there are M suitable sites for a home; and naturally, our animals come scurrying in to find one. We will assume (except in Exercise 2.23) that there are at least enough sites for all, i.e., $N + 1 \leq M$; hence $\sigma < 1$, where we define

$$(2.55) \qquad\qquad \sigma \;=\; \frac{N}{M}.$$

The sites, which we assume to be equally valuable to all animals, are randomly distributed across the habitat, however; and in each unit of time there is only probability ϵ that an animal will find a site. Furthermore, there is no guarantee that the site so found isn't already occupied by another animal; in which case, we shall refer to the occupier as the owner, and to the other animal as the intruder. The intruder can either attack the owner, in the hope of obtaining a site without further search, or surrender the site to its owner and look for another one. Likewise, if the intruder attacks, then the owner can either also attack, or else surrender the site to the intruder—who then becomes the owner—and search for another. What should an animal do in these circumstances—attack, surrender, or a bit of both?

To obtain an answer to this question—more precisely, to determine an ESS for the population—we make some rather bold assumptions. If two animals actually engage in a fight, then they are equally likely to win, and the loser is always injured before surrendering; but there is probability λ that the victor is also injured. Injured animals do not recover (within the time scale of the conflict), and are unable to fight or search for a site; and sites are worthless to an injured animal. Thus fights should not be lightly undertaken. Finally, we will assume that the animals cannot wait forever for a site; rather, to achieve their long-term objective of reproduction, they must secure a site within K units of time. We could imagine, perhaps, that the unit of time is an hour, that the new habitat becomes available at dawn, and that a site is required by dusk; in which case, a reasonable value for K might be 15, and we shall use this later for illustration. But the precise value of K will turn out to be unimportant, provided only that K is neither too large nor too small; see Exercises 2.15 and 2.16.

To help us keep track of the population, let us now define four possible states that an animal can occupy, namely:

(2.56)   1. Uninjured owner   3. Uninjured non-owner
           2. Injured owner   4. Injured non-owner

Thus an animal's objective is to be in state 1 at time K. We will analyze the conflict over territories as a game restricted to the following four pure strategies for uninjured animals:

(2.57)

HH: Attack if owner, attack if intruder
HD: Attack if owner, surrender if intruder
DH: Surrender if owner, attack if intruder
DD: Surrender if owner, surrender if intruder

where H stands for Hawk and D stands for Dove, as in §1.2; HH and DD are unconditional strategies, whereas HD and DH are conditional strategies.[8] Because we have assumed that injured animals are unable to search or fight, the question of strategy for an injured animal does not arise. If the injury is sustained in losing, then the animal remains in state 4 until time K; whereas if the injury is sustained in winning, then the animal remains on its site either until it is attacked by an (uninjured) intruder, in which case it surrenders the territory immediately, or until $t = K$ (if no one attacks). We will call this game Owners and Intruders. The strategy HD is often called Bourgeois [**132**, p. 94], because it respects and defends property rights; consequently, the strategy DH is also known as anti-Bourgeois.

Although Owners and Intruders is a game restricted to pure strategies, it will facilitate analysis if we first define appropriate mixed strategies (and later concentrate on the special case of interest). Accordingly, let u_1 be the (conditional) probability that an animal attacks if he is an owner, and u_2 the probability that an animal attacks if he is an intruder. Then an animal's strategy can be represented by

---

[8]This notation enables us to distinguish readily between strategies and tactics. In general, a *tactic* is an action that a player can take during a game, and a strategy is a rule for deploying tactics. There is no difference between strategies and tactics in the Hawk-Dove game of §2.3; but here, for example, H is one of two possible tactics, whereas HH is a strategy. Both are all Hawk, but they are not the same thing.

the vector $u = (u_1, u_2)$, with

(2.58) $\quad HH = (1,1), \quad HD = (1,0), \quad DH = (0,1), \quad DD = (0,0).$

It is important to note that the strategy $u = (u_1, u_2)$ differs from the mixed strategy we defined in §1.4, because the (conditional) probabilities u_1 and u_2 are completely independent, and so both can take any value between 0 and 1.[9] In other words, we allow (u_1, u_2) to be any point in the unit square; whereas, in §1.4, (u_1, u_2) was confined to the triangle (1.28).

As usual, to determine whether u is an ESS, we must obtain an expression for the reward $f(u, v)$ to a focal u-strategist, who is a potential mutant, in a population of v-strategists. Several steps are required to derive this expression (so let us be patient). To begin with, let the random variable $X_u(t)$ denote a u-strategist's state at time t, so that X_u has sample space $\{1, 2, 3, 4\}$. We suppose that animals can change their state only at discrete instants of time, say $t = k$, where k is a positive integer; and we denote the probability that a u-strategist is in state j at time k (i.e., immediately after time k) by $x_j(k)$. Thus

(2.59a) $\qquad x_j(k) \;=\; \text{Prob}(X_u(k) = j)$

for $j = 1, \ldots, 4$. Strictly, because x_j depends on u, we should use the notation $x_j(k, u)$—but that would be unnecessarily cumbersome. Instead, therefore, we shall use $x_j(k)$ for the probability that a u-strategist is in state j at time k, and

(2.59b) $\qquad y_j(k) \;=\; \text{Prob}(X_v(k) = j)$

(rather than $x_j(k, v)$) for the probability that a v-strategist is in state j at time k. Then, because any animal, whether u-strategist or v-strategist, must be in one of the four states defined by (2.56) at time k, we have

(2.60)
$$x_1(k) + x_2(k) + x_3(k) + x_4(k) \;=\; 1$$
$$y_1(k) + y_2(k) + y_3(k) + y_4(k) \;=\; 1$$

---

[9]Game theorists distinguish between mixed strategies, which randomize over pure strategies at the start of a game, and behavior or stochastic strategies, which randomize over tactics (defined in the previous footnote) during a game. Technically, u is a stochastic strategy, not a mixed one; but we have no use for the distinction, and so we will call either type a mixed strategy. For further discussion, see, e.g., [**206**, pp. 44-45].

for $0 \leq k \leq K$. Furthermore, if animals are injured only by fighting (animals injured by the hurricane fail to reach the new habitat), then all animals must be in state 3 initially; and so if $x(k)$ denotes the vector $(x_1(k), x_2(k), x_3(k), x_4(k))$ and $y(k)$ the vector $(y_1(k), y_2(k), y_3(k), y_4(k))$ for any $k \geq 0$, then

$$(2.61) \qquad\qquad x(0) \;=\; (0,0,1,0) \;=\; y(0).$$

Let $\phi_{ij}(k, u, v)$ be the conditional probability that a u-strategist in a population of v-strategists is in state j at time $k+1$, given that he is in state i at time k; that is, define

$$(2.62) \qquad \phi_{ij}(k, u, v) = \mathrm{Prob}\,(X_u(k+1) = j \mid X_u(k) = i)$$

for $1 \leq i, j \leq 4$, where $\mathrm{Prob}\,(Y \mid Z)$ is standard notation for the conditional probability of Y, given Z. This time, it really is necessary to use more explicit notation, because we have to distinguish between, on the one hand, the probability that a u-strategist goes from state i to state j, which is $\phi_{ij}(k, u, v)$; and, on the other hand, the probability that a v-strategist goes from state i to state j, which is

$$(2.63) \qquad \mathrm{Prob}\,(X_v(k+1) = j \mid X_v(k) = i) = \phi_{ij}(k, v, v).$$

We make at this point an additional assumption, namely, N is so large that every v-strategist is effectively playing only against v-strategists; although the population contains a single u-strategist, if N is very large then the probability that a v-strategist will meet that u-strategist is negligible. We now have

$$x_j(k+1) \;=\; \sum_{i=1}^{4} \mathrm{Prob}\,(X_u(k+1) = j \mid X_u(k) = i) \cdot \mathrm{Prob}\,(X_u(k) = i)$$

$$y_j(k+1) \;=\; \sum_{i=1}^{4} \mathrm{Prob}\,(X_v(k+1) = j \mid X_v(k) = i) \cdot \mathrm{Prob}\,(X_v(k) = i)$$

for $j = 1, ..., 4$ and $k = 0, \dots, K - 1$; or, on using (2.62)-(2.63),

$$(2.64a) \qquad\qquad x_j(k+1) \;=\; \sum_{i=1}^{4} \phi_{ij}(k, u, v) x_i(k)$$

$$(2.64b) \qquad\qquad y_j(k+1) \;=\; \sum_{i=1}^{4} \phi_{ij}(k, v, v) y_i(k).$$

Note that

$$(2.65) \qquad\qquad \sum_{j=1}^{4} \phi_{ij}(k, u, v) \;=\; 1, \quad 1 \leq i \leq 4, \; 0 \leq u, v \leq 1,$$

because if a u-strategist is in state i at time k, then at time $k + 1$ it must either remain in state i or enter one of the other states.

An animal's objective is to end the conflict as uninjured owner of a site, and so it is reasonable to regard

$$(2.66) \qquad F(u, v) = \begin{cases} 1 & \text{if } X_u(k) = 1 \\ 0 & \text{if } X_u(k) \geq 2 \end{cases}$$

as the payoff to a u-strategist among v-strategists. But F, so defined, is a random variable, which cannot itself be maximized; therefore, in the usual way, we assume instead that the reward to a u-strategist is

$$(2.67) \qquad f(u, v) = \mathrm{E}\,[F(u, v)]$$
$$= 1 \cdot \mathrm{Prob}\,(X_u(K) = 1) + 0 \cdot \mathrm{Prob}\,(X_u(K) \geq 2)$$
$$= x_1(K),$$

where E denotes expected value. Here, we have again taken liberties with notation, because the dependence of $f(u, v)$ on K is suppressed on the left-hand side of (2.67); whereas the dependence of $x_1(K)$ on u and v is suppressed on the right-hand side. But any other notation would be unnecessarily cumbersome.

We can deduce $x_1(K)$ from (2.61) and (2.64) if we first obtain an explicit expression for the 4×4 matrix $\phi(k, u, v)$, defined by (2.62). The key assumptions we make in this regard are that searching animals are equally likely to find any site; that at most one animal intrudes upon a site per unit of time; and that animals and sites are both so numerous that $N \to \infty$ and $M \to \infty$, but in such a way that σ defined by (2.55) is finite. The first assumption may be questionable, but it greatly simplifies analysis. On the other hand, the second assumption has been forced upon us by our earlier assumption that an animal can change state only at time $t = k$, where k is an integer; if we allowed multiple intrusions during the same unit of time, then we would also have to allow multiple changes of state. Similarly, the third assumption has been forced on us by the assumption we made to arrive at (2.64). How unreasonable in practice are these assumptions? See Exercise 2.17.

Granted these assumptions, let us first suppose the u-strategist to be in state 1 (uninjured owner) at time k. Then it can move into state 4 (injured non-owner) only if it is intruded upon, is attacked, attacks back, and then loses; whereas it can move into state 3 (uninjured non-owner) only if it is intruded upon, is attacked, and promptly surrenders; and it can move into state 2 (injured owner) only if it is intruded upon, is attacked, attacks back, and then wins, but sustains an injury. At time k, the expected number of uninjured non-owners is $Ny_3(k)$, because all of them are v-strategists; and each finds a site between time k and time $k + 1$ with probability ϵ. Because, as we have just assumed, they are all equally likely to find any site, the probability that one of them finds the particular site now occupied by our u-strategist is ϵ/M. The probability that one of them does not find the u-strategist is therefore $1 - \epsilon/M$; the probability that none of them finds the u-strategist is that number raised to the power of $Ny_3(k)$; and so the probability that at least one of them intrudes is

$$(2.68) \qquad 1 - \left(1 - \frac{\epsilon}{M}\right)^{Ny_3(k)} \equiv 1 - \left(1 - \frac{\epsilon}{M}\right)^{M\sigma y_3(k)},$$

by (2.55). Thus, in the limit as $M \to \infty$, the probability that at least one uninjured non-owner locates the u-strategist is $q(y_3(k))$, where

$$(2.69) \qquad q(y_3) = 1 - e^{-\epsilon\sigma y_3}$$

(and e denotes the exponential function). In view of our second assumption above, we interpret $q(y_3(k))$ as the probability that an uninjured non-owner intrudes during the interval between time k and time $k + 1$. (In the event that more than one animal locates the site, we must simply assume that the u-strategist interacts with the first, and that later arrivals ignore them both.) Then, because another animal attacks as intruder with probability v_2, and because the u-strategist attacks as owner with probability u_1, and loses with probability $\frac{1}{2}$ (conditional upon attacking), the probability that a non-owner intrudes *and* attacks, *and* that the u-strategist attacks *and* loses, is $q(y_3(k))$ times v_2 times u_1 times $\frac{1}{2}$; all conditional, of course, on the u-strategist being in state 1, and assuming that the intruder's attack probability is independent of the u-strategist's surrender probability. We have thus established that

$$(2.70a) \qquad \phi_{14}(k, u, v) = \frac{1}{2}u_1 v_2 q(y_3(k)).$$

Similarly, because the u-strategist surrenders as owner with probability $1 - u_1$, we have

$$(2.70\text{b}) \qquad \phi_{13}(k, u, v) = (1 - u_1)v_2 q(y_3(k));$$

and, because the u-strategist wins with probability $\frac{1}{2}$ but (conditional upon winning) sustains an injury with probability λ, we have

$$(2.70\text{c}) \qquad \phi_{12}(k, u, v) = \frac{1}{2}\lambda u_1 v_2 q(y_3(k)).$$

The probability that the u-strategist remains in state 1 is now readily deduced from (2.65) with $i = 1$:

$$(2.70\text{d}) \qquad \phi_{11}(k, u, v) = 1 - \left(1 - \frac{1}{2}\{1 - \lambda\}u_1\right)v_2 q(y_3(k)).$$

Let us next suppose that the u-strategist is in state 2 (injured owner). Then, because the u-strategist does not recover,

$$(2.71\text{a}) \qquad \phi_{21}(k, u, v) = 0 = \phi_{23}(k, u, v);$$

the u-strategist either remains an injured owner or becomes an injured non-owner. Given that it is already injured, it surrenders its site if it is intruded upon—which happens with probability $q(y_3(k))$—and attacked, which happens with probability v_2. Thus, on using (2.65) with $i = 2$, we have

$$(2.71\text{b}) \qquad \phi_{22}(k, u, v) = 1 - v_2 q(y_3(k)) = 1 - \phi_{24}(k, u, v).$$

Third, let us suppose that the u-strategist is in state 3 (uninjured non-owner). Then it will descend into state 4 (injured non-owner) only if it intrudes upon an uninjured owner, attacks, is attacked back, and then loses. At time k, the expected number of uninjured owners is $Ny_1(k)$, because every one of them is a v-strategist; whence, if the u-strategist finds a site, then the probability that it is occupied by an uninjured owner is $Ny_1(k)/M = \sigma y_1(k)$. The u-strategist finds a site between time k and time $k + 1$ with probability ϵ; and if it is occupied, then the u-strategist attacks as intruder with probability u_2, is attacked back by the owner with probability v_1, and subsequently loses with probability $\frac{1}{2}$. Thus, multiplying all the conditional probabilities together, we have

$$(2.72\text{a}) \qquad \phi_{34}(k, u, v) = \frac{1}{2}\sigma\epsilon u_2 v_1 y_1(k).$$

Similarly, because (conditional upon an intrusion and engagement) the u-strategist wins and is injured with probability $\frac{1}{2}\lambda$, we have

(2.72b) $$\phi_{32}(k, u, v) = \tfrac{1}{2}\lambda\sigma\epsilon u_2 v_1 y_1(k).$$

Just as the conditional probability that a site is occupied by an uninjured owner is $\sigma y_1(k)$, so the probability that a site is occupied by an injured owner is $\sigma y_2(k)$. Thus the probability that the u-strategist finds a site, whose occupier is either injured or uninjured, *and* does not attack is ϵ times $\sigma\{y_1(k) + y_2(k)\}$ times $1 - u_2$. The probability that the u-strategist does not find a site is $1 - \epsilon$. In either case, it remains in state 3. Thus

(2.72c) $$\phi_{33}(k, u, v) = 1 - \epsilon + \sigma\epsilon(1 - u_2)\{y_1(k) + y_2(k)\};$$

and (2.65) with $i = 3$ implies

(2.72d) $$\phi_{31}(k, u, v) = \epsilon(1 - \sigma(1 - u_2)\{y_1(k) + y_2(k)\}) - \tfrac{1}{2}\sigma\epsilon(1 + \lambda)u_2 v_1 y_1(k).$$

Finally, because an injured animal is unable to search, even for unoccupied sites, we have

(2.73) $$\phi_{4j}(k, u, v) = 0, \quad j = 1, 2, 3, \qquad \phi_{44}(k, u, v) = 1.$$

The matrix $\phi(k, u, v)$ has now been defined.

Substitution of (2.70)-(2.73) into (2.64) leads to a set of eight first-order, nonlinear recurrence equations of the form

(2.74a)
$$x_j(k + 1) = H_j(x_1(k), \dots, x_4(k), y_1(k), \dots, y_4(k)),$$
$$y_j(k + 1) = H_{j+4}(x_1(k), \dots, x_4(k), y_1(k), \dots, y_4(k)),$$

$j = 1, \dots, 4$, where $H_1, \dots, H_8$ are functions of $x_1, \dots, x_4, y_1, \dots, y_4$ (whose dependence on u and v has been suppressed by the notation); see Exercise 2.18. If we relabel y_1 as x_5, y_2 as x_6, y_3 as x_7 and y_4 as x_8, then (2.74a) has the form of (2.53) with $m = 8$ and $G = H$:

(2.74b) $$x(k + 1) = H(x(k)), \qquad k = 0, \dots, K - 1.$$

Note that although (2.74) describes short-term dynamics, and (2.52) describes long-term dynamics, the two dynamics have a common mathematical structure.

Table 2.1. $x(K)$ and $y(K)$ for $\epsilon = 0.4$, $\lambda = 0.4$, $\sigma = 0.7$, $x(0) = (0, 0, 1, 0) = y(0)$ and $(u, v) = (0.3, 0.7, 0.6, 0.4)$.

K	$x_1(K)$	$x_2(K)$	$x_3(K)$	$x_4(K)$	$y_1(K)$	$y_2(K)$	$y_3(K)$	$y_4(K)$
1	0.3000	0.0000	0.7000	0.0000	0.3000	0.0000	0.7000	0.0000
2	0.4689	0.0047	0.5147	0.0117	0.4627	0.0041	0.5230	0.0102
3	0.5757	0.0099	0.3891	0.0253	0.5645	0.0087	0.4046	0.0223
4	0.6477	0.0145	0.2995	0.0382	0.6335	0.0129	0.3198	0.0338
5	0.6984	0.0185	0.2337	0.0494	0.6828	0.0166	0.2565	0.0442
6	0.7350	0.0218	0.1841	0.0591	0.7192	0.0197	0.2079	0.0532
7	0.7621	0.0245	0.1462	0.0672	0.7468	0.0224	0.1698	0.0610
8	0.7824	0.0268	0.1169	0.0739	0.7683	0.0246	0.1395	0.0677
9	0.7979	0.0286	0.0940	0.0795	0.7851	0.0264	0.1151	0.0733
10	0.8098	0.0301	0.0760	0.0842	0.7985	0.0280	0.0954	0.0781
11	0.8190	0.0313	0.0616	0.0881	0.8093	0.0293	0.0792	0.0821
12	0.8262	0.0323	0.0502	0.0913	0.8181	0.0304	0.0660	0.0856
13	0.8319	0.0331	0.0410	0.0940	0.8252	0.0313	0.0550	0.0884
14	0.8365	0.0338	0.0336	0.0962	0.8311	0.0321	0.0460	0.0909
15	0.8401	0.0343	0.0276	0.0980	0.8359	0.0327	0.0385	0.0929

For any strategy combination (u, v), and for any value of K, the vectors $x(K)$ and $y(K)$ are now readily calculated from K successive iterations of (2.74). From (2.61) and (2.64) with $k = 0$ we have

$$(2.75) \qquad x(1) = (\epsilon, 0, 1 - \epsilon, 0) = y(1),$$

whence from (2.64) with $k = 1$:

$$(2.76) \quad x_1(2) = \epsilon(1 - \epsilon) - \sigma\epsilon^2(1 - \epsilon)\left\{1 - u_2 + \tfrac{1}{2}(1 + \lambda)u_2 v_1\right\}$$
$$- \epsilon\left\{1 - e^{-\sigma\epsilon(1-\epsilon)}\right\}\left\{1 - \tfrac{1}{2}(1 - \lambda)u_1\right\} v_2.$$

From (2.67), if $K = 2$ then (2.76) is an expression for $f(u, v)$. For larger values of K, however, calculating f is clearly a task for the computer. Suppose, e.g., that $\epsilon = 0.4$, $\lambda = 0.4$, $\sigma = 0.7$ and $(u, v) = (0.3, 0.7, 0.6, 0.4)$, so that the u-strategist attacks with probabililty 0.3 as owner and 0.7 as intruder, whereas the rest of the population attack with probability 0.6 as owner and 0.4 as intruder. Then you can readily verify, by using a computer to solve (2.74) recursively, that the vectors $x(K)$ and $y(K)$ have the values given in Table 2.1, for various values of K; e.g., if $K = 15$, then $f(u, v) = 0.8401$.

Having demonstrated how to calculate $f(u, v)$ for arbitrary (u, v), let us now return to Owners and Intruders. In keeping with §2.6, we relabel HH as strategy 1, HD as strategy 2, DH as strategy 3 and DD as strategy 4; and, for $1 \leq i, j \leq 4$, we denote the reward to strategy i against a population using strategy j by a_{ij}. Then, from

Table 2.2. Payoff matrix in Owners and Intruders for various λ when $K = 15, \epsilon = 0.1, \sigma = 0.7$ and $x(0) = (0, 0, 1, 0) = y(0)$.

$\lambda = 0$	$\lambda = 0.2$	$\lambda = 0.9$
0.621 0.704 0.685 0.794	0.592 0.686 0.665 0.794	0.504 0.623 0.602 0.794
0.594 0.671 0.575 0.671	0.579 0.671 0.557 0.671	0.531 0.671 0.502 0.671
0.603 0.704 0.673 0.794	0.589 0.686 0.673 0.794	0.544 0.623 0.673 0.794
0.571 0.671 0.543 0.671	0.571 0.671 0.543 0.671	0.571 0.671 0.543 0.671

(2.58), we obtain the *reward matrix*

$$(2.77) \quad A = \begin{bmatrix} f(1,1,1,1) & f(1,1,1,0) & f(1,1,0,1) & f(1,1,0,0) \\ f(1,0,1,1) & f(1,0,1,0) & f(1,0,0,1) & f(1,0,0,0) \\ f(0,1,1,1) & f(0,1,1,0) & f(0,1,0,1) & f(0,1,0,0) \\ f(0,0,1,1) & f(0,0,1,0) & f(0,0,0,1) & f(0,0,0,0) \end{bmatrix},$$

which is readily evaluated by computer—provided, of course, we have first agreed on the value of K. Note that A's dependence on K is suppressed by the notation.

A brief digression is now in order on the distinction we choose to observe henceforth between a payoff and a reward matrix. We have used reward for the expected value of a payoff, and the entries in A above are indeed all expected values, whereas only some of the entries in the payoff matrices for Crossroads (§1.1) and Four Ways (§1.4) are expected values; but that is not the distinguishing characteristic. Rather, we will use reward matrix and payoff matrix, respectively, to distinguish between the case in which a_{ij} is the reward to an i-strategist from interacting with a population of j-strategists (as here) and the case in which a_{ij} is the reward or payoff to an i-strategist from a pairwise interaction with a particular j-strategist, albeit one drawn randomly from the population (as in §§1.1 and 1.4). Note, however, that (2.37) defines an ESS in either case.[10]

Returning to our animals, suppose for illustration that $K = 15$, $\epsilon = 0.1$ and $\sigma = 0.7$. Then for various values of λ, the payoff matrix is recorded in Table 2.2. By direct application of (2.37), we see that HH is the only ESS when $\lambda = 0$. Thus, when the price of victory is zero, a population of inveterate Hawks is uninvadable. When $\lambda = 0.2$, on the other hand, the price of victory—a 20% risk of being unable to profit from the victory through injury—is sufficiently high that the

---

[10]In terms of Maynard Smith [**132**, p. 23], we use payoff and reward matrix to distinguish between pairwise contests and playing the field, respectively.

game has two evolutionarily stable strategies, namely, HH and DH. A population of inveterate Hawks is still uninvadable, but a population of animals who surrender as owners but attack as intruders is also uninvadable. Which of these strategies is ultimately adopted by the population must be determined by long-term dynamics, corresponding to those of (2.52), and will depend on the proportions using different strategies initially (and has nothing to do with $x(0)$ and $y(0)$, which merely define the state in which a particular game begins). Finally, when $\lambda = 0.9$, the price of victory is so high that the only ESS is DH, and a population of inveterate fighters is readily invaded. If DD were absent from the population, then HD would also be evolutionarily stable (delete the last row and column from the third matrix in Table 2.2). But HD is only collectively stable in the presence of DD, which means that DD cannot be prevented from drifting into a HD population and changing its composition to a mixture of HD and DD; see Exercise 2.22. If this mixed population is now infiltrated by HH or DH, however, then (Exercise 2.22) the proportion of HD will decrease and the proportion of DD will increase, because DD does better than HD against either HH or DH; in terms of §2.6, the mixed population is metastable. If either HH or DH were to infiltrate repeatedly, then the proportion of DD could become quite large (with a corresponding reduction in the proportion of HD). But DD is not even collectively stable, and is readily invaded by either HH or DH. Thus, as illustrated in Exercise 2.22, the population would eventually evolve from the metastable mixture of HD and DD to the sole ESS, namely, DH.[11]

Two points are illustrated by the above analysis. First, the evolutionary stability of a particular strategy can depend on one of the game's parameters, e.g., λ in Table 2.2. Further examples of this possibility arise in Chapters 5 and 6. Second, and more importantly, whether a strategy is an ESS is always a function of the other strategies present (more fundamentally, a model's predictions are always a function of the assumptions on which it is based). When $\lambda = 0.9$ in

---

[11] If the frequency of one of the strategies in a metastable mixture decreases over time without interference by strategies other than those already represented in the mixture, then the reduction is said to be caused by *random drift*; whereas, if the frequency of a metastable strategy decreases because it does less well against an infiltrator than other strategies in the mixture, then the reduction is said to be caused by *selection pressure*.

Owners and Intruders, for example, the presence or absence of inveterate doves (DD) is critical to the evolutionary stability of HD. The relevance of this second point will emerge again in §5.4.

How do ϵ and σ affect evolutionarily stability? See Exercise 2.19. For fixed ϵ, σ and λ, how sensitive are our results to the value of K? See Exercise 2.15. More generally, how good are the assumptions on which our model is based? See Exercise 2.17.

2.8. Spiders in a spin—a case of anti-Bourgeois?

It may seem surprising that a population would always be meek in the role of occupier but aggressive in the role of intruder, despite this rule's advantage—when universally obeyed—that nobody would ever fight. After all, if someone knocked on your door tonight and demanded your property, would you turn it over and run off to attack the neighbor? Your property is more valuable to you than your neighbor's property, of course; and we made the assumption in §2.7 that sites were equally valuable to all animals. But even in cases where this assumption is reasonable, it is perhaps surprising that strategy DH would ever prevail in nature. Indeed it is far commoner for territorial conflicts to be settled by behavior that approximates the strategy HD—see [**132**, p. 97]. An example of behavior that resembles DH is therefore of interest.

According to Burgess [**33**], the behavior of *Oecobius civitas*, a Mexican spider, features a curious combination of tolerance and avoidance. Burgess provides the following description:

> On the underside of the rock that shelters the spiders each individual weaves a small open-ended tube of silk that is its hiding place; around this retreat the spider constructs a thin, encircling alarm-system net close to the surface of the rock. The pair of structures makes up the spider's web, which is generally found in a hollow or a crevice of the rock. If a spider is disturbed and driven out of its retreat, it darts across the rock and, in the absence of a vacant crevice to hide in, may seek refuge in the hiding place of another spider of the same species. If the other spider is in residence when the intruder enters, it does not attack but darts out and seeks a new refuge of its own. Thus once the first spider is disturbed the process of sequential displacement from web to web may continue for several seconds, often causing a majority of the spiders in the aggregation to shift from their home refuge to an alien one. ... Moreover, within the local population the shift to another spider's shelter may be a semipermanent move. The reason is that when the spiders are undisturbed, they occupy a fixed web position for long periods.

Table 2.3. Payoff matrix in Owners and Intruders for $K = 15$, $\epsilon = 0.1$, $\sigma = 0.7$ and $x(0) = (0.99, 0, 0.01, 0) = y(0)$.

$\lambda = 0$				$\lambda = 0.9$			
0.9924	0.9952	0.9937	0.9979	0.9875	0.9927	0.9899	0.9979
0.9910	0.9937	0.9895	0.9937	0.9885	0.9937	0.9857	0.9937
0.9917	0.9952	0.9937	0.9979	0.9883	0.9927	0.9937	0.9979
0.9895	0.9937	0.9870	0.9937	0.9895	0.9937	0.9870	0.9937

Could this be an example of strategy DH at work in nature? It appears from Burgess's account that virtually all spiders are uninjured homeowners when a disturbance occurs; and so if Owners and Intruders is to provide even an approximate description of what happens subsequently, then we should at least start the game with virtually all players in state 1, and just a few in state 3. If, for example, 1% of owners are disturbed initially, then we should replace (2.61) by $x(0) = (0.99, 0, 0.01, 0) = y(0)$. But the payoff matrix (2.77) is still readily computed from successive iterations of the recurrence equation (2.74), and our conclusions are essentially the same. When $K = 15, \epsilon = 0.1$ and $\sigma = 0.7$, for example, the payoff matrices for $\lambda = 0$ and $\lambda = 0.9$ are as shown in Table 2.3. We see that HH is still the unique ESS when $\lambda = 0$, and that DH is still the unique ESS when $\lambda = 0.9$.

The values we have chosen for $\epsilon, \lambda, \sigma$ and K are totally arbitrary. But whether a strategy is an ESS is essentially unaffected by K; and its dependence on ϵ, λ and σ is quite robust, depending only on whether these parameters are small or large—see Exercises 2.15 and 2.19. We can therefore hazard a cautious guess as to why *O. civitas* behaves as it does. Let us posit that the supply of sites is adequate ($1 - \sigma$ sufficiently large); and that one spider's web is as good as another's in terms of snaring prey, so that sites are all equally valuable to all animals. Then our model suggests that the spider *O. civitas* may have evolved to DH because DD—inveterate Dove—has arisen too often as a mutant strategy for the less paradoxical HD to be evolutionarily stable; and because the probability of winning a fight without injury, i.e., $1 - \lambda$, is too low for HH to be evolutionarily stable. Of course, this is mere speculation—but we cannot explain observed behavior unless first of all we speculate about it!

On the other hand, even speculative models should at least be self-consistent, and in Exercise 2.17 you were asked to criticize some assumptions we made in §2.7. Perhaps you wondered what kind of animal could be so mobile as to be equally likely to reach all sites in a unit of time, in the limit as $M \to \infty$. We can rescue this assumption to some extent by supposing that the habitat is small—and Burgess does say that *O. civitas* lives under crowded conditions. But if the habitat is so small, can we really assume that at most one animal intrudes on a site per unit of time? We can also rescue this assumption by supposing that the unit of time is small—and Burgess's spiders exchange their homes on a time scale of several seconds (if only realtors were so efficient). Because K has essentially no effect on the ESS, decreasing the time unit (which, for a contest of given duration, is equivalent to increasing K) should leave our conclusions more or less intact. In practice, however, we would achieve the same end by choosing some common time unit—a second, say—and simply supposing that ϵ is small. Then $q(y_3)$ in (2.69) can be approximated by $\epsilon \sigma y_3$; see Exercise 2.20.

2.9. Commentary

In Chapter 2 we have introduced three criteria for distinguishing among Nash equilibria, due to Harsanyi and Selten (§2.1), Kalai and Samet (§2.2) and Maynard Smith (§2.3). Theirs are by no means the only criteria, but were chosen to illustrate both a diversity of ideas and a variety of predictions; and, in the case of Maynard Smith's criterion, because the concept of evolutionary stability is central to §§2.4-2.8 and Chapters 5-6. Alternative criteria—amply expounded in the monograph by van Damme [225] and critiqued in Chapter 5 of Kreps [116]—include Selten's concept of perfectness (insensitivity to arbitrary small random errors in choosing a strategy) and Myerson's concept of properness (modified perfectness that assigns lower probability to a more costly error than to a less costly error). In the symmetric version of Crossroads, the equilibrium selected by evolutionary stability is both perfect and proper (hence, for the sake of variety, our choice of persistence). Indeed there is at least some truth to the notion that perfectness and properness belong to the same conceptual flock as evolutionary stability; whereas the tracing procedure—described in Appendix A—is arguably a bird of a different feather. Furthermore, Harsanyi and Selten's criterion, augmented where necessary by their "logarithmic tracing procedure" [**93**], is the only criterion that guarantees uniqueness.

In §2.4, we transformed the symmetric version of Crossroads into a continuous population game to demonstrate that the concept of evolutionary stability is by no means restricted to bimatrix games. In §§2.5-2.6, we discussed population dynamics in terms of strategy composition, and we adapted our definition of evolutionary stability to games restricted to pure strategies—discrete population games. In §§2.7-2.8, we further applied the concept of evolutionary stability, this time to a territorial conflict in which players can change their roles.

There now exists a substantial research literature on ESSes and related concepts; see, e.g., [**4, 45, 86, 96, 118, 194, 229**], and further references in §5.11 and §6.9. More advanced texts include [**43, 99**] from the perspective of biology and [**197, 226, 234**] from that of economics. Much of this literature deals with bimatrix games, although—as we have seen in §2.4 and §2.7—the scope of Maynard Smith's solution concept is far more general; also, much of the literature assumes an infinite population, although some of it—e.g., [**42, 133, 198**]—considers a finite population. Infinite- versus finite-population effects will be studied in §§5.5-5.7 in the context of a discrete population game.

How much is lost in the restriction to pure strategies? If one takes the view that mixed strategies are merely proxies for unmodelled aspects of a conflict, then the answer in theory is nothing. An example will help to illustrate. Recall once more the game of Four Ways, and suppose that drivers do not have spinning arrows on their dashboards; rather, they select pure strategy C when their hearts are racing, pure strategy G when both their hearts are racing and they have sweaty palms, and otherwise strategy W. Nevertheless, the probabilities of being in these various states of agitation correspond to the probabilities of selecting G, W or C in the ESS of §2.3. Thus, when behavior is averaged over time, drivers appear to adopt a mixed strategy; yet in fact they adopt a conditional pure strategy that is state-dependent (just as strategies in Owners and Intruders were role-dependent). A conditional pure strategy can also be time-dependent; perhaps, for example, drivers select C when their palms are sweaty on Mondays, Wednesdays and Thursdays, but G if their palms are sweaty on Tuesdays and Fridays. More generally, a player's state can be defined to incorporate almost any physiological or environmental variable; and if decisions can depend on both state and time in a sufficiently general way, then it is arguable that mixed strategies are quite unnecessary. (In practice, of course, it may not be so easy to ensure that all relevant physiological or environmental variables are adequately represented.)

Here two remarks are in order. First, although the decision rule in such games is deterministic—in the sense that the functional relationship between state, time and action taken is deterministic—the state remains a random variable; thus the "internal" uncertainty that mixed strategies bring to a player's decisions is replaced by the "external" uncertainty of a player's state. Second, we must ultimately allow for mistakes by the players. But this uncertainty can also be externalized by considering purely deterministic (but conditional) decision rules of the form, "If you think that the state variable is that, then do this" (and then computing rewards by taking expectations over an appropriate probability distribution).

All things considered, state-dependent, dynamic games with conditional pure strategies are far more appealing as behavioral models than static games with mixed strategies; see Chapter 7 of Houston and McNamara [100], and references therein.[12] Analysis of such games is often extremely difficult, however; and so even if in theory one regards mixed strategies as merely a stopgap, one is often in practice still grateful to have them.

Exercises 2

1. Show that $u = 1$ is an uninvadable strategy of the symmetric version of Crossroads when $\theta > 1$. (Show that $f(1,1) > f(u,1)$ for all u such that $0 \leq u < 1$.) This appears to say that if both drivers are slow then both drivers should select G. Do you think

---

[12]Dynamic games rely heavily on stochastic dynamic programming, to which Clark and Mangel [38] is a gentler introduction (although it has little to say about dynamic games per se).

it would be better to replace "both drivers are slow" by "the junction is fast?" Comment.

2. (a) Show that any uninvadable strategy is a Nash-equilibrium strategy.

   (b) Show that if v^* is a strongly uninvadable strategy, then (v^*, v^*) must be a strong Nash equilibrium.

   (c) If (u^*, v^*) is a strong Nash equilibrium, are u^* and v^* strongly uninvadable?

3. Using (2.10) and (2.11), verify that $v^* = \theta$ is an ESS for the Hawk-Dove game when $\theta < 1$, whereas $v^* = 0$ and $v^* = 1$ are not; here θ is defined by (2.16).

4. Verify (2.19).

5. Show that if $\delta > \frac{1}{2}\tau$ in Four Ways, then the uninvadable strategy mixes all three pure strategies with positive probability, but that C should not be selected with greater probability than $\frac{1}{3}$.

6. Show that Bourgeois is an evolutionarily stable strategy in the Hawk-Dove-Bourgeois game defined in Exercise 1.24.

7. Find all evolutionarily stable strategies in the prisoner's dilemma game, defined by Exercise 1.25.

8. Show that $R = (0,0)$ and $\left(\frac{1}{2}, \frac{1}{2}\right)$, i.e., an equal mixture of H and D, are both evolutionarily stable strategies in the Hawk-Dove-Retaliator game defined by Exercise 1.27.

9. Show that $\left(\frac{1}{3}, \frac{1}{3}\right)$, in which each pure strategy is played with the same probability, is an ESS of the game defined by Exercise 1.28 if $\lambda > 0$, but that there is no ESS if $\lambda < 0$.

10. (a) Find the rational reaction set for Crossroads II with slow drivers, thus verifying Figure 2.5 and (2.31).

   (b) Show that the average delay in Crossroads II decreases with respect to η.

   (c) Generalize Crossroads II to allow for an arbitrary nonuniform distribution of lateness, i.e., for any g such that $g(\xi) \geq 0$ and $\int_0^1 g(\xi)\, d\xi = 1$ in place of (2.23). Assume $\delta > \frac{1}{2}\tau$.

11. In §2.3 we showed that strategy θ in Crossroads could not be invaded by strategy 0 or strategy 1 in isolation. Reconcile this conclusion with dynamical model (2.35) in §2.5.

12. Show that if (2.38) is satisfied, then the only evolutionarily stable strategies—pure or mixed—of the symmetric game with payoff matrix $A = \left[\begin{smallmatrix} a_{11} & a_{12} \\ a_{21} & a_{22} \end{smallmatrix}\right]$ are the two pure strategies.

13. Verify (2.43).

14. **(a)** Establish (2.49).

  (b) Show that (2.48) and (2.52) imply $x_1(n) + x_2(n) + \ldots + x_m(n) = 1, 0 \le n < \infty$.

15. Demonstrate that the precise value of K in Owners and Intruders is unimportant, provided it is neither too large nor too small.

16. Suppose that K in Owners and Intruders is a random variable whose distribution is known. How must the analysis be modified? How are the conclusions affected?

17. How consistent are our assumptions in §2.7 that searching animals are equally likely to find any site, that at most one animal intrudes upon a site per unit of time, and that animals and sites are infinitely numerous, but in such a way that σ defined by (2.55) is finite?

18. Obtain $H_1, \ldots, H_8$ in equation (2.74a) as functions of $x_1, \ldots, x_4$, $y_1, \ldots, y_4$.

19. How does the ESS in Owners and Intruders depend on ϵ and σ?

20. Use Taylor expansion of (2.69) to show that if ϵ is small, then $q(y_3)$ in Owners and Intruders can be replaced by $\epsilon \sigma y_3$. How good is this approximation in practice?

21. In the context of discrete population games (§2.6):

  (a) Show that any ESS is stable against pure infiltration.

  (b) Show that a strong ESS is stable against mixed infiltration.

  (c) Show that a dominant strategy is also a strong ESS.

22. In this exercise, you will use a calculator or computer to solve (2.52) with Table 2.2's payoff matrix for $\lambda = 0.9$.

  (a) Take $x(0) = (\delta_1, \alpha, \delta_2, \delta_3)$, where δ_1, δ_2 and δ_3 are small positive numbers and $\alpha = 1 - \delta_1 - \delta_2 - \delta_3$, and solve (2.52) for various values of α close to 1 (e.g., in the range $0.9 < \alpha < 1$). Show that $x(n) \to (0, c, 0, 1 - c)$ as $n \to \infty$, where c is close to 1 but $c < \alpha$.

  (b) For each value of c in **(a)**, solve (2.52) for various values of $x(0) = (\delta_1, c - \delta_1, \delta_2, 1 - c - \delta_2)$, where δ_1 and δ_2 are small positive numbers. Show that $x(n) \to (0, c_1, 0, 1 - c_1)$ as

$n \to \infty$, where c_1 is close to c but $c_1 < c$. Together, **(a)** and **(b)** illustrate how infiltration by HH or DH can steadily reduce the proportion of HD and increase the proportion of DD.

(c) Now solve (2.52) with $x(0) = \frac{1}{3}(1-\alpha, 3\alpha, 1-\alpha, 1-\alpha)$ for various values of $\alpha \in (0, 1)$. Show that there is a critical value of α, say α_c, such that if $\alpha > \alpha_c$ then $x(n) \to (0, c, 0, 1-c)$ as $n \to \infty$, where $c < \alpha$; but if $\alpha < \alpha_c$, then $x(n) \to (0, 0, 1, 0)$ as $n \to \infty$. These dynamics illustrate that repeated infiltration by HH and DH will ultimately cause the population to evolve to DH. What is the value of α_c?[13]

(d) Criticize the use of (2.52) with (2.50) and (2.77) for the long-term dynamics of Owners and Intruders.

23. Analyze the game of Owners and Intruders for $\sigma > 1$, where σ is defined by (2.55). How must (2.72) be modified?

24. In Owners and Intruders we assumed that if two animals actually engage in a fight then they are equally likely to win, and that sites are worthless to an injured animal. Analyze a game in which these assumptions are relaxed.

25. Find all evolutionarily stable strategies for the game defined in Exercise 1.32. Given that there is more than one ESS, which ESS will establish itself? Does it matter whether $\tau_1 = \tau_2$ or $\tau_1 \neq \tau_2$?

---

[13]$x(0)$ in this exercise is the initial composition of the population in §2.6; it has nothing to do with the initial distribution of states for a play of Owners and Intruders.

Chapter 3

Cooperative Games in Strategic Form

Consider a game between two players, and let each choose a single scalar variable—u for the first player, v for the second. Then to every feasible strategy combination (u, v), i.e., to every (u, v) in the decision set D, there corresponds a vector of rewards $f = (f_1, f_2)$; and the equations $f_1 = f_1(u, v)$, $f_2 = f_2(u, v)$ define a vector-valued function, f, from the decision set D, which is a subset of the u-v plane, into the f_1-f_2 plane. In Crossroads, for example, f_1 and f_2 are defined by (1.13)–(1.14). The range of the function f, i.e., the subset of the f_1-f_2 plane onto which f maps D, is known as the *reward set*; it contains all reward vectors that are achievable by some combination of strategies in D. We denote the reward set by $\overline{F}$, or $f(D)$. Note that f is not in general invertible; i.e., the equations $f_1 = f_1(u, v)$, $f_2 = f_2(u, v)$ do not define a (single-valued) function from $\overline{F}$ onto D. For example, we saw in Exercise 1.19 that when both drivers in Crossroads are fast, the joint max-min strategy combination $(\tilde{u}, \tilde{v})$ and the Nash-equilibrium strategy combination (θ_1, θ_2) are mapped to the same point in $\overline{F}$ by f; however, $(\tilde{u}, \tilde{v}) \neq (\theta_1, \theta_2)$.

The concept of reward set is readily generalized to games among n players, the k-th of whom selects an s_k-dimensional strategy vector; f is then a vector-valued function from a space of dimension $s_1 + s_2 + \ldots + s_k$ into a space of dimension n. But the concept is most useful for $n = 2$. For $n = 3$ the reward set is difficult to sketch, and for $n > 3$ it is difficult even to visualize.

For games between specific individuals who are able to make binding agreements with one another if it benefits them to do so, Nash's concept of noncooperative equilibrium loses its appeal to various cooperative solution concepts (the most enduring of which is still due to Nash). In understanding why, we will find that a picture of the reward set is worth several thousand words. Accordingly, an analysis of the reward set for the game of Crossroads will serve as our springboard to cooperative games. In this chapter, we refer to such games as cooperative games in strategic form—merely to distinguish them from cooperative games in characteristic-function form, which are the subject of Chapter 4.

We begin, as promised, by returning to Chapter 1's game of Crossroads. For the sake of simplicity, we assume throughout that Nan and San are both fast drivers. Therefore

$$(3.1) \qquad 2\delta \; > \; 2\epsilon \; > \; \max(\tau_1, \tau_2).$$

3.1. Unimprovability: group rationality

For Crossroads, the reward vector f is defined by (1.13)–(1.14), i.e.,

$$(3.2a) \quad f_1(u, v) \; = \; (\epsilon + \tfrac{1}{2}\tau_2 - \{\delta + \epsilon\}v)u + (\epsilon - \tfrac{1}{2}\tau_2)v - \epsilon - \tfrac{1}{2}\tau_2$$

$$(3.2b) \quad f_2(u, v) \; = \; (\epsilon + \tfrac{1}{2}\tau_1 - \{\delta + \epsilon\}u)v + (\epsilon - \tfrac{1}{2}\tau_1)u - \epsilon - \tfrac{1}{2}\tau_1.$$

Suppose, for example, that the game is symmetric with

$$(3.3) \qquad \delta = 3, \qquad \epsilon = 2, \qquad \tau_1 = 2 = \tau_2.$$

Then

$$(3.4) \qquad f_1 = (3 - 5v)u + v - 3, \qquad f_2 = u - 3 + (3 - 5u)v.$$

The reward set $\overline{F}$ is the image of the decision set

$$(3.5) \qquad D \; = \; \{(u, v) \mid 0 \le u, v \le 1\}$$

under the mapping defined by (3.4). To obtain $\overline{F}$, imagine that D is covered by an infinity of line segments parallel to the v-axis. On each of these line segments u is constant, but v increases from 0 to 1. Let us first obtain the image under (3.4) of one of these line segments,

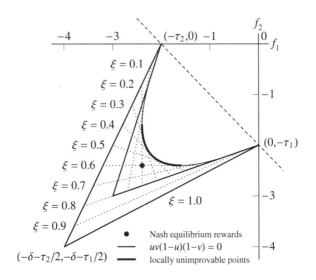

Figure 3.1. The reward set for Crossroads, where $(0, -\tau_1)$ and $(-\tau_2, 0)$ are globally unimprovable. The dashed line on which these points both lie has equation $\tau_1 f_1 + \tau_2 f_2 + \tau_1 \tau_2 = 0$.

then allow u to vary between 0 and 1. The images of all the line segments together (with some duplication) will constitute $\overline{F}$.

Consider, therefore, the line segment

$$(3.6) \qquad L(\xi) = \{(\xi, v) \mid 0 \le v \le 1\}$$

on which $u = \xi = $ constant (and $0 \le \xi \le 1$). Its image under the mapping f, $f(L(\xi))$, is the set of all points (f_1, f_2) such that

$$(3.7) \qquad f_1 = 3(\xi - 1) + (1 - 5\xi)v, \qquad f_2 = \xi - 3 + (3 - 5\xi)v$$

for some $v \in [0, 1]$. Eliminating v, we see that $f(L(\xi))$ is part of the line in the f_1-f_2 plane with equation

$$(3.8) \qquad (3 - 5\xi)f_1 + (5\xi - 1)f_2 + 10\xi^2 - 8\xi + 6 = 0;$$

see Exercise 3.1. But $f(L(\xi))$ is not the whole of this line, because $0 \le v \le 1$; rather, it is that part which extends in the f_1-f_2 plane from the point $(3\xi - 3, \xi - 3)$, corresponding to $v = 0$, to the point $(-2 - 2\xi, -4\xi)$, corresponding to $v = 1$. The line segment $f(L(\xi))$ is sketched in Figure 3.1 for values of ξ at increments of 0.1 between 0 and 1. For example, $f(L(0))$, which is the image of side $u = 0$ of the unit square, is that part of the line $3f_1 - f_2 + 6 = 0$ which stretches from $(-3, -3)$ to $(-2, 0)$. It is marked in Figure 3.1 as the upper prong of the smaller V-shape, the lower prong of which is the image of side $v = 0$ of the unit square and satisfies $f_1 - 3f_2 - 6 = 0$. Again, $f(L(1))$, which is the image of side $u = 1$ of the unit square, is that part of the line $f_1 - 2f_2 - 4 = 0$ which stretches from $(-4, -4)$ to $(0, -2)$. In Figure 3.1 it is the lower prong of the larger V-shape, the upper prong of which is the image of side $v = 1$ of the unit square and satisfies $2f_1 - f_2 + 4 = 0$. And so on (Exercise 3.1).

A glance at Figure 3.1 now reveals that the reward set is a curvilinear triangle, which is reminiscent of the open mouth of a fledgling (and the picture is similar for all other values of δ, ϵ, τ_1 and τ_2 such that (3.1) is satisfied). The straight edges of the triangle have equations $2f_1 - f_2 + 4 = 0 = f_1 - 2f_2 - 4$. The curved edge of the triangle is the curve to which every line segment in the family

$$(3.9) \qquad \qquad \{f(L(\xi)) \mid 0 \le \xi \le \tfrac{4}{5}\}$$

is a tangent somewhere; we call this curve the *envelope* of the family. The envelope has two straight segments. The first runs between $\left(-\frac{11}{5}, -\frac{3}{5}\right)$ and $(-2, 0)$; it corresponds to limiting member $L(0)$ of family (3.9), and hence has equation $3f_1 - f_2 + 6 = 0$. The second straight segment of the envelope runs between $\left(-\frac{3}{5}, -\frac{11}{5}\right)$ and $(0, -2)$; it corresponds to limiting member $L(4/5)$ of family (3.9), and so has equation $f_1 - 3f_2 - 6 = 0$.

To find the equation of the curved part of the envelope between $\left(-\frac{11}{5}, -\frac{3}{5}\right)$ and $\left(-\frac{3}{5}, -\frac{11}{5}\right)$, suppose that the line segment $f(L(\xi))$ has equation $\psi(f_1, f_2, \xi) = 0$, so that ψ is defined by the left-hand side of (3.8); and that $f(L(\xi))$ touches the envelope at the point with coordinates $(F_1(\xi), F_2(\xi))$, so that the parametric equations of the envelope are $f_1 = F_1(\xi), f_2 = F_2(\xi), 0 \le \xi \le 4/5$. Then the vector normal to $f(L(\xi))$ has direction $(\partial\psi/\partial f_1, \partial\psi/\partial f_2)$; and the tangent vector to the envelope has direction $(F_1'(\xi), F_2'(\xi))$, where a prime denotes differentiation. Because these two vectors are perpendicular at the point $(F_1(\xi), F_2(\xi))$, we have $\partial\psi/\partial f_1 \cdot F_1'(\xi) + \partial\psi/\partial f_2 \cdot F_2'(\xi) = 0$. But $(F_1(\xi), F_2(\xi))$ must lie on $f(L(\xi))$, i.e., $\psi(F_1(\xi), F_2(\xi), \xi) = 0$. Differentiating this equation with respect to ξ, we obtain $\partial\psi/\partial f_1 \cdot F_1'(\xi) +$

$\partial \psi / \partial f_2 \cdot F_2'(\xi) + \partial \psi / \partial \xi = 0$; therefore, $\partial \psi / \partial \xi = 0$ at $(F_1(\xi), F_2(\xi))$. Thus

$$(3.10) \qquad \psi(f_1, f_2, \xi) \;=\; 0 \;=\; \frac{\partial \psi}{\partial \xi}(f_1, f_2, \xi)$$

must be satisfied by all points (f_1, f_2) on the envelope. By eliminating ξ between these equations, we obtain the equation of the envelope:

$$(3.11) \qquad 25\,(f_1 - f_2)^2 - 40(f_1 + f_2) - 176 \;=\; 0$$

(Exercise 3.2). Thus the curved edge of $\overline{F}$ has equation

$$(3.12a) \qquad\qquad 3f_1 - f_2 + 6 = 0 \quad \text{if } -\tfrac{3}{5} \le f_2 \le 0$$

$$(3.12b) \quad 25\,(f_1 - f_2)^2 - 40(f_1 + f_2) = 176 \quad \text{if } -\tfrac{12}{5} \le f_1, f_2 \le -\tfrac{3}{5}$$

$$(3.12c) \qquad\qquad f_1 - 3f_2 - 6 = 0 \quad \text{if } -\tfrac{3}{5} \le f_1 \le 0.$$

Note that (3.10) would yield the parametric equations of the envelope of (3.9) even if $\psi = 0$ were not a straight line.

Now, if Nan and San agree to cooperate, then which point of $\overline{F}$ will be most agreeable to them? Suppose that they somehow pick a tentative point. Then, because Nan wants f_1 to be as large as possible and San wants f_2 to be as large as possible, points just above or to the right of that point will always be at least as agreeable to them, no matter where in the reward set the tentative point lies. Therefore, Nan and San will revise their tentative point. Proceeding in this manner, the two players will quickly eliminate all points either in the interior of the reward set or on the southern or western edge, because all such points can be improved upon by moving infinitesimally upwards or to the right. Thus agreeable points must lie on (3.12).

But even part of this curved edge can be eliminated. Implicit differentiation of (3.11) yields

$$(3.13) \qquad\qquad \frac{df_2}{df_1} \;=\; \frac{5(f_1 - f_2) - 4}{5(f_1 - f_2) + 4},$$

whence the tangent to the curve (3.12) is parallel to the f_1-axis at $\left(-\tfrac{8}{5}, -\tfrac{12}{5}\right)$ and to the f_2-axis at $\left(-\tfrac{12}{5}, -\tfrac{8}{5}\right)$; see Exercise 3.3. Between $\left(-\tfrac{8}{5}, -\tfrac{12}{5}\right)$ and $\left(-\tfrac{3}{5}, -\tfrac{11}{5}\right)$, the slope increases from 0 to $\tfrac{1}{3}$; thereafter it is constant. Hence both f_1 and f_2 are greater at every point between $\left(-\tfrac{8}{5}, -\tfrac{12}{5}\right)$ and $(0, -2)$ than they are at $\left(-\tfrac{8}{5}, -\tfrac{12}{5}\right)$. All such points, with the exception of $(0, -2)$, are improvable, and so Nan and San would eliminate them. Likewise, f_1 and f_2 are greater at every point between $\left(-\frac{12}{5}, -\frac{8}{5}\right)$ and $(-2,0)$ than they are at $\left(-\frac{12}{5}, -\frac{8}{5}\right)$; all such points, with the exception of $(-2,0)$, are improvable and would be eliminated. Thus the only remaining points on (3.12) are $(-2,0)$, $(0,-2)$ and points that lie between $\left(-\frac{8}{5}, -\frac{12}{5}\right)$ and $\left(-\frac{12}{5}, -\frac{8}{5}\right)$. If one of these points has been reached, then it is impossible to move to a *neighboring* point in the reward set—i.e., a point in the reward set that is arbitrarily close but nevertheless distinct—without making at least one player worse off. Such points are therefore said to be *locally unimprovable*, or locally Pareto-optimal (after the sociologist Vilfredo Pareto) or locally *group rational*. But among these points, $(-2,0)$ and $(0,-2)$ possess an even stronger measure of unimprovability: it is impossible to move from either one to *any* point in the reward set without making at least one player worse off. Such points are said to be *globally* unimprovable; and clearly, global unimprovability implies local unimprovability. Locally unimprovable points of $\overline{F}$ are marked in Figure 3.1 by a solid curve from $\left(-\frac{8}{5}, -\frac{12}{5}\right)$ to $\left(-\frac{12}{5}, -\frac{8}{5}\right)$. Note, however, that the endpoints are excluded.

More formally, the strategy combination (u,v) is locally unimprovable if there exists *no* neighboring point $(\overline{u}, \overline{v})$ in D such that

$$
(3.14) \quad
\begin{array}{llll}
\text{EITHER} & f_1(\overline{u},\overline{v}) > f_1(u,v) & \text{AND} & f_2(\overline{u},\overline{v}) \geq f_2(u,v) \\
\text{OR} & f_1(\overline{u},\overline{v}) \geq f_1(u,v) & \text{AND} & f_2(\overline{u},\overline{v}) > f_2(u,v);
\end{array}
$$

and (u,v) is globally unimprovable if there exists no $(\overline{u}, \overline{v})$ *anywhere* in D such that (3.14) is satisfied. Note that if (u,v) and $(\overline{u}, \overline{v})$ are neighboring points in D, then $(f_1(u,v), f_2(u,v))$ and $(f_1(\overline{u},\overline{v}), f_2(\overline{u},\overline{v}))$ are neighboring points in $\overline{F}$, because f_1 and f_2 are continuous functions.

These definitions are adequate for our purposes, and would hold even if Player 1 had a vector u of strategies, and Player 2 a vector v of strategies (as in §1.4). Nevertheless, we note that the concept of unimprovability is readily generalized to games among an arbitrary number of players—say n—as follows. Let $N = \{1, 2, \ldots, n\}$ denote the set of players, and for each $k \in N$ let Player k have an s_k-dimensional vector of strategies, which we denote by w^k; thus $w^1 = u, w^2 = v$ in (3.14). For each $k \in N$, let Player k's reward be $f_k = f_k(w)$, where w denotes the strategy combination $(w^1, w^2, \ldots, w^n)$; and let D be the the decision set, i.e., the set of all feasible w. Then the strategy combination w is locally unimprovable if there exists *no* other neighboring point $\overline{w} = (\overline{w}^1, \overline{w}^2, \dots, \overline{w}^n)$ in D such that

$$
\begin{aligned}
f_k(\overline{w}) &\geq f_k(w) \quad \text{for } all \ k \in N \\
\text{AND} \quad f_i(\overline{w}) &> f_i(w) \quad \text{for } some \ i \in N;
\end{aligned}
$$

(3.15)

and w is globally unimprovable if there exists no point $\overline{w}$ anywhere in D such that (3.15) is satisfied.

Clearly, provided the players have agreed to cooperate, any strategy combination that is not locally unimprovable would be irrational, because the players could agree to a neighboring combination that would yield no less a reward for all of them and a somewhat greater reward for at least one of them. It is not so clear that a combination would necessarily be irrational if it were not globally unimprovable, however, because the only globally unimprovable points in Figure 3.1 are $(0, -2)$ and $(-2, 0)$; and although these points represent best possible outcomes for Nan or San as individuals, it is difficult to see how both as a group could agree to either. Therefore, we shall proceed on the assumption that the best cooperative strategy combination in a cooperative game must be at least locally unimprovable, but not necessarily globally unimprovable; and if only a single combination were locally unimprovable, then we would not hesitate to regard it as the solution of the game. These special circumstances almost never arise, however; rather, many strategy combinations are locally unimprovable. Accordingly, we denote the set of all locally unimprovable strategy combinations by P (for Pareto), and the set of all globally unimprovable strategy combinations by P_G. Of course, $P_G \subset P$: a strategy cannot be globally unimprovable unless first of all it is locally unimprovable. Note that $P_G \subset P$ allows for the possibility that $P_G = P$, as in Figure 3.3.

To obtain P for the game defined by (3.3), we substitute (3.4) into (3.12b); after simplification, we obtain (Exercise 3.3)

(3.16) $$\{10(u + v) - 8\}^2 = 0.$$

But in Figure 3.1 the thick solid curve corresponds to values of u between $\frac{1}{5}$ and $\frac{3}{5}$. Thus P consists of the line segment in the u-v plane that joins $\left(\frac{1}{5}, \frac{3}{5}\right)$ to $\left(\frac{3}{5}, \frac{1}{5}\right)$, together with $(0, 1)$ and $(1, 0)$:

(3.17) $$P = \{0, 1\} \cup \left\{(u, v) \mid u + v = \tfrac{4}{5}, \ \tfrac{1}{5} < u < \tfrac{3}{5}\right\} \cup \{1, 0\}.$$

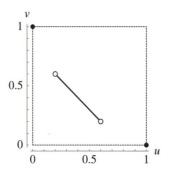

Figure 3.2. The bargaining set for Crossroads with $\delta = 3$, ϵ $= 2$ and $\tau_1 = 2 = \tau_2$.

Also, clearly,

$$(3.18) \qquad\qquad P_G = \{0,1\} \cup \{1,0\}.$$

The set P is sketched in Figure 3.2. Note that P is disconnected—it consists of three separate pieces—and excludes $\left(\frac{1}{5}, \frac{3}{5}\right)$ and $\left(\frac{3}{5}, \frac{1}{5}\right)$, as indicated by the open circles. Furthermore, although every locally unimprovable reward vector lies on the boundary of $\overline{F}$, all but two locally unimprovable strategy combinations lie in the interior of D.

If you are familiar with the implicit function theorem, then you will recognize that (3.16) could also have been obtained from the vanishing of the Jacobian of the mapping defined by (3.4), i.e., from

$$(3.19) \qquad\qquad \begin{vmatrix} \frac{\partial f_1}{\partial u} & \frac{\partial f_1}{\partial v} \\ \frac{\partial f_2}{\partial u} & \frac{\partial f_2}{\partial v} \end{vmatrix} = 0;$$

see Exercise 3.4. This happens because, on the one hand, the mapping f is locally invertible wherever (3.19) does not vanish; and, on the other hand, if you think of the unit square in the u-v plane as a sheet of rubber that must be stretched or shrunk by the mapping f until it corresponds to the reward set, then the envelope lies where the sheet gets folded back on itself by f. The mapping is not locally invertible here because, for points in $\overline{F}$ that are arbitrarily close to the envelope, it is impossible to say whether the inverse image in the unit square should correspond to the upper of the two folds or to the lower one.

If P contains more than a single strategy combination, then which element of P is the solution of the game? We will consider this question in §3.3. It is already clear from Figure 3.1 that the question is worth answering, however, because both Nan and San prefer *any* strategy combination in P to the Nash equilibrium $\left(\frac{3}{5}, \frac{3}{5}\right)$, which gives each a reward of only $-\frac{12}{5}$. On the other hand, P may be too large a set from which to select a solution, and in this regard it will be convenient to introduce some new terminology. For $k \in N$, let $\tilde{f}_k$ denote Player k's max-min reward; i.e., define $\tilde{f}_k = m_k(\tilde{w}^k)$, where m_k is the minimizing function defined by (1.74) and $\tilde{w}^k$ is a max-min strategy for Player k. If $f_k(w) \geq \tilde{f}_k$, i.e., if w gives Player k at least her max-min reward, then we say that w is *individually rational* for player k. Let

$$(3.20) \qquad D^* = \left\{ w \in D \mid f_k(w) \geq \tilde{f}_k \text{ for all } k \in N \right\}$$

denote the set of strategy combinations that are individually rational for all players, and let

$$(3.21) \qquad P^* = \left\{ w \in P \mid f_k(w) \geq \tilde{f}_k \text{ for } all \ k \in N \right\} = P \cap D^*$$

denote the set of all locally unimprovable strategy combinations that are individually rational for all players. We call P^* the *bargaining set*. Now, it would make no sense for Player k to accept the reward $f_k(w)$ if $f_k(w) < \tilde{f}_k$, because she could obtain at least $\tilde{f}_k$ without any cooperation at all. Accordingly, we seek cooperative solutions that lie not only in P, but also in P^*.

Note, however, that P^* often coincides with P, as illustrated by Figure 3.1 and Figure 3.3.

3.2. Necessary conditions for unimprovability

To determine whether a strategy combination $w \in D$ is locally improvable, we compare the rewards at w with those at a neighboring point, say $w + \lambda h$, where the vector $h = (h_1, h_2, \ldots, h_n)$ yields the direction of movement from w and λ is a small positive number; h must not be the zero vector. If there exists $\lambda > 0$, no matter how small, such that $f_k(w+\lambda h) \geq f_k(w)$ for all $k \in N$ and $f_i(w+\lambda h) > f_i(w)$ for some $i \in N$, then w is not (either locally or globally) unimprovable, because $w + \lambda h$ is an improved strategy combination—provided, of

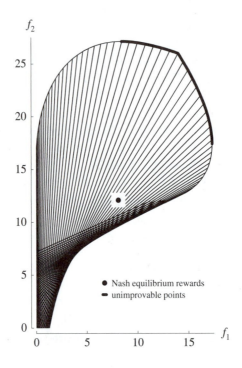

Figure 3.3. The reward set for Store Wars in §1.5 with $\alpha =$ 10 and $c = 1$. The hexagonal decision set D in Figure 1.8 can be covered by a family of straight lines parallel to the axis of symmetry, $u = v$; and the images of these straight lines under the mapping $f_1 = f_1(u, v), f_2 = f_1(u, v)$—which are also straight lines—trace out the reward set $\overline{F} = f(D)$, as indicated in the diagram. The corner in the interior of the set of unimprovable points corresponds to the Nash bargaining solution (§3.3). Note that the max-min rewards are $\tilde{f}_1 = 32/25$ and $\tilde{f}_2 = 12$. For further details, see Exercise 3.11.

course, that $w + \lambda h \in D$. Accordingly, we define h to be an *admissible direction* at w if $w + \lambda h \in D$ for sufficiently small $\lambda > 0$. Clearly, any h is admissible for w in the interior of D, whereas only vectors that point into D are admissible for w on the boundary of D. Suppose, for example, that $u = w_1, v = w_2$ and that D is the unit square $\{(u, v) \mid 0 \le u, v \le 1\}$. Then $h = (h_1, h_2)$ must satisfy $h_1 \ge 0$ to be admissible on the side $u = 0$; however, h_2 is unrestricted on $u = 0$ (except at the points $(0,0)$, where we require $h_2 \geq 0$, and $(0,1)$, where we require $h_2 \leq 0$). Similar considerations apply to the other three sides of the square (although any h is admissible in the interior).

Now we can say that if, at w, there exists an admissible direction h and a number $\lambda > 0$, no matter how small, such that $f_k(w + \lambda h) - f_k(w) \geq 0$ for all $k \in N$ and $f_i(w + \lambda h) - f_i(w) > 0$ for some $i \in N$, then $w \notin P$ (and hence $w \notin P^*$). In principle, by applying this test to each $w \in D$ in turn, we could systematically eliminate all locally improvable strategy combinations. The test is not practicable, however, because D contains infinitely many points.

To devise a practicable test for unimprovability, it is necessary to make assumptions about the nature of the functions f_k, principally, that f_k is (at least once) differentiable for all $k \in N$. From little more than this assumption, it is possible to derive powerful necessary and sufficient conditions for unimprovability; see, e.g., [**230**]. To follow this approach in its full generality would greatly distract us from our purpose, however, and so we assume instead that each player controls only a single scalar variable. Then, because Player k's strategy, w^k, is no longer a vector, we prefer to denote it by w_k; and the joint strategy combination w becomes an n-dimensional row vector, namely, $w = (w_1, w_2, \dots, w_n)$. Furthermore, we shall restrict our attention to *necessary conditions* for unimprovability; which, as we shall see, eliminate most—but not all—improvable strategy combinations.

Let us now recall that if f_k is differentiable with respect to w_k for all $k \in N$, then from Taylor's theorem for functions of several variables we have

$$(3.22) \qquad f_k(w + \lambda h) - f_k(w) = \lambda \frac{\partial f_k}{\partial w} h^T + o(\lambda),$$

where $\frac{\partial f_k}{\partial w}$ denotes the gradient vector $\left(\frac{\partial f_k}{\partial w_1}, \frac{\partial f_k}{\partial w_2}, \dots, \frac{\partial f_k}{\partial w_n} \right)$, h^T is the transpose of h, $\lambda > 0$, and $o(\lambda)$ denotes terms so small that you can divide them by λ and the result will still tend to zero as $\lambda \to 0$. The first term on the right hand side of (3.22) dominates $o(\lambda)$ for sufficiently small λ if $\frac{\partial f_k}{\partial w} h^T \neq 0$. Therefore, if

$$(3.23) \qquad \frac{\partial f_k}{\partial w} h^T > 0, \qquad k = 1, 2, \dots, n$$

for *any* admissible h, then w is not (either locally or globally) unimprovable; because then (3.22) and (3.23) imply

$$(3.24) \qquad f_k(w + \lambda h) - f_k(w) \; > \; 0$$

for all $k \in N$ for sufficiently small λ (> 0), so that $w + \lambda h$ is an improved strategy combination. Strategy combinations that do not satisfy (3.23) for any admissible h are candidates for unimprovability; however, we cannot be sure that they are unimprovable, even locally. Accordingly, let us denote by P_{nec} the set of all $w \in D$ that do not satisfy (3.23) for any admissible h, and by P_{nec}^* the set of all $w \in P_{nec}$ that are individually rational for all players. Then $P \subset P_{nec}$ and $P^* \subset P_{nec}^*$; but $P \neq P_{nec}$ and $P^* \neq P_{nec}^*$, at least in general.

Suppose, for example, that $n = 2$ and set $u = w_1, v = w_2$. Then (3.23) requires us to eliminate $(u, v) \in D$ if BOTH

$$(3.25) \qquad \frac{\partial f_1}{\partial u} h_1 + \frac{\partial f_1}{\partial v} h_2 > 0 \qquad \text{AND} \qquad \frac{\partial f_2}{\partial u} h_1 + \frac{\partial f_2}{\partial v} h_2 > 0$$

for any admissible direction $h = (h_1, h_2)$. To be quite specific, let us consider the version of Crossroads defined by (3.3). Then, from (3.4) and (3.25), (u, v) is improvable if

$$(3.26) \qquad \begin{array}{c} (3 - 5v)h_1 + (1 - 5u)h_2 > 0 \\ \text{AND} \qquad (1 - 5v)h_1 + (3 - 5u)h_2 > 0 \end{array}$$

for any admissible $h = (h_1, h_2)$. Now, $h = (1, 0)$ is an admissible direction everywhere in the unit square except on the side $u = 1$. Accordingly, from points in the square such that $0 \leq u < 1, 0 \leq v \leq 1$ we must exclude those where $3 - 5v > 0$ and $1 - 5v > 0$, i.e., points where $v < \frac{1}{5}$. Similarly, because $h = (-1, 0)$ is an admissible direction everywhere in the square except on the side $u = 0$, from points in the unit square such that $0 < u \leq 1, 0 \leq v \leq 1$ we must exclude those where $3 - 5v < 0$ and $1 - 5v < 0$, i.e., points where $v > \frac{3}{5}$. Continuing in this manner, we find that choosing $h = (0, 1)$ excludes points where $u < \frac{1}{5}$, and that choosing $h = (0, -1)$ excludes points where $u > \frac{1}{5}$. Now the only points remaining are $(0, 1)$, $(1, 0)$ and those which satisfy $\frac{1}{5} \leq u, v \leq \frac{3}{5}$. This set of candidates is sketched in Figure 3.4(a).

We already know that $(0, 1)$ and $(1, 0)$ are globally unimprovable, from §3.1. So we concentrate on the region $\{(u, v) \mid \frac{1}{5} \leq u, v \leq \frac{3}{5}\}$. Because this square lies totally within the interior of D, any h is

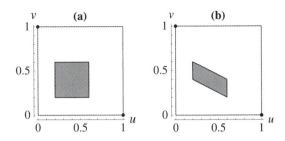

Figure 3.4. Sets that contain the set of locally unimprovable strategy combinations.

admissible. We can choose, for example, $h = (2, 1)$. Then (3.26) requires us to exclude (u, v) such that $5u + 10v < 7$ and $5u + 10v < 5$, i.e., points where $u + 2v < 1$. Similarly, choosing $h = (-2, -1)$ requires us to exclude points such that $5u + 10v > 7$ and $5u + 10v > 5$, i.e., points where $u + 2v > \frac{7}{5}$. We have now reduced our set of candidates to the parallelogram sketched in Figure 3.4(b).

We could eliminate improvable points indefinitely in this manner, but let us not prolong our agony. From (3.26), $h = (1, 1)$ rules out points in the shaded region of Figure 3.4(b) where $4 - 5(u + v) > 0$, and $h = (-1, -1)$ rules out points in that region where $4 - 5(u + v) < 0$; therefore, all locally unimprovable points must satisfy $u + v = \frac{4}{5}$. We have therefore shown that

$$(3.27) \quad P_{nec} = \{0, 1\} \cup \left\{ (u, v) \mid u + v = \tfrac{4}{5}, \ \tfrac{1}{5} \le u \le \tfrac{3}{5} \right\} \cup \{1, 0\}.$$

Note, however, that (3.26) cannot eliminate $\left(\frac{1}{5}, \frac{3}{5} \right)$ and $\left(\frac{3}{5}, \frac{1}{5} \right)$ as candidates, although we already know from §3.1 that neither point is unimprovable. Thus $P \subset P_{nec}$, but $P \ne P_{nec}$.

The guesswork involved in choosing an admissible direction that yields useful information is clearly unsatisfactory, and we can eliminate much of it by having recourse to the so-called "Theorem of the Alternative." A special case of this theorem, which will suffice for our purposes, is the following: if J is an $n \times n$ matrix and K is an $m \times n$ matrix, then *either* there exists a $1 \times n$ (row) vector h such that

$$(3.28) \qquad J h^T > 0^n \qquad \text{AND} \qquad K h^T \ge 0^m;$$

or there exists a $1 \times n$ vector η with nonnegative components, at least one of which is strictly positive, and a $1 \times m$ vector μ with nonnegative components, such that

$$(3.29) \qquad \eta J + \mu K = 0_n;$$

but *never both*. Here 0^n stands for the $n \times 1$ zero vector, 0_n stands for the $1 \times n$ zero vector, and a vector inequality $w > 0$ (or $w \geq 0$) means that every component of the vector w must be positive (or nonnegative); thus, in the statement of the theorem, $\eta \geq 0_n, \eta \neq 0_n, \mu \geq 0_m$. For a proof of the theorem, see Mangasarian [124].

Now, under our scalar-strategy assumption, constraints on the admissibility of h at a given point can always be written in the form $Kh^T \geq 0^m$ for suitable K and m, and the n inequalities in (3.23) are equivalent to $Jh^T > 0$, where J is the Jacobian matrix

$$(3.30) \qquad J(w) = \begin{bmatrix} \frac{\partial f_1}{\partial w_1} & \frac{\partial f_1}{\partial w_2} & \cdots & \frac{\partial f_1}{\partial w_n} \\ \frac{\partial f_2}{\partial w_1} & \frac{\partial f_2}{\partial w_2} & \cdots & \frac{\partial f_2}{\partial w_n} \\ \cdot & \cdot & \cdots & \cdot \\ \cdot & \cdot & \cdots & \cdot \\ \cdot & \cdot & \cdots & \cdot \\ \frac{\partial f_n}{\partial w_1} & \frac{\partial f_n}{\partial w_2} & \cdots & \frac{\partial f_n}{\partial w_n} \end{bmatrix}$$

for the particular w whose unimprovability is being tested. But if w is unimprovable, then (3.23) precludes alternative (3.28); therefore, we must instead have (3.29). In particular, for w in the interior of D (where any h is admissible), because we can set K equal to the $m \times n$ zero matrix (for any value of m), and because η must not be the zero vector, we deduce from (3.29) that $\eta J(w) = 0_n, \eta \neq 0_n$. Thus it follows immediately from the theory of linear algebra that if w is an interior, unimprovable point then

$$(3.31) \qquad |J(w)| = 0,$$

i.e., the determinant of the Jacobian must vanish at w.[1] The result does not imply that every interior point w satisfying (3.31) is a candidate for unimprovability, however, because more than just $|J| = 0$

---

[1] Note that (3.19) is a special case of this result.

is implied by $\eta J = 0_n, \eta \neq 0_n$. When $n = 2$, for example, $\eta J = 0_n$ becomes

$$(3.32) \qquad \eta_1 \frac{\partial f_1}{\partial u} + \eta_2 \frac{\partial f_2}{\partial u} = 0, \qquad \eta_1 \frac{\partial f_1}{\partial v} + \eta_2 \frac{\partial f_2}{\partial v} = 0,$$

where $\eta = (\eta_1, \eta_2)$. With $(\eta_1, \eta_2) \geq (0, 0)$ and $(\eta_1, \eta_2) \neq (0, 0)$ there are three possibilities for η, namely, that $\eta_1 = 0$ and $\eta_2 > 0$, that $\eta_1 > 0$ and $\eta_2 = 0$, or that $\eta_1 > 0$ and $\eta_2 > 0$. From (3.32), the first case requires $\partial f_2/\partial u = 0 = \partial f_2/\partial v$; the second case requires $\partial f_1/\partial u = 0 = \partial f_1/\partial v$; and the third requires $\partial f_1/\partial u \cdot \partial f_2/\partial u < 0$ and $\partial f_1/\partial v \cdot \partial f_2/\partial v < 0$. For the version of Crossroads defined by (3.3), you can easily verify that these restrictions on interior unimprovable points correspond to (3.27).

We can also apply (3.29) to boundary points. Consider, for example, $(1, 0)$. The restrictions on h at that point are $h_1 \leq 0, h_2 \geq 0$, implying $m = 2$, $K = \begin{bmatrix} -1 & 0 \\ 0 & 1 \end{bmatrix}$ and $J = J((1, 0)) = \begin{bmatrix} 3 & -4 \\ 1 & -2 \end{bmatrix}$. So $\eta J + \mu K = 0_n = 0_2$ becomes $\begin{bmatrix} \eta_1 & \eta_2 \end{bmatrix} \begin{bmatrix} 3 & -4 \\ 1 & -2 \end{bmatrix} + \begin{bmatrix} \mu_1 & \mu_2 \end{bmatrix} \begin{bmatrix} -1 & 0 \\ 0 & 1 \end{bmatrix} = \begin{bmatrix} 0 & 0 \end{bmatrix}$ or $3\eta_1 + \eta_2 - \mu_1 = 0, -4\eta_1 - 2\eta_2 + \mu_2 = 0$; and these equalities are easily satisfied with $\eta \geq 0_2, \eta \neq 0_2, \mu \geq 0_2$. Therefore $(1, 0)$ is a candidate for unimprovability. On the other hand, for the point $(0, 0)$, where the restrictions on h are $h \geq 0_2$, so that K is the 2×2 identity matrix, a similar analysis yields the equalities $3\eta_1 + \eta_2 + \mu_1 = 0$ and $\eta_1 + 3\eta_2 + \mu_2 = 0$. Now $\mu \geq 0_2$ implies $3\eta_1 + \eta_2 \leq 0, \eta_1 + 3\eta_2 \leq 0$, which contradicts $\eta \geq 0_2, \eta \neq 0_2$. Thus $(0, 0)$ is improvable. Again, at points other than $(1, 0)$ and $(1, 1)$ on the side of D where $u = 1$, the only restriction on h is $h_1 \leq 0$, so that $m = 1$, and $K = \begin{bmatrix} -1 & 0 \end{bmatrix}$. Now $\eta J + \mu K = 0_2$ yields $(3 - 5v)\eta_1 + (1 - 5v)\eta_2 - \mu_1 = 0, 4\eta_1 + 2\eta_2 = 0$, the second of which contradicts $\eta \geq 0_2, \eta \neq 0_2$, regardless of the value of v. Continuing in this manner, the remaining improvable points on the boundary of D are readily eliminated; see Exercise 3.6. Then try Exercise 3.7.

3.3. The Nash bargaining solution

Despite the Theorem of the Alternative, eliminating improvable points is rarely a straightforward exercise, even if $n = 2$. On the other hand, bargaining points—i.e., strategy combinations in the bargaining sets—are almost never unique; so that, even after the strenuous labor of calculating P^*, we are still faced with the problem of deciding which w in P^*, say $w = \hat{w}$, should be the solution of the game. If it were somehow possible to determine $\hat{w}$ without first calculating P^*, then clearly we could save ourselves a great deal of trouble. One such approach to determining $\hat{w}$ is provided by Nash's ([**163**]) bargaining solution, which we now describe.

Because Player k benefits from cooperation with the other players only if she obtains a reward in excess of the max-min reward she can guarantee for herself, we will define

$$(3.33) \qquad\qquad \overline{x}_k \;=\; f_k(w) - \tilde{f}_k$$

to be Player k's benefit of cooperation from the joint strategy combination $w \in D$; for $w \in P^*, \overline{x}_k \geq 0$. Because, for each $k \in N$, Player k wants $\overline{x}_k$ to be as large as possible (and certainly nonnegative), an agreeable choice of the vector

$$(3.34) \qquad\qquad \overline{x} \;=\; (\overline{x}_1, \overline{x}_2, \ldots, \overline{x}_n)$$

should lie as far as possible from the origin (and in the nonnegative orthant) of the n-dimensional space of cooperation benefits. But how should one measure distance $\overline{d}$ from the origin? Should one use

$$(3.35) \qquad\qquad \overline{d} \;=\; \overline{x}_1 + \overline{x}_2 + \cdots + \overline{x}_n \,?$$

Or

$$(3.36) \qquad\qquad \overline{d} \;=\; \overline{x}_1 \cdot \overline{x}_2 \cdots \overline{x}_n \,?$$

Or yet another formula expressing the idea that all players prefer a larger benefit to a smaller one, i.e., $\partial \overline{d}/\partial \overline{x} > 0_n$ for all $\overline{x} > 0_n$? Clearly, there is no limit to the number of such formulae!

Now, each player measures her cooperation benefit according to her own scale of merit, i.e., subjectively. Suppose, however, that there exists some "objective" scale of merit, against which a supreme arbitrator could assess all players' subjective valuations; and let one unit of objective merit equal γ_k units of Player k's subjective merit, for all $k \in N$. Then the objective distances corresponding to (3.35) and (3.36), which the supreme arbitrator could perhaps supply, are

$$(3.37) \qquad\qquad \overline{d} \;=\; \frac{\overline{x}_1}{\gamma_1} + \frac{\overline{x}_2}{\gamma_2} + \cdots + \frac{\overline{x}_n}{\gamma_n}$$

and

$$(3.38) \qquad \overline{d} \; = \; \frac{\overline{x}_1}{\gamma_1} \cdot \frac{\overline{x}_2}{\gamma_2} \cdots \frac{\overline{x}_n}{\gamma_n} \; = \; \frac{\overline{x}_1 \cdot \overline{x}_2 \cdots \overline{x}_n}{\gamma_1 \cdot \gamma_2 \cdots \gamma_n},$$

respectively. But who is this supreme arbitrator? Who is this person who is capable of assigning values to the numbers $\gamma_1, \gamma_2, \ldots, \gamma_n$—or, as game theorists prefer to say, of making *interpersonal comparisons of utility*? Suppose there is no such person. If distance is measured according to (3.35), then this is a most unfortunate circumstance, because we cannot maximize the corresponding objective distance (3.37) until we know the values of $\gamma_1, \gamma_2, \ldots, \gamma_n$. If distance is measured according to (3.36), however, then it matters not a whit—maximizing (3.36) and maximizing (3.38) are one and the same thing, for any values of $\gamma_1, \gamma_2, \ldots, \gamma_n$. Thus formula (3.36) has a very desirable property that (3.35) and other formulae do not possess. We will say that $\hat{w} \in P^*$ is a *Nash bargaining solution* if $\hat{w}$ maximizes

$$(3.39) \qquad \overline{d}(w) \; = \; \{f_1(w) - \tilde{f}_1\} \cdot \{f_2(w) - \tilde{f}_2\} \cdots \{f_n(w) - \tilde{f}_n\}$$

over P^*; and if $\hat{w}$ is unique—i.e., if no other $w \in P^*$ satisfies $\overline{d}(w) = \overline{d}(\hat{w})$—then we will regard $\hat{w}$ as the solution of our cooperative game.

In practice, we can often obtain $\hat{w}$ by maximizing $\overline{d}(w)$, not over P^*, but rather over the whole of D, and then checking that $\hat{w}$ is individually rational for all players, i.e., $\hat{w} \in D^*$. It is clear that $\hat{w}$ so found must be globally (and hence also locally) unimprovable; for if $\hat{w}$ were improvable, then the improved strategy combination would yield a larger value of $\overline{d}$. If, of course, the w that maximizes $\overline{d}(w)$ over D does not belong to D^*, then we must maximize $\overline{d}(w)$ over D^* instead. Either way, we can find $\hat{w}$ that maximizes $\overline{d}(w)$ over P^* without actually calculating P^*. Both cases are illustrated by the following two examples.

First, let us calculate the Nash bargaining solution for Store Wars II from §1.6. From Exercise 1.20, the max-min rewards for Nan, Van and Zan are, respectively, $\tilde{f}_1 = ac\pi/9, \tilde{f}_2 = ac\pi/16$ and $\tilde{f}_3 = 25ac\pi/144$. Therefore, from (1.63), we have

$$(3.40a) \qquad f_1(u, v, z) - \tilde{f}_1 \; = \; 8ac\pi \left\{ u \left(\tfrac{1}{3} + v - 2u + z \right) - \tfrac{1}{72} \right\}$$

$$(3.40b) \qquad f_2(u, v, z) - \tilde{f}_2 \; = \; 8ac\pi \left\{ v \left(\tfrac{1}{4} + u - 2v + z \right) - \tfrac{1}{128} \right\}$$

$$(3.40c) \qquad f_3(u, v, z) - \tilde{f}_3 \; = \; 8ac\pi \left\{ z \left(\tfrac{5}{12} + u - 2z + v \right) - \tfrac{25}{1152} \right\};$$

and on using (3.39) with $w = (u, v, z)$ we obtain

(3.41) $\overline{d}(u, v, z) = \frac{(ac\pi)^3}{20736}\hat{x}_1(u, v, z)\hat{x}_2(u, v, z)\hat{x}_3(u, v, z)$

with $\hat{x}_1$ defined by $\hat{x}_1(u, v, z) = 24u(1+3v-6u+3z)-1$, $\hat{x}_2$ defined by $\hat{x}_2(u, v, z) = 32v(1 + 4u - 8v + 4z) - 1$ and $\hat{x}_3$ defined by $\hat{x}_3(u, v, z) = 96z(5 + 12u - 24z + 12v) - 25$.

To obtain $(\hat{u}, \hat{v}, \hat{z})$ we maximize $\overline{d}$ over the decision set D defined by (1.64). For the sake of definiteness, let us suppose that the value of α, which determines the maximum price in (1.59), is $\alpha = 10$. Then D consists of all (u, v, z) such that

(3.42) $0 \le u, v, z \le 10,\quad |u - v| \le \frac{1}{12},\quad |v - z| \le \frac{1}{6},\quad |u - z| \le \frac{1}{4}.$

Note that if u, v and z were not constrained by a price ceiling, i.e., if $\alpha \to \infty$ in (1.59), then $\overline{d}$ would be unbounded on D (because, e.g., $\overline{d}$ would increase without bound as $t \to \infty$ along the line defined by $u = v = z = t$ if it weren't for the constraints $u \le 10, v \le 10, z \le 10$). Thus collusion among storekeepers is bad for the consumer. It enables one of the players—from Figure 1.12, clearly Zan—to set her price at the ceiling $8\pi ac\alpha$, with the others not far behind; whereas competition in §1.6 kept prices in rein.

Maximizing (3.41) subject to (3.42) is a problem in constrained nonlinear programming, a discussion of which would take us far beyond our brief. Therefore, we simply observe that packages for solving such problems are now widely available even on desktop computers.[2] By using such a package, we discover that the maximum of (3.41) on D occurs where $u = 9.98$, $v = 9.95$ and $z = 10$: for all practical purposes, the prices are at their ceiling. The corresponding rewards are $f_1 = 25.8ac\pi(> \tilde{f}_1), f_2 = 26.3ac\pi(> \tilde{f}_2)$ and $f_3 = 27.7ac\pi(> \tilde{f}_3)$, confirming that $(9.98, 9.95, 10) \in D^*$ (which, however, we have not had to calculate). If we compare the Nash bargaining rewards to the Nash equilibrium rewards $f_1 = 0.44ac\pi, f_2 = 0.36ac\pi$ and $f_3 = 0.54ac\pi$, obtained in §1.6, then we see how much the players

---

[2]For a discussion of constrained nonlinear programming see, e.g., Chapters 10, 12 and 13 of [**123**] or Chapters 9 and 10 of [**106**]. The first edition of this book [**140**] describes how to find the maximum with a Fortran subroutine. If you accept on intuitive grounds that the maximum must occur where Zan's price is at its ceiling, however, then the result can be obtained more easily by using numerical optimization routines supplied with standard desktop mathematical software (see, e.g., [**239**]) to calculate the unconstrained maximum of $\overline{d}(u, v, 10)$ and then checking that (3.42) is satisfied anyway; it is necessary to supply a suitable initial guess (e.g., $u = 10 = v$).

benefit by collusion. Note that Van's reward is greater than Nan's under cooperation, whereas Nan's reward is greater than Van's under competition; in both cases, however, Van's price is lower than Nan's.

The $\hat{w}$ that maximizes $\overline{d}$ need not be unique, however, as our second example—Crossroads with fast drivers—will illustrate. From Exercise 1.19, when $2\epsilon > \max(\tau_1, \tau_2)$ the max-min rewards are $\tilde{f}_1 = m_1(\tilde{u}) = -(\delta + \frac{1}{2}\tau_2)\theta_2$ and $\tilde{f}_2 = m_2(\tilde{v}) = -(\delta + \frac{1}{2}\tau_1)\theta_1$, where $\theta_k = \frac{2\epsilon + \tau_k}{2\epsilon + 2\delta}$ is defined by (1.22) and m_1, m_2 are defined by (1.75). Thus, from (3.2), we have

$$(3.43a) \qquad f_1(u,v) - \tilde{f}_1 = (\epsilon - \tau_2/2 - \{\delta + \epsilon\}u)(v - \theta_2),$$

$$(3.43b) \qquad f_2(u,v) - \tilde{f}_2 = (\epsilon - \tau_1/2 - \{\delta + \epsilon\}v)(u - \theta_1);$$

and (3.39) yields

$$(3.44) \qquad \overline{d}(u,v) = (\delta + \epsilon)^2(u - \theta_1)(\xi_1 - u)(v - \theta_2)(\xi_2 - v)$$

where ξ_1 and ξ_2 are defined by

$$(3.45) \qquad (\delta + \epsilon)\xi_1 = \epsilon - \tfrac{1}{2}\tau_2, \qquad (\delta + \epsilon)\xi_2 = \epsilon - \tfrac{1}{2}\tau_1.$$

In the special case of (3.3), we have

$$(3.46) \qquad \overline{d}(u,v) = \tfrac{1}{25}(5u - 3)(1 - 5u)(5v - 3)(1 - 5v).$$

The maximum of (3.46) over $D = \{(u,v) \mid 0 \le u, v \le 1\}$ occurs at $(u,v) = (1,1)$, but $(1,1) \notin D^*$. Accordingly, we must maximize $\overline{d}$ over D^* instead; and, from Exercise 3.8, we find that the maximum, $\frac{24}{25}$, occurs at both $(1,0)$ and $(0,1)$. It appears that there are two Nash bargaining solutions, each of which is the best possible outcome for one of the players; and, as we have remarked already in §3.1, it is difficult to see how they could agree to either.

Here two remarks are in order. First, the game of Crossroads defined by (3.3) is symmetric ($\tau_1 = \tau_2$). There is no basis for distinguishing between the players, and so any cooperative solution should also be symmetric, i.e., satisfy $u = v$. But P^* in Figure 3.2 contains only a single, symmetric strategy combination, namely, $(u,v) = \left(\frac{2}{5}, \frac{2}{5}\right)$. We therefore propose that the solution of this cooperative game should be neither $(1,0)$ nor $(0,1)$, but rather $\left(\frac{2}{5}, \frac{2}{5}\right)$, the center of the bargaining set in Figure 3.2.[3]

---

[3]Which corresponds to a *local* maximum on D^* of $\overline{d}$ defined by (3.46).

Second, the function $\bar{d}$ defined by (3.46) achieves its maximum twice on D^* only because the game is symmetric. In general, if Crossroads is asymmetric, i.e., if $\tau_1 \neq \tau_2$, then the Nash bargaining solution is again unique. Suppose, for example, that

$$(3.47) \qquad \delta = 5, \quad \epsilon = 3, \quad \tau_1 = 4, \quad \tau_2 = 2.$$

Then (Exercise 3.8) the unique Nash bargaining solution is $(\tilde{u}, \tilde{v}) = (0, 1)$, i.e., Nan always waits and San always goes. This solution is intuitively attractive because, although both drivers are fast, San is considerably faster; and so it makes good sense for Nan to let her whip across the junction before she dawdles into gear herself.

More fundamentally, there is no unique Nash bargaining solution for the game defined by (3.3) because $\overline{F}$ is not *convex*, i.e., it isn't possible to join any two points in $\overline{F}$ by a straight line segment that never leaves $\overline{F}$ (consider, for example, $(1, 0)$ and $(0, 1)$). If $\overline{F}$ is convex, then Nash's bargaining solution is always unique; see, for example, Exercise 3.10. More generally, Nash's bargaining solution is unique whenever a convex subset of $\overline{F}$ contains its unimprovable boundary, regardless of whether $\overline{F}$ itself is convex, as illustrated by Figure 3.3.

3.4. Independent versus correlated strategies

The difference between what is rational for an individual and what is rational for a group is exemplified by comparing the noncooperative Nash-equilibrium solution for the game of Crossroads defined by (3.47) with the ad hoc cooperative solution derived at the end of the previous section. Nan and San's Nash bargaining rewards are $-\frac{11}{5}$; whereas their Nash-equilibrium rewards are only $-\frac{12}{5}$. Clearly, each prefers $-\frac{11}{5}$ to $-\frac{12}{5}$, and so it is in their mutual interest to select their cooperative strategies $u = \frac{2}{5} = v$ over their Nash equilibrium strategies $u^* = \frac{3}{5} = v^*$; in other words, to be less aggressive and assume right of way less often. If, however, Nan were to select $u = \frac{2}{5}$ without first making her intentions clear to San, then from Figure 1.5 (with $\theta_1 = \frac{3}{5} = \theta_2$) the rational thing for San to do would be to select $v = 1$ (always go), because $\left(\frac{2}{5}, 1\right) \in R_2$. It would be irrational for San to select $v = \frac{2}{5}$, because $\left(\frac{2}{5}, \frac{2}{5}\right) \notin R_2$. Likewise, if Nan knew that San would play $v = \frac{2}{5}$ then the rational thing for Nan to do, from a selfish point of view, would be to welch on San and play $u = 1$, because

$\left(1, \frac{2}{5}\right) \in R_1$. Thus cooperative solutions are rational only if there is a gentlewoman's agreement among the players—however enforced, whether voluntarily or by compulsion—to abide by their bargaining strategies. Indeed in theory, it is merely the existence or absence of such an agreement that determines whether a game is cooperative or noncooperative. In practice, however, what often determines whether a game is cooperative or noncooperative is whether or not the game is among specific individuals who meet repeatedly in similar circumstances, recognize one another, and have the ability to communicate; for rational beings will avail themselves of any device for maximizing rewards—including, if they have the means, cooperation.

Provided the players can trust one another, however, then the potential exists for even greater benefits from cooperation than are possible when strategies are chosen independently. Consider, for example, Crossroads. As in §1.3, we can imagine that each driver has a spinning arrow on her dashboard, which determines whether to go or wait in any particular confrontation. Now, if Nan's arrow always came to rest over the shaded sector of her disk when San's arrow came to rest over the unshaded part of her disk, and vice versa, then the two players would never waste any time wondering who should go after WW or who should back down after GG, because the only pure strategy combinations ever selected would be WG or GW. But, of course, WG and GW are not the only pure strategy combinations selected, because the players spin their arrows independently, and all angles of rest between 0 and 2π are equally likely.

If the players have already agreed to cooperate, however, then they can further reduce delays by agreeing to correlate strategies, as follows. Naturally, there is potential for such collaboration only when Crossroads is played between two specific individuals—a particular Nan and a particular San.

Let us imagine that, after agreeing to cooperate, this Nan and San discard their individual spinning arrows and replace them by a large arrow and disk at the junction itself (which can be started by remote control). This disk is divided into four sectors, which subtend angles $2\pi\omega_{11}, 2\pi\omega_{12}, 2\pi\omega_{21}$ and $2\pi\omega_{22}$, respectively. If the arrow comes to rest in the first sector, then Nan and San will both go (GG); if in the second sector, then Nan will go but San will wait (GW); if in the third sector, then San will go but Nan will wait (WG); and if in the fourth sector, then Nan and San will both wait (WW). The device is equivalent to selecting pure strategy combination (i, j) with probability ω_{ij}, for $i, j = 1, 2$. Thus

$$(3.48) \qquad \omega_{11} + \omega_{12} + \omega_{21} + \omega_{22} = 1;$$

and Nan's reward f_1, the expected value of her payoff F_1, is

$$(3.49) \qquad \begin{aligned} f_1 &= -\left(\delta + \tfrac{1}{2}\tau_2\right) \cdot \mathrm{Prob}(GG) + 0 \cdot \mathrm{Prob}(GW) \\ &\quad - \tau_2 \cdot \mathrm{Prob}(WG) - \left(\epsilon + \tfrac{1}{2}\tau_2\right) \cdot \mathrm{Prob}(WW) \\ &= -\left(\delta + \tfrac{1}{2}\tau_2\right)\omega_{11} - \tau_2\omega_{21} - \left(\epsilon + \tfrac{1}{2}\tau_2\right)\omega_{22} \end{aligned}$$

from Table 1.1. Similarly, San's reward is

$$(3.50) \qquad f_2 = -\left(\delta + \tfrac{1}{2}\tau_1\right)\omega_{11} - \tau_1\omega_{12} - \left(\epsilon + \tfrac{1}{2}\tau_1\right)\omega_{22}.$$

We will refer to the vector $\omega = (\omega_{11}, \omega_{12}, \omega_{21}, \omega_{22})$ as a *correlated strategy*.

When Nan and San had two separate arrows and chose independent strategies u and v, we had $\omega_{11} = uv$, $\omega_{12} = u(1 - v)$, $\omega_{21} = (1 - u)v$ and $\omega_{22} = (1 - u)(1 - v)$, from §1.3 (p. 11). Thus maximization of f_1 and f_2 when strategies are selected independently is equivalent to maximization of f_1 and f_2 under the constraints

$$(3.51) \qquad \omega_{11} + \omega_{12} = u, \qquad \omega_{21} + \omega_{22} = v$$

and, of course, (3.48); whereas maximization of f_1 and f_2 when strategies are correlated is subject only to (3.48). Because there are (two) fewer constraints, we can expect maximum rewards to be larger.

Now, in designing their disk, Nan and San do not hesitate to set $\omega_{11} = 0 = \omega_{22}$, which increases both f_1 and f_2; as we have remarked already, it is obvious that nothing can be gained under correlated strategies by selecting GG or WW. Therefore, from (3.48)-(3.50), Nan and San reduce their task to selecting ω_{12} and ω_{21} such that $\omega_{12} + \omega_{21} = 1$, with rewards $f_1 = -\tau_2\omega_{21}$ and $f_2 = -\tau_1\omega_{12}$. Thus, regardless of which ω_{12} (and hence ω_{21}) the players choose, the corresponding rewards will satisfy

$$(3.52) \qquad \tau_1 f_1 + \tau_2 f_2 + \tau_1\tau_2 = 0, \qquad f_1, f_2 \le 0.$$

Their reward pair will therefore lie in the the f_1-f_2 plane on the line segment that joins $(-\tau_2, 0)$ and $(0, -\tau_1)$; and every point on this line segment is achievable by some choice of correlated strategies. Furthermore, from Figure 3.1 (or the equivalent diagram for other values of the parameters δ, ϵ, τ_1 and τ_2), this line segment lies on or above the unimprovable boundary of $\overline{F}$ at every point; in particular, in the case where $\delta = 3, \epsilon = 2$ and $\tau_1 = 2 = \tau_2$, it lies above the point that corresponds to the ad hoc cooperative solution we found at the end of §3.3. It thus appears that if you are going to cooperate, then you might as well correlate.

There is, however, a price to be paid—which reward pair on this line segment yields the solution of the game? Intuition suggests that we should select the "fair" solution

$$(3.53) \qquad f_1 = -\frac{\tau_1 \tau_2}{\tau_1 + \tau_2} = f_2,$$

achieved by the correlated strategy ω for which $\omega_{11} = 0 = \omega_{22}$ and

$$(3.54) \qquad \omega_{12} = \frac{\tau_2}{\tau_1 + \tau_2}, \qquad \omega_{21} = \frac{\tau_1}{\tau_1 + \tau_2}$$

(but see Exercise 3.10). If we accept this solution, however, then tacitly we have made an interpersonal comparison of utilities: we have assumed that a minute of San's time is worth a minute of Nan's, so that if they can save a minute together, then each should reap 30 seconds of the benefits. But who is to say whether Nan's time and San's time are equally valuable? What if Nan is a brain surgeon and San a cashier—then couldn't one argue that San's time is more valuable than Nan's, if they meet in the morning on their way to work (because if San is five minutes late for work she may lose her job, whereas if Nan is five minutes late for work—well, do you really expect a brain surgeon to be on time?) Nevertheless, there are many circumstances in which interpersonal comparisons of utility are quite acceptable—for example, Nan and San may both be brain surgeons or cashiers—and if the price is small then we shall happily pay it.

If we agree to correlate strategies, however, and if we agree to make interpersonal comparisons of utility (my time's worth as much as your time, my dollar's just as good as your dollar, etc.), then solving a game loses much of its strategic interest. It becomes instead a matter of seeking a fair distribution among the players of some benefit of cooperation (e.g., time saved, money saved), of which there exists a definite total amount. Or, if you prefer, there's a finite pie to be distributed among the players, and we need to establish how big a slice is a player's just desert (dessert?). But already we are talking about characteristic function games, and these are the subject of the following chapter.

3.5. Commentary

In Chapter 3 we have introduced the most important concepts of co-operative games in strategic form, namely, unimprovability or Pareto-optimality (§§3.1-3.2), Nash bargaining solutions (§3.3) and independent versus correlated strategies (§3.4). In §3.3 we introduced the Nash bargaining solution in the context of independent strategies. Usually, however, the solution is applied in the context of correlated strategies. The reward set is then always convex (so that local unimprovability and global unimprovability are equivalent); and, as remarked at the end of §3.3, a unique Nash bargaining solution is then guaranteed. For a proof of this fact and further properties of Nash's bargaining solution see, e.g., Owen [**173**]. Note, finally, that although Nash's bargaining solution appears to be the most enduring solution concept for cooperative games in strategic form, it is by no means the only one; others are described by, e.g., Shubik [**207**, pp. 196-200].

Exercises 3

1. (a) Verify (3.8).
    (b) Show that the image of the line segment $\{(u, 0)|0 \leq u \leq 1\}$ under the mapping f defined by (3.4) is that part of the line $f_1 - 3f_2 - 6 = 0$ which extends from $(-3, -3)$ to $(0, -2)$. Note that this image is largely coincident with $f(L(4/5))$, but $f(L(4/5))$ does not extend from $\left(-\frac{3}{5}, -\frac{11}{5}\right)$ to $(0, -2)$.
    (c) Show that the image of the line segment $\{(u, 1)|0 \leq u \leq 1\}$ under the mapping f defined by (3.4) is that part of the line $2f_1 - f_2 + 4 = 0$ which extends from $(-4, -4)$ to $(-2, 0)$.
2. (a) Show that (3.11) is the envelope of the family of lines $\psi = 0$, where ψ is defined by the left hand side of (3.8).

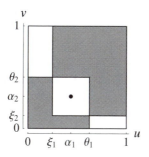

Figure 3.5

(b) Show that this envelope meets (touches) the line (3.12a) at the point $\left(-\frac{11}{5}, -\frac{3}{5}\right)$, and crosses the f_1-axis where $f_1 > -2$; and that it meets the line (3.12c) at the point $\left(-\frac{3}{5}, -\frac{11}{5}\right)$, and crosses the f_2-axis where $f_2 > -2$. Why are the two small curvilinear triangles enclosed by the envelope, the lines and the axes not part of the reward set?

3. (a) Deduce from (3.13) that the tangent to the curved edge of $\overline{F}$ is parallel to the f_1-axis at $\left(-\frac{8}{5}, -\frac{12}{5}\right)$, and to the f_2-axis at $\left(-\frac{12}{5}, -\frac{8}{5}\right)$.

   (b) Which strategy combinations correspond to these points?

   (c) Verify (3.16)-(3.18).

4. Verify that (3.19) vanishes where $u + v = \frac{4}{5}$.

5. Using the method of §3.1, find the bargaining sets for Crossroads when $\delta = 5, \epsilon = 3, \tau_1 = 4$ and $\tau_2 = 2$. Sketch the reward set, marking in particular all locally or globally unimprovable points.

6. Use the Theorem of the Alternative (§3.2) to show that no points other than $(1, 0)$ and $(0, 1)$ on the boundary of the unit square are unimprovable in Crossroads when $\delta = 3, \epsilon = 2$ and $\tau_1 = 2 = \tau_2$.

7. Use the Theorem of the Alternative to find P_{nec}^* for Crossroads when $\delta = 5, \epsilon = 3, \tau_1 = 4$ and $\tau_2 = 2$. Verify that your results agree with those you obtained in Exercise 3.5.

8. (a) For Chapter 1's original (asymmetric) game of Crossroads between fast drivers, show that D^* is the unshaded region in Figure 3.5, where ξ_1 and ξ_2 are defined by (3.45), θ_1 and

θ_2 are defined by (1.22) and the point marked by a dot has coordinates $(\alpha_1, \alpha_2) = \frac{\epsilon}{\delta + \epsilon}(1, 1) + \frac{\{\tau_1 - \tau_2\}}{4\{\delta + \epsilon\}}(1, -1)$.

(b) Describe the Nash bargaining solution.

(c) With regard to the discussion at the end of §3.3, does the Nash bargaining solution invariably lack uniqueness when the game is symmetric?

9. For the game of the prisoner's dilemma defined by Exercise 1.25:

   (a) Show that $P_G = P$, and find P^*

   (b) Find the Nash bargaining solution.

10. Find the Nash bargaining solution under correlated strategies of Chapter 1's original (asymmetric) game of Crossroads. Is $f_1 = f_2$ at this solution?[4]

11. Find parametric equations for the boundary of the Store Wars reward set when $\alpha = 10$ and $c = 1$. (This reward set is sketched in Figure 3.3).[5]

12. Use your results from Exercise 3.11 to find the bargaining set and the Nash bargaining solution for Store Wars when $\alpha = 10$.

---

[4]From (3.49) and (3.50), the vector $f = (f_1, f_2)$ from the origin to the point with coordinates (f_1, f_2) is a "convex linear combination" of the vectors $(-\delta - \tau_2/2, -\delta - \tau_2/2), (0, -\tau_1), (-\tau_2, 0)$ and $(-\epsilon - \tau_2/2, -\epsilon - \tau_2/2)$, i.e., a linear combination with nonnegative coefficients that sum to 1, by (3.48). The point (f_1, f_2) must therefore lie somewhere in the triangle with vertices $(-\delta - \tau_2/2, -\delta - \tau_2/2), (0, -\tau_1)$ and $(-\tau_2, 0)$. Thus finding $\hat{\omega}$ corresponds to maximizing the area of a rectangle, one corner of which is constrained to lie on the line with equation (3.52).

[5]Any line segment $L(\xi)$ with parametric equations $u = \xi + t, v = t, 0 \leq t \leq \alpha - \xi$ is parallel to $u = v$ in Figure 1.8. For $1 \leq \xi \leq 6$ such line segments cover D_A, and for $0 \leq \xi \leq 1$ they cover the part of D_B that lies below $u = v$. The remainder of D is covered by line segments of the form $u = t, v = \xi + t, 0 \leq t \leq \alpha - \xi$; for $1 \leq \xi \leq 6$ they cover D_C, and for $0 \leq \xi \leq 1$ they cover the rest of D_B.

Chapter 4

Characteristic Function Games

A characteristic function game or CFG is a purely cooperative game among n players who seek a fair distribution for a benefit that is freely transferable. It is assumed that all players would like as much as possible of the benefit, and that one unit of the benefit is worth the same to all players; thus, in terms of §3.4, characteristic function games imply interpersonal utility comparisons. Usually, but not necessarily (see, e.g., §4.1), the benefit to be shared is money. In this chapter, we introduce two solution concepts for characteristic function games, the nucleolus and the Shapley value.

The fairness of a distribution is assumed to depend on the bargaining strengths of the various coalitions that could possibly form among some, but not all, of the players. Nevertheless, and at first sight paradoxically, the fundamental assumption of a CFG is that all players are cooperating. In other words, a *grand coalition* of all n players has formed—perhaps voluntarily, but the grand coalition may also have been enforced by the action of some external agent or circumstance. Thus coalitions of fewer than n players can use as bargaining leverage the strength they *would* have had without the others, *if* the others weren't there; but the grand coalition can never actually dissolve (or the theory dissolves with it).

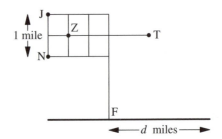

Figure 4.1. Jed, Ned, Ted and Zed's stomping ground.

It will be convenient in this chapter to regard each player as a fictitious coalition of one person; however, characteristic function games are interesting only when there could exist true coalitions of less than all the players. Therefore, we shall assume throughout that $n \geq 3$.[1]

As usual, we introduce solution concepts by means of examples.

4.1. Characteristic functions and reasonable sets

Jed, Ned and Ted are neighbors. Their houses are marked J, N and T, respectively, in Figure 4.1. They work in the same office at the same times on the same days, and in order to save money they would like to form a car pool. They must first agree on how to share the costs of this cooperative venture, however; or, which is the view we prefer to adopt, on how to share the car pool's benefits. We explore the matter here. Later, in §4.4, we shall consider adding a fourth neighbor, Zed, to the car pool.

Let's suppose that Jed, Ned and Ted drive identical cars, and that the cost of driving to work, *including depreciation*, is k dollars per mile. Because depreciation is included, it doesn't matter in principle whose car is used (though in practice they might take turns). Let the distance to work from point F in the diagram, where the road through their neighborhood crosses a freeway, be d miles. Then, assuming each player selects the shortest route, Jed lives $4 + d$ miles from work, whereas Ned and Ted are both $3 + d$ miles away.

---

[1] For $n = 2$, both the nucleolus and the Shapley value give the players equal shares of the benefit to be shared; see Exercises 4.7 and 4.17.

Let Jed be Player 1, Ned Player 2 and Ted Player 3; and let $c(\{i\})$ denote Player i's cost in dollars of driving to work alone. Then $c(\{1\}) = (4+d)k$ and $c(\{2\}) = (3+d)k = c(\{3\})$. Round-trip travel costs are just twice these amounts. Let $c(\{1,2,3\})$ denote the cost in dollars if all three players drive to work in a single car, assuming that the shortest route is adopted. Then, clearly, $c(\{1,2,3\}) = (7+d)k$, regardless of whose car is used. If Jed and Ned were to form a car pool without Ted, then the cost would be $c(\{1,2\}) = (4+d)k$, again assuming the shortest route—and, of course, that they would use Jed's car. Similarly, the costs of car pools that excluded Ned or Jed would be $c(\{1,3\}) = (6+d)k = c(\{2,3\})$; it would not matter whose car Ned and Ted used, although Jed and Ted would have to use Jed's. We have now calculated the travel costs associated with each of the seven coalitions that three players could form, namely, $\{1\}$, $\{2\}$, $\{3\}$, $\{1,2\}$, $\{1,3\}$, $\{2,3\}$ and $\{1,2,3\}$. More generally, among n players, $2^n - 1$ coalitions are possible.

Let $\overline{v}(S)$ denote the benefit of cooperation associated with the coalition S. Then

$$(4.1) \qquad \overline{v}(S) \;=\; \sum_{i \in S} c(\{i\}) - c(S),$$

i.e., $\overline{v}(S)$ is the difference between the sum of the costs that the individual members of S would have to bear if they did not cooperate and the cost when they club together. For example, the benefit associated with the grand coalition $\{1,2,3\}$ among all three players is saving

$$
\begin{aligned}
(4.2) \qquad \overline{v}(\{1,2,3\}) \;&=\; c(\{1\}) + c(\{2\}) + c(\{3\}) - c(\{1,2,3\}) \\
&=\; (3+2d)k
\end{aligned}
$$

dollars. A car pool without Ted would save Jed and Ned

$$(4.3) \qquad \overline{v}(\{1,2\}) \;=\; c(\{1\}) + c(\{2\}) - c(\{1,2\}) \;=\; (3+d)k$$

dollars, whereas the benefits of cooperations that excluded Ned or Jed would be

$$(4.4) \qquad \overline{v}(\{1,3\}) \;=\; c(\{1\}) + c(\{3\}) - c(\{1,3\}) \;=\; (1+d)k$$

and

$$(4.5) \qquad \overline{v}(\{2,3\}) \;=\; c(\{2\}) + c(\{3\}) - c(\{2,3\}) \;=\; dk,$$

respectively. Of course, the benefit of cooperation associated with not cooperating is precisely zero; that is, from (4.1) with $S = \{i\}$,

$$(4.6) \qquad\qquad \bar{\nu}(\{i\}) = 0$$

for any value of i.

For the sake of definiteness, let's imagine that the car pool $\{1, 2, 3\}$ will always use Jed's car, so that it is Jed who actually foots the bills. How much should Ned and Ted pay Jed for each one-way trip? Let us suppose that Jed receives fraction x_1 of the grand car pool's benefit $(3 + 2d)k$, and that Ned and Ted receive fractions x_2 and x_3, respectively, where

$$(4.7) \qquad\qquad 0 \le x_1, x_2, x_3 \le 1, \quad x_1 + x_2 + x_3 = 1.$$

Then Ned or Ted should pay Jed the amount

$$(4.8) \qquad\qquad c(\{i\}) - (3 + 2d)kx_i$$

dollars per trip, where $i = 2$ for Ned and $i = 3$ for Ted; and (4.8) with $i = 1$ is the part of the bill that Jed must pay himself. Our task is therefore to determine the fractions x_1, x_2 and x_3.

Now, (4.2)-(4.6) define a function $\bar{\nu}$ from the set of all coalitions among three players into the real numbers. We call $\bar{\nu}$ the characteristic function. For set-theoretic purposes, it is convenient to suppose that the set of all coalitions of three players contains, in addition to $\{1\}$, $\{2\}$, $\{3\}$, $\{1,2\}$, $\{1,3\}$, $\{2,3\}$ and $\{1,2,3\}$, the "empty coalition" $\varnothing$, which, because it contains no players, cannot benefit from cooperation. For completeness, therefore, we append

$$(4.9) \qquad\qquad \bar{\nu}(\varnothing) = 0.$$

Note that (4.9) is not a gratuitous appendage; if in any doubt, compute the right-hand side of (4.15) with $T = \{i\}$.

The number $\bar{\nu}(S)$, the benefit that the players in S can obtain if they cooperate with each other but not with the players outside S, is a measure of the bargaining strength of the coalition S. It is convenient, however, to express this strength as a fraction of the strength of the grand coalition $\{1, 2, 3\}$, i.e., as $\nu(S) = \bar{\nu}(S)/\bar{\nu}(\{1,2,3\})$, where S is any coalition. Thus $\nu(\{i\}) = 0$, $i = 1, 2, 3$,

$$(4.10) \qquad \nu(\{1,2\}) = \tfrac{3+d}{3+2d}, \quad \nu(\{1,3\}) = \tfrac{1+d}{3+2d}, \quad \nu(\{2,3\}) = \tfrac{d}{3+2d}$$

and $\nu(\{1,2,3\}) = 1$. More generally, for a game among n players, we define the *normalized characteristic function*, ν, by

$$(4.11) \qquad \nu(S) = \frac{\overline{\nu}(S)}{\overline{\nu}(N)},$$

where N denotes the set of all players or grand coalition, i.e., $N = \{1, 2, ..., n\}$; and S is any of the 2^n coalitions (including $\varnothing$) or, which is the same thing, S is any of the 2^n subsets of N. Note that (4.11) does not make sense unless $\overline{\nu}(N) > 0$, that is, unless the benefits of cooperation are positive, and we shall assume throughout that this condition is satisfied. A CFG such that $\overline{\nu}(N) > 0$ is said to be *essential*, whereas a CFG such that $\overline{\nu}(N) = 0$ is *inessential*. Thus we restrict attention to the essential variety.[2]

Now, with regard to the car pool, our task is to determine a 3-dimensional vector $x = (x_1, x_2, x_3)$ satisfying (4.7) that stipulates how the benefits of cooperation are to be distributed among the grand coalition $\{1, 2, 3\}$. We will refer to x as an *imputation*, and to the i-th component of x, namely x_i, as Player i's *allocation* at x. Furthermore, for each coalition S, we will refer to the sum $\sum_{i \in S} x_i$ of allocations to players in S as the coalition's allocation at x (or as the amount that x allocates to S).

More generally, an imputation of a CFG among n players is an n-dimensional vector $x = (x_1, x_2, \ldots, x_n)$ such that

$$(4.12a) \qquad x_i \geq 0, \quad i = 1, 2, \ldots, n$$

$$(4.12b) \qquad x_1 + x_2 + \ldots + x_n = 1.$$

The i-th component of x, namely x_i, is Player i's allocation at x. We will denote the set of all imputations by X. Thus $x \in X$ if and only if (4.12) is satisfied. Obviously, if $x_i \geq 0$ for all $i \in N$, then (4.12b) implies $x_i \leq 1$ for all $i \in N$. If x satisfies (4.12b), then x is said to be unimprovable or *group rational*, because it is impossible to increase one player's allocation without decreasing that of another. If

---

[2]If benefits are first defined such that $\overline{\nu}(\{i\}) \neq 0, i = 1, 2, \ldots, n$, as would have happened in the car pool if we had first defined $\overline{\nu} = -c$, then the benefits associated with acting alone must first be subtracted out (as in (4.1)) before the benefits of cooperation are properly defined; in which case, (4.11) must be replaced by

$$\nu(S) = \frac{\overline{\nu}(S) - \sum_{i \in S} \overline{\nu}(\{i\})}{\overline{\nu}(N) - \sum_{i=1}^{n} \overline{\nu}(\{i\})}.$$

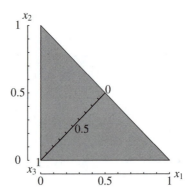

Figure 4.2. The set of imputations for a 3-player CFG.

x satisfies (4.12a), then x is said to be *individually rational*, because each player is at least as well off in the grand coalition as he would have been all by himself. Thus imputations are vectors that are both individually and group rational.

For $n = 3$, because $x_3 \geq 0$ implies $0 \leq x_1 + x_2 \leq 1$, X can be represented in two dimensions as a right-angled isosceles triangle. From Figure 4.2, where X is shaded, we see that x_1 increases to the east and x_2 to the north, whereas x_3 increases to the southwest. We should bear in mind, however, that this representation is a distortion, albeit a convenient one, of the true picture, which is that X is a 2-dimensional equilateral triangle imbedded in 3-dimensional space. More generally, for a CFG among n players, X is an $(n-1)$-dimensional "hypertriangle"—more commonly known as a *simplex*—in n-dimensional space.

Which of the infinitely many imputations in X can be regarded as a fair distribution of the benefits of cooperation? We can attempt to answer this question by considering the order in which the grand coalition of all n players might actually form. Suppose, for example, that $n \geq 4$ and that Player a is first to join, Player b second, Player c third and Player i fourth, where a, b, c and i are any positive integers between 1 and n. Let $T = \{a, b, c, i\}$. Then, because the players have joined in the given order, it is reasonable to say that Player a has contributed $\nu(\{a\}) - \nu(\varnothing) = 0$ to the grand coalition, that Player

b has added $\nu(\{a, b\}) - \nu(\{a\}) = \nu(\{a, b\})$, that Player c has added $\nu(\{a, b, c\}) - \nu(\{a, b\})$, and that Player i has added $\nu(T) - \nu(\{a, b, c\})$; in which case, $\nu(T) - \nu(\{a, b, c\})$ is Player i's fair allocation. But $\{a, b, c\} = T - \{i\}$, where we define the difference of two sets to consist of all elements in the first but not the second, i.e.,

$$(4.13) \qquad A - B = \{x \mid x \in A, x \notin B\}.$$

Thus $\nu(T) - \nu(T - \{i\})$ is Player i's fair allocation.

More generally, if Player i is the j-th player to join and the $j - 1$ players who have joined already are $T - \{i\}$, then $\nu(T) - \nu(T - \{i\})$ is a fair allocation for i. Unfortunately, we do not know the identity of T: there are 2^{n-1} coalitions containing i (see Exercise 4.16), and T could be any one of them. What we can be sure about, however, is that if we calculate $\nu(T) - \nu(T - \{i\})$ for every coalition T containing i, i.e., for every coalition in the set Π^i defined by

$$(4.14) \qquad \Pi^i = \{S \mid i \in S \text{ and } S \subset N\},$$

then a fair allocation for Player i should not exceed the maximum of those 2^{n-1} numbers. Therefore, the set of imputations such that

$$(4.15) \qquad x_i \leq \max_{T \in \Pi^i} \{\nu(T) - \nu(T - \{i\})\}$$

for all $i = 1, 2, ..., n$ is called the *reasonable set*.

Returning now to the car pool, let us suppose, for the sake of definiteness, that the office where Jed, Ned and Ted all work is 9 miles from the point marked F in Figure 4.1. Then $d = 9$ in (4.10). For Jed, the maximum in (4.15) is taken over all coalitions T that contain $\{1\}$, i.e., over all $T \in \Pi^1 = \{\{1\}, \{1, 2\}, \{1, 3\}, \{1, 2, 3\}\}$. Thus if x belongs to the reasonable set we must, in particular, have

$$x_1 \leq \max\Big\{\nu(\{1, 2, 3\}) - \nu(\{2, 3\}), \ \nu(\{1, 2\}) - \nu(\{2\}),$$
$$\nu(\{1, 3\}) - \nu(\{3\}), \ \nu(\{1\}) - \nu(\varnothing)\Big\}$$

$= \max\big\{1 - \frac{3}{7}, \frac{4}{7} - 0, \frac{10}{21} - 0, 0 - 0\big\} = \frac{4}{7}$. Similarly, on taking $i = 2$ and then $i = 3$ in (4.15), we must also have $x_2 \leq \max\{\frac{11}{21}, \frac{3}{7}, \frac{4}{7}\} = \frac{4}{7}$ and $x_3 \leq \max\{\frac{3}{7}, \frac{10}{21}, \frac{3}{7}\} = \frac{10}{21}$ (Exercise 4.1). The reasonable set consists of points that satisfy both these inequalities and (4.7). Because $x_3 = 1 - x_1 - x_2$, the inequality $x_3 \leq \frac{10}{21}$ is equivalent to $x_1 + x_2 \geq \frac{11}{21}$; and so

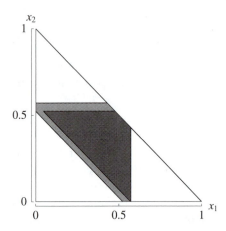

Figure 4.3. The reasonable set (light and dark shading) and
the core (dark shading) for the 3-person car pool with $d = 9$.

the reasonable set is the hexagon depicted by the shaded regions (both
light and dark) in Figure 4.3. We see that, although (4.15) excludes
imputations near the corners of X—it would not be reasonable to give
most of the benefits of car pooling to a single player—the concept of
reasonable set provides minimal constraints on what we may consider
a fair imputation. Nevertheless, the thinking that led to (4.15), when
suitably refined, is capable of yielding an attractive solution concept,
and we shall give it our attention in §4.7.

Before proceeding, a word about notation. For most purposes, it
is preferable to regard imputations of a 3-player game as vectors in 3-
dimensional space; but X is 2-dimensional, and for graphical purposes
it is preferable to regard the imputations as points in two dimensions.
Thus the corner $(1, 0)$ of the triangle in Figure 4.3 represents the
(unreasonable) imputation $(1, 0, 0)$, which would give all the benefits
of cooperation to Jed; the corner $(0, 1)$ represents the imputation
$(0, 1, 0)$, which would give all the benefits of cooperation to Ned; and
the corner $(0, 0)$ represents the imputation $(0, 0, 1)$, which would give
all the benefits of cooperation to Ted. More generally, from (4.7), the
point (x_1, x_2) of the triangle in Figure 4.3 represents the imputation
$(x_1, x_2, 1 - x_1 - x_2)$.

4.2. Core-related concepts

The concept of reasonable set has enabled us to exclude the most unreasonable points from the set of imputations for the car pool. But infinitely many imputations still remain. If we are serious about helping Jed, Ned and Ted reach agreement over their car pool, then we had better find a way to exclude more of X.

A concept that is useful in this regard is that of *excess*. For each imputation $x \in X$ in an n-player CFG, the excess of the coalition S at x, denoted by $e(S, x)$, is the difference between the fraction of the benefits of cooperation that S can obtain for itself, even if it does not cooperate with players outside S, and the fraction of the benefits of cooperation that x allocates to S:

$$(4.16) \qquad e(S, x) \;=\; \nu(S) - \sum_{i \in S} x_i.$$

For example, for any CFG, it follows from (4.16) that

$$(4.17\text{a}) \qquad e(\{i\}, x) \;=\; -x_i, \quad i = 1, 2, \ldots, n$$

$$(4.17\text{b}) \qquad e(\varnothing, x) \;=\; 0 \;=\; e(N, x);$$

and for the car pool of §4.1 with $d = 9$ we have

$$(4.18\text{a}) \qquad e(\{1, 2\}, x) \;=\; \tfrac{4}{7} - x_1 - x_2$$

$$(4.18\text{b}) \qquad e(\{1, 3\}, x) \;=\; \tfrac{10}{21} - x_1 - x_3 \;=\; x_2 - \tfrac{11}{21}$$

$$(4.18\text{c}) \qquad e(\{2, 3\}, x) \;=\; \tfrac{3}{7} - x_2 - x_3 \;=\; x_1 - \tfrac{4}{7}.$$

If $e(S, x) > 0$, then the players in S will regard the imputation x as unfair, because they would receive greater benefits if they did not have to form the grand coalition. It therefore seems sensible, if possible, to exclude imputations such that $e(S, x) > 0$ for some coalition S (even if only one). The imputations that remain, if any, are said to form the *core* of the game. Here, and in the following section, we will assume that the core exists; and for reasons about to emerge, we will denote it by $C^+(0)$. In set-theoretic notation:

$$(4.19) \qquad C^+(0) \;=\; \{x \in X \mid e(S, x) \leq 0 \text{ for } \textit{all} \text{ coalitions } S\}.$$

For example, from (4.18), the core of the car-pool game with $d = 9$ is the dark quadrilateral in Figure 4.3.[3] If the core exists, then it must be a subset of the reasonable set (Exercise 4.26), possibly the whole of it (Exercise 4.10). But a game may have no core (or, if you prefer, $C^+(0) = \varnothing$), and we shall consider this possibility in §4.4. We note in passing that a sufficient condition[4] for the core to exist is that the game be *convex*, i.e., that

$$(4.20) \qquad \nu(S \cup T) \geq \nu(S) + \nu(T) - \nu(S \cap T)$$

for all coalitions S and T. But convexity is by no means a necessary condition, and often fails to hold; for example, the car-pooling game defined by (4.10) is never convex but always has a core (Exercise 4.9).

By excluding imputations that lie inside the reasonable set but outside the core (the lighter shaded region in Figure 4.3), we move nearer to a car pool agreement. But infinitely many imputations still remain. Then which of them represents the fairest distribution of the benefits of cooperation?

From Jed's point of view, the best points in the dark quadrilateral of Figure 4.3 lie on the boundary where $x_1 = \frac{4}{7}$; from Ned's point of view, they lie on the boundary where $x_2 = \frac{11}{21}$; and from Ted's point of view they lie on the boundary where $x_1 + x_2 = \frac{4}{7}$. It thus appears that the fairest compromise would be somehow to locate the "center" of the core. But how do we find an imputation that we can reasonably interpret as the center? One approach would be to move the walls of the boundary inward, all at the same speed, until they coalesce in a point. Suppose, for example, that we move the walls inward at the rate of one unit every 21 seconds. Then, after $\frac{1}{21}$ seconds, we will have shrunk the boundary to the inner of the two quadrilaterals in Figure 4.4. If we continue at the same rate, then after $\frac{2}{21}$ seconds we will have shrunk the boundary to the outer of the two triangles, and after $\frac{1}{7}$ seconds to the inner triangle. Ultimately, after $\frac{11}{63}$ seconds, the boundary will collapse to the point $\left(\frac{25}{63}, \frac{22}{63}\right)$, which is marked by a dot in Figure 4.4. We might therefore propose that the imputation $x^* = \left(\frac{25}{63}, \frac{22}{63}, \frac{16}{63}\right)$ is the fair solution of the game; in which case, from (4.8), Jed and Ned should each pay $\frac{14}{3}k$ dollars of the cost per trip,

---

[3] The core of a 3-player CFG need not be a quadrilateral; it can be a point, a line segment, a triangle, a pentagon (as in Exercise 4.10) or a hexagon.

[4] See [**205**].

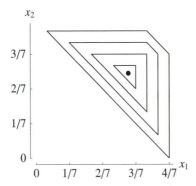

Figure 4.4. Rational ϵ-core boundaries for $\epsilon = 0$ (outer quadrilateral), $\epsilon = -1/21$, $\epsilon = -2/21$, $\epsilon = -1/7$ and $\epsilon = -11/63$ (dot) when $d = 9$ in the car-pool game of §4.1.

whereas Ted should pay $\frac{20}{3}k$ dollars. Ted should pay most because Jed must go out of his way to drive him home.

With a view to later developments, it will be convenient to denote by Σ^0 the set of all coalitions that are neither the empty coalition nor the grand coalition:

$$(4.21) \qquad \Sigma^0 = \{S \mid S \subset N, S \neq \varnothing, S \neq N\}.$$

We can now generalize the ideas that yielded our fair solution to an arbitrary n-person CFG by defining the *rational ϵ-core*, denoted by $C^+(\epsilon)$, as the set of all imputations at which no coalition other than $\varnothing$ or N has a greater excess than ϵ. That is,

$$(4.22) \qquad C^+(\epsilon) = \{x \in X \mid e(S, x) \leq \epsilon \text{ for all } S \in \Sigma^0\}.$$

Thus the core (if it exists) is the rational 0-core; and for $\epsilon < 0$, $C^+(\epsilon)$ is the set to which the core has shrunk after 1 second, when its walls are moved inward at $|\epsilon|$ units per second (again, if it exists). At imputation x, however, $e(S, x) \leq \epsilon$ for all $S \in \Sigma^0$ if and only if the maximum of $e(S, x)$—taken over all coalitions in Σ^0—is less than or equal to ϵ. Accordingly, and again with a view to later developments, let us define a function ϕ_0 from X to the real numbers by

$$(4.23) \qquad \phi_0(x) = \max_{S \in \Sigma^0} e(S, x).$$

We can now define the rational ϵ-core more succinctly as

$$(4.24) \qquad C^+(\epsilon) \;=\; \{x \in X \mid \phi_0(x) \le \epsilon\}.$$

Reducing ϵ shrinks $C^+(\epsilon)$; but if ϵ is too small, then there are no imputations such that $e(S, x) \le \epsilon$ for all $S \in \Sigma^0$. So there is a least value of ϵ for which $C^+(\epsilon)$ exists. Indeed, if we denote this value by ϵ_1, then it is clear that

$$(4.25) \qquad\qquad \epsilon_1 \;=\; \min_{x \in X} \phi_0(x),$$

because reducing ϵ causes more and more imputations to violate the condition $\phi_0(x) \le \epsilon$, until all that remain finally are the imputations for which ϕ_0 attains its minimum on X. We shall refer to $C^+(\epsilon_1)$, the set over which ϕ_0 attains its minimum, as the *least rational core*. With a view to later developments, however, it will be convenient to have an alternative notation for the least rational core, namely, X^1. Thus $C^+(\epsilon_1)$ and X^1 are the same subset of X.

Returning to the car pool for illustration, we find that x belongs to $C^+(\epsilon)$ if, on using (4.18),

$$(4.26a) \qquad \tfrac{4}{7} - x_1 - x_2 \le \epsilon, \;\; x_2 - \tfrac{11}{21} \le \epsilon, \;\; x_1 - \tfrac{4}{7} \le \epsilon$$

and, on using (4.17a),[5]

$$(4.26b) \qquad\qquad -x_1 \le \epsilon, \;\; -x_2 \le \epsilon, \;\; -x_3 \le \epsilon;$$

or, which is the same thing (verify), if

$$(4.27) \qquad \begin{aligned} -\epsilon \le x_1 \le \tfrac{4}{7} + \epsilon, \;\; -\epsilon \le x_2 \le \tfrac{11}{21} + \epsilon \\ \tfrac{4}{7} - \epsilon \le x_1 + x_2 \le 1 + \epsilon. \end{aligned}$$

The boundary of the region corresponding to (4.27) is sketched in Figure 4.4 for $\epsilon = 0$ (outer quadrilateral), $\epsilon = -\tfrac{1}{21}$ (inner quadrilateral), $\epsilon = -\tfrac{2}{21}$ (outer triangle) and $\epsilon = -\tfrac{1}{7}$ (inner triangle); these triangles, and the open curves consisting of the longest three sides of the quadrilaterals, are $\phi_0 = \epsilon$ contours of the function defined by (4.23). The dot corresponds to the least rational core, which is

$$(4.28) \qquad X^1 \;=\; \{x^*\} \;=\; \left\{\left(\tfrac{25}{63}, \tfrac{22}{63}, \tfrac{16}{63}\right)\right\}.$$

---

[5]Note that if $\epsilon > 0$ then (4.26b) is superseded by condition (4.12a) that allocations be nonnegative.

Note that the value $\epsilon_1 = -\frac{11}{63}$ can be found analytically, but we postpone this matter until §4.3.

Now $e(S, x)$ is a measure of coalition S's dissatisfaction with the imputation x. If $e(S, x) > 0$ then S's allocation from x would be less than the benefit it could obtain for itself, and the players in S would rather dissolve the grand coalition than accept x (but they cannot dissolve it, because a fundamental assumption of CFG analysis is that the grand coalition has formed). If $e(S, x) = 0$ then the players in S would be indifferent between maintaining or dissolving the grand coalition. On the other hand, if $e(S, x) < 0$, then the players in S would prefer to remain in the grand coalition (even if it were possible to dissolve it). Regardless of whether $e(S, x)$ is positive or negative, however, the lower the value of $e(S, x)$, the lower the dissatisfaction of the players in S with imputation x (or, if you prefer, the higher their satisfaction). If $x \in C^+(\epsilon)$, then no coalition's dissatisfaction exceeds ϵ (or no coalition's satisfaction is less than $-\epsilon$); and the lower the value of ϵ, the lower the value of the maximum dissatisfaction among all coalitions that could possibly form. Thus if $X^1 = C^+(\epsilon_1)$ contains a single imputation, say x^*, then x^* minimizes maximum dissatisfaction; and to the extent that minimizing maximum dissatisfaction is fair—which we assume henceforward until §4.7—x^* is the fair solution of the game.

Although $X^1 = C^+(\epsilon_1)$ always exists, it may contain infinitely many imputations. Then which should be regarded as the fair solution? We will address this matter in §4.5. Meanwhile, the least rational core will be an adequate solution concept, because in §§4.3-4.4 we consider games for which X^1 contains only a single imputation, x^*. Before proceeding, however, we pause to note that—although reducing ϵ always shrinks $C^+(\epsilon)$, until eventually at $\epsilon = \epsilon_1$ it disappears—it is not quite true that increasing ϵ always expands $C^+(\epsilon)$, because $C^+(\epsilon)$ is a subset of X. Rather, there is a maximum value of ϵ, say ϵ_0, beyond which the inequalities $e(S, x) \leq \epsilon$ are all superseded by $x \in X$. Thus if $\epsilon \geq \epsilon_0$ then $C^+(\epsilon) = X$; or, if you prefer, X is the greatest rational core. For example, by moving the walls outward in Figure 4.4 we see that $\epsilon_0 = \frac{3}{7}$ for the 3-person car pool with $d = 9$.

4.3. A four-person car pool

The purpose of the present section is twofold: to present an example of a 4-player game, and to show how to calculate X^1. Accordingly, let Jed, Ned and Ted have a fourth neighbor, Zed, whose house is marked Z in Figure 4.1. He works in Jed, Ned and Ted's office at the same times on the same days, owns the same kind of small car, and in order to save money would like to join their car pool. Now, in practice it might happen that the existing car pool would bargain with Zed as a unit, so that the bargaining would reduce to a 2-player game; but we wish to consider a 4-person game. Let us therefore assume, at least until the end of the section, that Zed is a good friend of all the others, and that they work out the costs of the 4-person car pool from scratch. As stated in the previous section, we also assume that a fair distribution of the benefits of cooperation is one that minimizes maximum dissatisfaction. Thus our task is to calculate the least rational core of a 4-player game.

With $n = 4$, and hence $N = \{1, 2, 3, 4\}$, there are 15 coalitions excluding $\varnothing$. Let us assume that Zed is Player 4, and that Jed, Ned and Ted are Players 1, 2 and 3, as before. Now, it is clear from §4.1 that the value of k has no effect on the outcome; therefore, we may as well express the costs in units of k dollars. Then from Figure 4.1 we readily find that the costs of the car pools that could possibly form (predicated on travel by the shortest route) are as follows:

$$c(\{1\}) = 4 + d, \ c(\{2\}) = 3 + d$$
$$c(\{3\}) = 3 + d, \ c(\{4\}) = 3 + d$$
$$(4.29) \qquad c(\{1,2\}) = 4 + d, \ c(\{1,3\}) = 6 + d, \ c(\{1,4\}) = 4 + d$$
$$c(\{2,3\}) = 6 + d, \ c(\{2,4\}) = 4 + d, \ c(\{3,4\}) = 5 + d$$
$$c(\{1,2,3\}) = 7 + d, \ c(\{1,2,4\}) = 5 + d, \ c(\{1,3,4\}) = 6 + d$$
$$c(\{2,3,4\}) = 6 + d, \ c(\{1,2,3,4\}) = 7 + d.$$

From (4.1), we deduce the associated benefits of cooperation:

$$\overline{v}(\{i\}) = 0, \ i = 1, 2, 3, 4$$
$$\overline{v}(\{1,2\}) = 3 + d, \ \overline{v}(\{1,3\}) = 1 + d, \ \overline{v}(\{1,4\}) = 3 + d$$
$$(4.30) \qquad \overline{v}(\{2,3\}) = d, \ \overline{v}(\{2,4\}) = 2 + d, \ \overline{v}(\{3,4\}) = 1 + d$$
$$\overline{v}(\{1,2,3\}) = 3 + 2d, \ \overline{v}(\{1,2,4\}) = 5 + 2d$$
$$\overline{v}(\{1,3,4\}) = 4 + 2d, \ \overline{v}(\{2,3,4\}) = 3 + 2d$$
$$\overline{v}(\{1,2,3,4\}) = 6 + 3d.$$

For the sake of definiteness, let us suppose that $d = 2$ (you may consider this value of d to be unrealistic, but higher values are considered in Exercises 4.5 and 4.6). Then, from (4.11) and (4.30), the normalized characteristic function is defined by

$$
\begin{gathered}
\nu(\{i\}) = 0, \quad i = 1, 2, 3, 4 \\
\nu(\{1,2\}) = \tfrac{5}{12}, \quad \nu(\{1,3\}) = \tfrac{1}{4}, \quad \nu(\{1,4\}) = \tfrac{5}{12} \\
\nu(\{2,3\}) = \tfrac{1}{6}, \quad \nu(\{2,4\}) = \tfrac{1}{3}, \quad \nu(\{3,4\}) = \tfrac{1}{4} \\
\nu(\{1,2,3\}) = \tfrac{7}{12}, \quad \nu(\{1,2,4\}) = \tfrac{3}{4}, \quad \nu(\{1,3,4\}) = \tfrac{2}{3} \\
\nu(\{2,3,4\}) = \tfrac{7}{12}, \quad \nu(\{1,2,3,4\}) = 1.
\end{gathered}
$$
(4.31)

Thus the rational ϵ-core consists of imputations $x = (x_1, x_2, x_3, x_4)$ that satisfy

(4.32a)
$$
-x_1 \le \epsilon, \quad -x_2 \le \epsilon, \quad -x_3 \le \epsilon, \quad -x_4 \le \epsilon
$$

(4.32b)
$$
\tfrac{5}{12} - x_1 - x_2 \le \epsilon, \quad \tfrac{1}{4} - x_1 - x_3 \le \epsilon, \quad \tfrac{5}{12} - x_1 - x_4 \le \epsilon
$$
$$
\tfrac{1}{6} - x_2 - x_3 \le \epsilon, \quad \tfrac{1}{3} - x_2 - x_4 \le \epsilon, \quad \tfrac{1}{4} - x_3 - x_4 \le \epsilon
$$

(4.32c)
$$
\tfrac{7}{12} - x_1 - x_2 - x_3 \le \epsilon, \quad \tfrac{3}{4} - x_1 - x_2 - x_4 \le \epsilon
$$
$$
\tfrac{2}{3} - x_1 - x_3 - x_4 \le \epsilon, \quad \tfrac{7}{12} - x_2 - x_3 - x_4 \le \epsilon
$$

(4.33)
$$
x_1 + x_2 + x_3 + x_4 = 1.
$$

Of course, if $\epsilon > 0$, then (4.32a) is superseded by condition (4.12a) that allocations be nonnegative.

On using (4.33) to eliminate x_4 from (4.32), we find that $C^+(\epsilon)$ is the 3-dimensional simplex of vectors $x \in X$ such that

(4.34a)
$$
-\epsilon \le x_1 \le \tfrac{5}{12} + \epsilon, \quad -\epsilon \le x_2 \le \tfrac{1}{3} + \epsilon, \quad -\epsilon \le x_3 \le \tfrac{1}{4} + \epsilon
$$

(4.34b)
$$
\tfrac{5}{12} - \epsilon \le x_1 + x_2 \le \tfrac{3}{4} + \epsilon, \quad \tfrac{1}{4} - \epsilon \le x_1 + x_3 \le \tfrac{2}{3} + \epsilon
$$
$$
\tfrac{1}{6} - \epsilon \le x_2 + x_3 \le \tfrac{7}{12} + \epsilon
$$

(4.34c)
$$
\tfrac{7}{12} - \epsilon \le x_1 + x_2 + x_3 \le 1 + \epsilon.
$$

You should verify for yourself that (4.32) and (4.34) are equivalent; for example, the second of (4.34a) combines the second of (4.32a) with the third of (4.32c), whereas the third of (4.34b) combines the third and fourth of (4.32b).

We can now determine ϵ_1. From the first of (4.34a) we have $-\epsilon \le \tfrac{5}{12} + \epsilon$, or $\epsilon \ge -\tfrac{5}{24}$. Similarly, from the second and third of (4.34a), we have $\epsilon \ge -\tfrac{1}{6}$ and $\epsilon \ge -\tfrac{1}{8}$; from (4.34b) we have $\epsilon \ge -\tfrac{1}{6}$

and $\epsilon \geq -\frac{5}{24}$ (twice); and from (4.34c) we again have $\epsilon \geq -\frac{5}{24}$. We have now obtained three lower bounds on ϵ, namely, $-\frac{5}{24}$, $-\frac{1}{6}$ and $-\frac{1}{8}$, and the greatest of these lower bounds is $-\frac{1}{8}$. We have therefore established that $\epsilon_1 \geq -\frac{1}{8}$. Can we improve this lower bound? Adding the first two inequalities of (4.34a) yields

$$(4.35) \qquad\qquad -2\epsilon \leq x_1 + x_2 \leq \tfrac{3}{4} + 2\epsilon,$$

which in conjunction with the first of (4.34b) yields

$$(4.36) \qquad\qquad \tfrac{5}{12} - \epsilon \leq \tfrac{3}{4} + 2\epsilon, \quad -2\epsilon \leq \tfrac{3}{4} + \epsilon$$

or $\epsilon \geq -\frac{1}{9}$ and $\epsilon \geq -\frac{1}{4}$, the second of which is superseded by the first. Hence $\epsilon_1 \geq -\frac{1}{9}$, which supersedes $\epsilon_1 \geq -\frac{1}{8}$. Can we further improve this lower bound? Adding all three of the inequalities in (4.34a) yields

$$(4.37) \qquad\qquad -3\epsilon \leq x_1 + x_2 + x_3 \leq 1 + 3\epsilon,$$

which in conjunction with (4.34c) yields

$$(4.38) \qquad\qquad \tfrac{7}{12} - \epsilon \leq 1 + 3\epsilon, \quad -3\epsilon \leq 1 + \epsilon$$

or $\epsilon \geq -\frac{5}{48}$ and $\epsilon \geq -\frac{1}{4}$, the second of which is again superseded by the first. Thus $\epsilon_1 \geq -\frac{5}{48}$, an improved lower bound. Continuing in this manner, we obtain four more lower bounds on ϵ, but none of them supersedes $\epsilon \geq -\frac{5}{48}$. No further restrictions on ϵ are implied by (4.34); ϵ can be as small as it pleases, as long as it satisfies $\epsilon \geq -\frac{5}{48}$. We conclude that

$$(4.39) \qquad\qquad \epsilon_1 = -\tfrac{5}{48}.$$

With $\epsilon = -\frac{5}{48}$, the first inequality of (4.34c) and the second of (4.37) together imply that

$$(4.40) \qquad\qquad \tfrac{11}{16} \leq x_1 + x_2 + x_3 \leq \tfrac{11}{16}$$

and hence, of course, that $x_1 + x_2 + x_3 = \frac{11}{16}$. We can now eliminate x_3 from (4.34); after simplification, the least rational core consists of imputations satisfying

$$(4.41) \qquad \tfrac{5}{24} \leq x_1 \leq \tfrac{5}{16}, \quad \tfrac{1}{8} \leq x_2 \leq \tfrac{11}{48}, \quad \tfrac{13}{24} \leq x_1 + x_2 \leq \tfrac{7}{12}.$$

These inequalities are satisfied only if $x_1 = \frac{5}{16}$ and $x_2 = \frac{11}{48}$. Hence, from (4.40) and (4.33), the least rational core is

$$(4.42a) \qquad X^1 = C^+\left(-\tfrac{5}{48}\right) = \left\{\left(\tfrac{5}{16}, \tfrac{11}{48}, \tfrac{7}{48}, \tfrac{5}{16}\right)\right\}.$$

In other words, the fairest distribution of the benefits of cooperation (according to our agreed criterion) is given by the imputation

(4.42b) $$x^* = \left(\tfrac{5}{16}, \tfrac{11}{48}, \tfrac{7}{48}, \tfrac{5}{16} \right).$$

By analogy with (4.8), it follows from (4.30) that Player i should contribute

(4.43) $$c(\{i\}) - (6 + 3d)x_i = c(\{i\}) - 12x_i$$

to the cost per trip (which, you will recall, is measured in units of k dollars). Thus Jed and Ned should each cover 2.25 units of the total cost of 9 units; whereas Ted should contribute 3.25 units, and Zed only 1.25. Again, Ted is penalized for being the outlier.

Now, it could be argued that this "fair" solution is more than fair to the newcomer Zed, and hence less than fair to the other three. If Jed, Ned and Ted were to bargain as a unit then we would have, in effect, a 2-player CFG with costs

(4.44) $$c(\overline{1}) = 9, \quad c(\overline{2}) = 5, \quad c(\overline{1} \cup \overline{2}) = 9$$

and hence benefits

(4.45) $$\overline{\nu}(\overline{1}) = 0 = \overline{\nu}(\overline{2}), \quad \overline{\nu}(\overline{1} \cup \overline{2}) = 5$$

where $\overline{1} = \{1, 2, 3\}$ and $\overline{2} = \{4\}$. From Exercise 4.7 the fair solution of this modified CFG is to share the benefits of cooperation equally between $\overline{1}$ and $\overline{2}$. Thus Zed would have to contribute 2.5 units—twice as much as previously—whereas Jed, Ned and Ted would have reduced their total costs from 7.75 units to 6.5. How would they distribute the extra benefit of 1.25 units among themselves? No two of them could secure part of this benefit by acting without the other; the benefit would accrue only if all three of them agreed to gang up on Zed. Thus Jed, Ned and Ted would be led to a CFG with normalized characteristic function defined by

(4.46) $$\nu(S) = 0 \text{ if } S \neq \{1, 2, 3\}, \quad \nu(\{1, 2, 3\}) = 1.$$

Again (Exercise 4.7), the fair solution of this game is to share the extra benefit equally. Thus Jed, Ned and Ted would each reduce their cost per trip by $\tfrac{5}{12}$ of a unit, so that Jed and Ned would cover $\tfrac{11}{6}$ units of the total cost of 9 units, whereas Ted would contribute $\tfrac{17}{6}$

units (and Zed $\frac{5}{2}$). Whether this or the previous solution ultimately emerged would depend on details of how the car pool formed.

We remark, finally, that the method we used to determine $C^+(\epsilon_1)$ for a 4-player game applies even more readily to 3-player games. In particular, it can be used to deduce (4.28) from (4.27); see Exercise 4.2. For further examples, see Exercises 4.3-4.6 and 4.10.

4.4. Log hauling: a coreless game

Many years ago when they were young and sprightly, Zan and Zed felled enough trees to make 150 8-foot logs, which they neatly stacked in their yard. They had intended to turn the logs into a cabin; but years went by, and never did they find the time. Now Zan and Zed are too old to lift the logs, and no longer need the cabin. They have therefore decided to sell the logs for a dollar each to the first person or persons who will haul them away, and they have placed an advertisement in the local newspaper.

In response to this advertisement, three people and their pickup trucks arrive simultaneously at Zan and Zed's at the appointed time. These eager beavers, who would all like to haul as many of the bargain logs as possible, are our old friends Jed (Player 1), Ned (Player 2) and Ted (Player 3). Jed's truck can haul 45 logs, Ned's can haul 60 and Ted's can haul 75. If there were 180 logs, then all three could leave with a full load. Unfortunately, however, there are only 150 logs; and because our friends all arrived simultaneously, no one can claim to have arrived first. Moreover, because the logs are too heavy for one person to lift, the players cannot resort to a scramble; they must cooperate. Then how do they divvy up the logs? We must solve another CFG.

Let $\bar{\nu}(S)$, the benefit of cooperation that accrues to coalition S, be the number of logs that the players in S can haul without the help of players outside S. Then, clearly:

$$\bar{\nu}(\{i\}) = 0, \ i = 1, 2, 3$$

(4.47) $\qquad \bar{\nu}(\{1,2\}) = 105, \ \bar{\nu}(\{1,3\}) = 120, \ \bar{\nu}(\{2,3\}) = 135$

$$\bar{\nu}(\{1,2,3\}) = 150.$$

The characteristic function is therefore defined by

(4.48) $\qquad \nu(\{1,2\}) = \frac{7}{10}, \ \nu(\{1,3\}) = \frac{4}{5}, \ \nu(\{2,3\}) = \frac{9}{10}$

(and, of course, $\nu(\{i\}) = 0$, $i = 1, 2, 3$, $\nu(N) = 1$). It follows readily (Exercise 4.8) that the imputation $x = (x_1, x_2, x_3)$ belongs to $C^+(\epsilon)$ if and only if

$$(4.49) \qquad \begin{gathered} -\epsilon \le x_1 \le \tfrac{1}{10} + \epsilon, \quad -\epsilon \le x_2 \le \tfrac{1}{5} + \epsilon \\ \tfrac{7}{10} - \epsilon \le x_1 + x_2 \le 1 + \epsilon. \end{gathered}$$

The sum of the second and fourth inequalities is consistent with the fifth if and only if $\tfrac{7}{10} - \epsilon \le \tfrac{3}{10} + 2\epsilon$, or $\epsilon \ge \tfrac{2}{15}$. Thus $\epsilon_1 = \tfrac{2}{15}$. We conclude immediately that the game has no core.

Whenever a game is coreless or, which is the same thing, whenever $\epsilon_1 > 0$ (if $\epsilon_1 = 0$, then the core contains a single imputation), the maximum dissatisfaction must be positive; no matter which imputation is selected, at least one coalition (and in this particular case, every coalition) will have a lower allocation than if the grand coalition did not form. Thus Jed, Ned and Ted all wish that one of the other two had not arrived, or had arrived late; but he did arrive, and the three of them must reach an agreement. Characteristic function games assume that the grand coalition has formed, but have nothing to say about how it formed. If the game has a core, then it is probable that the grand coalition formed voluntarily; the more players the merrier (up to a point), because everyone benefits from the cooperation. If, on the other hand, the game has no core, then the players were probably coerced into forming the grand coalition; the fewer players the better (down to a point), but no one can be barred from playing. To put it another way, games with cores are about sharing a surplus of benefits, whereas coreless games are about rationing a shortage of benefits; but in either case the existence of the grand coalition is assumed, and the least rational core, if it contains a single imputation, offers a solution that's fair in the sense of minimizing maximum dissatisfaction.

With $\epsilon_1 = \tfrac{2}{15}$, (4.49) can be satisfied only if $x_1 = \tfrac{7}{30}$ and $x_2 = \tfrac{1}{3}$ (Exercise 4.8), and so the least rational core is

$$(4.50) \qquad X^1 \;=\; C^+\left(\tfrac{2}{15}\right) \;=\; \left\{\left(\tfrac{7}{30}, \tfrac{1}{3}, \tfrac{13}{30}\right)\right\}.$$

Our fair solution gives 35 logs to Jed, 50 to Ned and 65 to Ted. Hence it distributes the 30-log shortage equally among the players.[6]

---

[6]There remains the problem of who should put the logs on the trucks (assume that they all have family to help at home). Because it takes two to lift a log, and

Another way to reach the same conclusion is to say that 15 logs are attributable to Jed regardless of any fairness considerations, because 15 logs would be left over even if Ned and Ted had both filled their trucks. Similarly, 30 logs are attributable to Jed, and 45 logs are attributable to Ted. So only 60 of the 150 logs are not attributable without invoking considerations of fairness. If we divide these 60 logs equally among the players, then we recover the least rational core.

The preceding argument readily generalizes. Consider the 3-player game defined by

$$(4.51) \qquad \nu(\{2,3\}) = a, \quad \nu(\{1,3\}) = b, \quad \nu(\{1,2\}) = c$$

with

$$(4.52) \qquad\qquad m = \max(a, b, c) \leq 1.$$

Because Players 2 and 3 can obtain only a of total savings by themselves, let $1-a$ be attributed to Player 1; and likewise, let $1-b$ and $1-c$ be attributed to Players 2 and 3, respectively. Then "nonattributable savings" are $1 - (1 - a) - (1 - b) - (1 - c) = a + b + c - 2$, which is positive whenever no core exists (Exercise 4.9). If players receive the savings thus attributed to them plus equal shares of nonattributable savings, then the game's solution is the imputation x^T defined by

$$(4.53) \qquad x^T = \tfrac{1}{3}(1 - 2a + b + c, 1 + a - 2b + c, 1 + a + b - 2c).$$

This solution is of considerable historical interest, because it was used by engineers and economists of the Tennessee Valley Authority in the 1930s, long before game theory emerged as a formal field of study, to allocate joint costs of dam systems among participating users fairly [218]. The solution foreshadowed later developments in game theory, because it agrees with the "nucleolus"—which we are about to introduce—whenever

$$(4.54) \qquad\qquad \min(a + b + c, 2a + 2b + 2c - 3m) \geq 1$$

(Exercise 4.27). Note that (4.54) is satisfied by (4.48).

Whether the above is the conclusion that Jed, Ned and Ted would actually reach, however, is an open question. Would it be fairer to

---

because the player's allocations are in the ratio 7:10:13, it seems fairest for Jed to lift 70 logs, Ned to lift 100 and Ted 130. For example, Ted could lift all 35 of Jed's logs, 12 with Jed and 23 with Ned; Ned could lift all 50 of his own logs, 20 with Jed and 30 with Ted; and Ted could lift all 65 of his own logs, 27 with Ned and 38 with Jed.

give each player 50 logs in the first instance, but then transfer 5 of Jed's logs to Ned or Ted because Jed can haul only 45? If so, how many of the extra 5 logs should Ted get? ... Well, what do you think?

4.5. Antique dealing. The nucleolus

The boundary of the core does not always collapse to a single point when we move its walls inward at a uniform rate; that is, the least rational core may contain more than one imputation. Then what is the center of the core in such cases? An answer is provided by the nucleolus, which we introduce in this section.[7]

Let us begin by returning to the 3-person car pool of §4.1. If $d = 1$ then, from (4.10), the characteristic function is defined by

$$(4.55) \qquad \nu(\{1,2\}) = \tfrac{4}{5}, \ \nu(\{1,3\}) = \tfrac{2}{5}, \ \nu(\{2,3\}) = \tfrac{1}{5}$$

(and, of course, $\nu(N) = 1$, $\nu(\{i\}) = 0$, $i = 1, 2, 3$). It is questionable whether a car pool would really form for such a low value of d; but the value is convenient for our present purpose, and higher values of d are considered in the exercises.

From Exercise 4.2, the rational ϵ-core is

$$(4.56) \quad C^+(\epsilon) = \left\{ x \in X \left| \ -\epsilon \leq x_1 \leq \tfrac{4}{5} + \epsilon, \right. \right.$$
$$\left. -\epsilon \leq x_2 \leq \tfrac{3}{5} + \epsilon, \tfrac{4}{5} - \epsilon \leq x_1 + x_2 \leq 1 + \epsilon \right\},$$

where X is defined by (4.7); for example, $C^+(0)$ and $C^+(-1/20)$ are the two quadrilaterals in Figure 4.5. By the method of §4.3, and because (4.56) implies $\tfrac{4}{5} - \epsilon \leq 1 + \epsilon$, it is readily found that $\epsilon_1 = -\tfrac{1}{10}$, and so the least rational core is

$$(4.57) \quad X^1 = \left\{ x \in X \mid \tfrac{2}{5} \leq x_1 \leq \tfrac{7}{10}, \tfrac{1}{5} \leq x_2 \leq \tfrac{1}{2}, x_1 + x_2 = \tfrac{9}{10} \right\}.$$

It no longer contains a single imputation; rather, it contains all imputations corresponding to points of the line $x_1 + x_2 = \tfrac{9}{10}$ between $(x_1, x_2) = \left(\tfrac{2}{5}, \tfrac{1}{2} \right)$ and $(x_1, x_2) = \left(\tfrac{7}{10}, \tfrac{1}{5} \right)$ in Figure 4.5. Which of all these imputations is the fair solution of the game?

To answer this question, let us continue to assume that fairness means minimizing maximum dissatisfaction. Now, by construction,

---

[7]The concept of nucleolus is due to Schmeidler [**201**]. Strictly, the concept we introduce is Maschler, Peleg and Shapley's lexicographic center—which, however, coincides with the nucleolus. See [**128**, pp. 331-336].

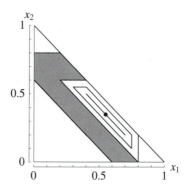

Figure 4.5. Reasonable set (hexagon) and rational ϵ-cores
for $\epsilon = 0$ (outer quadrilateral), $\epsilon = -1/20$ and $\epsilon = -1/10$
(line segment) when $d = 1$ in the 3-person car pool. The
quadrilaterals are $\phi_0 = \epsilon$ contours, where ϕ_0 is defined by
(4.23). The dot denotes the nucleolus. The shaded region is
the difference between the reasonable set and the core.

for $x \in X^1$ the maximum dissatisfaction of any coalition is $\epsilon_1 = -\frac{1}{10}$.
That is, every coalition obtains at least $\frac{1}{10}$ more of the benefits of
cooperation than it could obtain if the grand coalition had not formed,
which we can verify by using (4.16) and (4.55) to compute the excesses
at $x \in X^1$:

$$e(\{1\}, x) = -x_1, \quad e(\{2\}, x) = -x_2$$
$$(4.58) \qquad e(\{1,3\}, x) = x_2 - \tfrac{3}{5}, \quad e(\{2,3\}, x) = x_1 - \tfrac{4}{5}$$
$$e(\{1,2\}, x) = -\tfrac{1}{10} = e(\{3\}, x).$$

Because $\frac{2}{5} \le x_1 \le \frac{7}{10}$ and $\frac{1}{5} \le x_2 \le \frac{1}{2}$, none of these excesses
exceeds $-\frac{1}{10}$, which is the maximum dissatisfaction. But it is also the
minimum possible maximum dissatisfaction.

Because $e(S, x)$ is independent of x for the two most dissatisfied
coalitions, namely, $\{1, 2\}$ and $\{3\}$, we can vary x within X^1 without affecting their excesses. For the other four coalitions, however,
$e(S, x)$ remains a function of x; and so, by varying x, we can seek
to minimize their maximum excess, i.e., minimize the maximum of
$e(\{1,3\}, x), e(\{2,3\}, x), e(\{1\}, x)$ and $e(\{2\}, x)$. In other words, coalitions $\{1, 2\}$ and $\{3\}$, being indifferent among all imputations in X^1,

are already as satisfied as it is possible to make them; so we exclude them from further reckoning, and concentrate instead on reducing the dissatisfaction of the remaining coalitions below $-\frac{1}{10}$. What this means in practice is that Ted has been allocated a tenth of the benefits, and Jed and Ned have together been allocated nine tenths, but it is not yet clear how to divvy it up between them.

With a view to generalization, let Σ^1 denote the set of coalitions whose excess can be reduced below ϵ_1 by an imputation in X^1. That is, define

$$(4.59) \qquad \Sigma^1 = \left\{ S \in \Sigma^0 \mid e(S,x) < \epsilon_1 \text{ for some } x \in X^1 \right\},$$

(so that $\Sigma^1 = \{\{1\}, \{2\}, \{1,3\}, \{2,3\}\}$ for our 3-person car pool with $d = 1$). Still with a view to generalization, for $x \in X^1$ let

$$(4.60) \qquad \phi_1(x) = \max_{S \in \Sigma^1} e(S,x)$$

be the maximum excess at imputation x of the remaining coalitions; and let ϵ_2 be its minimum, i.e.,

$$(4.61) \qquad \epsilon_2 = \min_{x \in X^1} \phi_1(x).$$

Furthermore, let X^2 consist of all imputations at which ϕ_1 achieves its minimum, i.e., define

$$(4.62) \qquad X^2 = \left\{ x \in X^1 \mid \phi_1(x) = \epsilon_2 \right\}.$$

Then, in the case of the car pool, because $x_1 + x_2 = \frac{9}{10}$ for $x \in X^1$, ϕ_1 can be regarded as function of x_1 alone. In fact, from (4.58)-(4.60), we have

$$(4.63) \quad
\begin{aligned}
\phi_1(x) &= \max \left(x_2 - \tfrac{3}{5}, x_1 - \tfrac{4}{5}, -x_1, -x_2 \right) \\
&= \max \left(\tfrac{3}{10} - x_1, x_1 - \tfrac{4}{5}, -x_1, x_1 - \tfrac{9}{10} \right)
\end{aligned}$$

for $\frac{2}{5} \le x_1 \le \frac{7}{10}$.

The graph of ϕ_1 is the solid vee in Figure 4.6. Note that ϕ_1 assumes its minimum of $-\frac{1}{4}$ where $x_1 = \frac{11}{20}$. Thus $\epsilon_2 = -\frac{1}{4}$ and

$$(4.64a) \qquad X^2 = \left\{ \left(\tfrac{11}{20}, \tfrac{7}{20}, \tfrac{1}{10} \right) \right\}.$$

From (4.58), the coalitions for which $-\frac{1}{4}$ is the actual excess at x^* are $\{1,3\}$ and $\{2,3\}$; for $\{1\}$ and $\{2\}$, of course, the excesses at x^* are

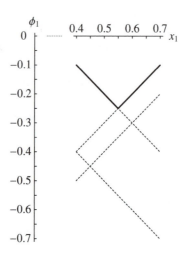

Figure 4.6. Graph of ϕ_1 defined by (4.63).

$-\frac{11}{20}$ and $-\frac{7}{20}$, respectively. Because X^2 contains a single imputation

$$(4.64b) \qquad\qquad x^* = \left(\tfrac{11}{20}, \tfrac{7}{20}, \tfrac{1}{10}\right),$$

no further variation of x is possible. We have reduced unfairness as much as possible. Thus x^* is the solution of the CFG. From (4.8), Jed, Ned and Ted's fair (fare?) shares of the cost per trip are $2.25k$, $2.25k$ and $3.5k$, respectively.

The ideas we have just developed to reach this solution are readily generalized to games with more than three players, for which the least rational core need be neither a point nor a line segment.[8] Consider, for example, the following four-player variation of Exercise 4.10. Jed, Ned, Ted and Zed are antique dealers, who conduct their businesses in separate but adjoining rooms of a common premises. Their advertised office hours are shown in Figure 4.7; for example, Ted's advertised hours are from 10:00 a.m. until 4:00 p.m. Because the dealers have other jobs and the store is never so busy that one guy

---

[8] Even if the least rational core is a line segment, X^2 need not be its mid-point. Although the walls of the least rational core are moved inward at the same speed, the components of the velocities of inward movement along the line segment corresponding to the least core need not be equal. For an example, see Exercise 4.12.

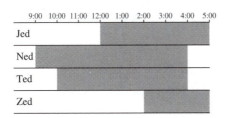

Figure 4.7. Antique dealers' business hours.

couldn't take care of everyone's customers, it is in the dealers' interests to pool their time in minding the store: there is no need for two people between 10:00 a.m. and noon or between 4:00 p.m. and 5:00 p.m, for three people between noon and 2:00 p.m., or for four people between 2:00 p.m. and 4:00 p.m. Cooperation will enable dealers to leave earlier than their advertised closing hours, or arrive later than their advertised opening hours, or both. But how much earlier or later? What are fair allocations of store-minding duty?

Let Jed, Ned, Ted and Zed be Players 1, 2, 3 and 4, respectively, and let the benefit of cooperation to coalition S be the number of dealer-hours that S saves by pooling time. Then $\bar{\nu}(i) = 0$, $i = 1, 2, 3, 4$, because no time is saved if no one cooperates; $\bar{\nu}(\{1, 2\}) = 4$, because either Jed or Ned can mind the other's business between noon and 4:00; $\bar{\nu}(N) = 13$, where $N = \{1, 2, 3, 4\}$, because only 8 of the 21 dealer hours currently advertised are actually necessary; etc. Thus

$$
\begin{aligned}
\nu(\{i\}) &= 0, \ i = 1, 2, 3, 4 \\
\nu(\{1, 2\}) &= \tfrac{4}{13}, \ \nu(\{1, 3\}) = \tfrac{4}{13}, \ \nu(\{1, 4\}) = \tfrac{3}{13} \\
\nu(\{2, 3\}) &= \tfrac{6}{13}, \ \nu(\{2, 4\}) = \tfrac{2}{13}, \ \nu(\{3, 4\}) = \tfrac{2}{13} \\
\nu(\{1, 2, 3\}) &= \tfrac{10}{13}, \ \nu(\{1, 2, 4\}) = \tfrac{7}{13}, \ \nu(\{1, 3, 4\}) = \tfrac{7}{13} \\
\nu(\{2, 3, 4\}) &= \tfrac{8}{13}, \ \nu(\{1, 2, 3, 4\}) = 1
\end{aligned}
$$

(4.65)

and it follows readily that the rational ϵ-core contains all vectors $x = (x_1, x_2, x_3, x_4) \in X$ such that

(4.66a) $\quad -\epsilon \le x_1 \le \tfrac{5}{13} + \epsilon, \ -\epsilon \le x_2 \le \tfrac{6}{13} + \epsilon, \ -\epsilon \le x_3 \le \tfrac{6}{13} + \epsilon$

(4.66b) $\quad \tfrac{4}{13} - \epsilon \le x_1 + x_2 \le \tfrac{11}{13} + \epsilon, \ \tfrac{4}{13} - \epsilon \le x_1 + x_3 \le \tfrac{11}{13} + \epsilon$

$$\tfrac{6}{13} - \epsilon \le x_2 + x_3 \le \tfrac{10}{13} + \epsilon$$

(4.66c) $\quad \tfrac{10}{13} - \epsilon \le x_1 + x_2 + x_3 \le 1 + \epsilon$

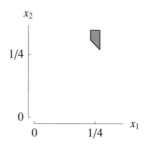

Figure 4.8. Least rational core of the antique-minding game. For greater detail, see Figure 4.9.

(Exercise 4.11). By the method of §4.3, we find that

$$(4.67) \qquad\qquad \epsilon_1 \; = \; -\tfrac{3}{26}$$

(note, in particular, that (4.66c) implies $2\epsilon \geq -3/13$; and so it follows from (4.66c) that

$$(4.68) \qquad\qquad x_1 + x_2 + x_3 \; = \; \tfrac{23}{26}$$

for $x \in X^1$. Using this equality to eliminate x_3, we find after simplification (Exercise 4.11) that

$$(4.69) \quad X^1 \; = \; \{x \in X \mid \tfrac{3}{13} \leq x_1 \leq \tfrac{7}{26}, x_1 + x_2 \geq \tfrac{7}{13}, x_2 \leq \tfrac{9}{26}\}.$$

Thus the least rational core is a quadrilateral, which is represented by the shaded region in Figure 4.8. We remind ourselves that this picture distorts the true quadrilateral, which, by virtue of (4.68), is imbedded in the hyperplane

$$(4.70) \qquad\qquad x_4 \; = \; \tfrac{3}{26}.$$

By construction, for $x \in X^1$ the maximum dissatisfaction of any coalition is $-\tfrac{3}{26}$; that is, every coalition obtains at least $-\tfrac{3}{26}$ more of the benefits of cooperation than it could obtain if the grand coalition had not formed. Two coalitions, namely, $\{1,2,3\}$ and $\{4\}$, obtain an excess of precisely $-\tfrac{3}{26}$, because $e(\{1,2,3\}) = \tfrac{10}{13} - x_1 - x_2 - x_3 = -\tfrac{3}{26}$ and $e(\{4\}) = -x_4 = -\tfrac{3}{26}$ for all $x \in X^1$, on using (4.16), (4.65), (4.68) and (4.70). These are the coalitions who cannot be allocated less dissatisfaction (or, if you prefer, greater satisfaction) than they receive from any imputation in the least rational core.[9] On the other hand, the excesses in X^1 of $\{1\}$, $\{2\}$, $\{3\}$, $\{1,2\}$, $\{1,3\}$, $\{1,4\}$, $\{2,3\}$, $\{2,4\}$, $\{3,4\}$, $\{1,2,4\}$, $\{1,3,4\}$, $\{2,3,4\}$ still vary with x, and are strictly less than $-\frac{3}{26}$ at least somewhere. So we exclude $\{1,2,3\}$ and $\{4\}$ from further reckoning and focus on eliminating unfairness among the twelve remaining coalitions, which constitute Σ^1 defined by (4.59).

With the help of (4.16), (4.65), (4.68) and (4.70), and using $e_{ij...}$ as a convenient and obvious shorthand for $e(\{i, j, \dots\}, x)$, the excesses of these coalitions at $x \in X^1$ are readily found to be

$$e_1 = -x_1 \qquad e_2 = -x_2 \qquad e_3 = x_1 + x_2 - \frac{23}{26}$$

$$e_{12} = \frac{4}{13} - x_1 - x_2 \qquad e_{13} = x_2 - \frac{15}{26} \qquad e_{14} = \frac{3}{26} - x_1$$

$$e_{23} = x_1 - \frac{11}{26} \qquad e_{24} = \frac{1}{26} - x_2 \qquad e_{34} = x_1 + x_2 - \frac{11}{13}$$

$$e_{124} = \frac{11}{26} - x_1 - x_2 \qquad e_{134} = x_2 - \frac{6}{13} \qquad e_{234} = x_1 - \frac{5}{13}.$$

It is straightforward to show that none of $e_1, e_2, e_3, e_{12}, e_{13}, e_{24}$ or e_{34} can exceed $-\frac{3}{13}$ on X^1; whereas none of the remaining five excesses can be less than $-\frac{5}{26}$. Moreover, $e_{234} > e_{23}$. Therefore, the maximum in definition (4.60) of ϕ_1 is effectively taken, not over the whole of Σ^1, but rather over the four coalitions $\{1,4\}, \{1,2,4\}, \{1,3,4\}$ and $\{2,3,4\}$ with excesses e_{14}, e_{124}, e_{134} and e_{234}, respectively; that is,

$$(4.71) \qquad \phi_1(x) = \max\left(\frac{3}{26} - x_1, \frac{11}{26} - x_1 - x_2, x_2 - \frac{6}{13}, x_1 - \frac{5}{13}\right)$$

for $x \in X^1$. Because ϕ_1 is now a function of two variables, its graph is three-dimensional, and we can no longer plot it in two dimensions as we did in Figure 4.6. Instead, therefore, in Figure 4.9 we sketch its contour map, from which we see that the graph of ϕ_1 is shaped like an asymmetric boat or inverted roof, the lowest points of which lie on the contour where $\phi_1 = -\frac{7}{52}$. It follows that

$$(4.72) \qquad \epsilon_2 = \min_{x \in X^1} \phi_1(x) = -\frac{7}{52}$$

and

$$(4.73) \qquad X^2 = \{x \in X \mid \phi_1(x) = \epsilon_2\}$$
$$= \left\{x \in X \mid x_1 = \tfrac{1}{4}, \tfrac{4}{13} \le x_2 \le \tfrac{7}{52}, x_4 = \tfrac{3}{26}\right\}.$$

---

[9]What this means in practice, of course, is that Zed has been allocated $\frac{3}{26}$ of the benefits of cooperation; whereas Jed, Ned and Ted have together been allocated $\frac{23}{26}$, but it is not yet clear how to divvy it up between them.

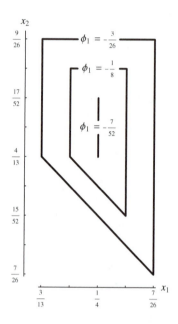

Figure 4.9. Contour map of function ϕ_1 defined over shaded region in Figure 4.8.

We can think of X^2—which has been obtained, in effect, by moving the walls of X^1 inward at a rate of one unit every 104 seconds—as the second-order least rational core.

By construction, for $x \in X^2$ the maximum dissatisfaction of any coalition is $-\frac{7}{52}$. Two coalitions, namely, $\{1,4\}$ and $\{2,3,4\}$, obtain an excess of precisely $-\frac{7}{52}$, because $e_{14} = 0 = e_{234}$ for all $x \in X^2$ from (4.73) and the expressions on p. 153. These coalitions cannot be allocated less dissatisfaction than they receive from any imputation in X^2. On the other hand, the excesses in X^2 of $\{1\}$, $\{2\}$, $\{3\}$, $\{1,2\}$, $\{1,3\}$, $\{2,3\}$, $\{2,4\}$, $\{3,4\}$, $\{1,2,4\}$ and $\{1,3,4\}$ still vary with x, and are strictly less than $-\frac{7}{52}$ at least somewhere. So we exclude $\{1,4\}$ and $\{2,3,4\}$ from further reckoning and concentrate our attention on eliminating unfairness among the the ten remaining coalitions, which constitute

(4.74) $$\Sigma^2 = \left\{ S \in \Sigma^1 \mid e(S, x) < \epsilon_2 \text{ for some } x \in X^2 \right\}.$$

By now it should be clear how we proceed. By analogy with (4.60)-(4.62), for $x \in X^2$ let

(4.75) $$\phi_2(x) = \max_{S \in \Sigma^2} e(S, x)$$

be the maximum excess at imputation x of the remaining coalitions; and let

(4.76) $$\epsilon_3 = \min_{x \in X^2} \phi_2(x)$$

be its minimum. Furthermore, let X^3 consist of all imputations at which ϕ_2 achieves its minimum, i.e., define

(4.77) $$X^3 = \left\{ x \in X^2 \mid \phi_2(x) = \epsilon_3 \right\}.$$

We can think of X^3 as the third-order least rational core. Because $x_1 = \frac{1}{4}$ for $x \in X^2$, ϕ_2 can be regarded as function of x_2 alone. Moreover, none of $e_1, e_2, e_3, e_{12}, e_{13}, e_{23}, e_{24}$ or e_{34} can exceed $-\frac{9}{52}$ on X^2, and neither of the other excesses can be less than $-\frac{2}{13}$. So the maximum in definition (4.75) of ϕ_2 is effectively taken, not over the whole of Σ^2, but only over coalitions $\{1, 2, 4\}$ and $\{1, 3, 4\}$. That is,

(4.78) $$\phi_2(x) = \max\left(\tfrac{9}{52} - x_2, x_2 - \tfrac{6}{13} \right)$$

for $\frac{4}{13} \leq x_2 \leq \frac{17}{52}$. The graph of ϕ_2 is similar to Figure 4.6, and its minimum occurs where $\frac{9}{52} - x_2 = x_2 - \frac{6}{13}$, or $x_2 = \frac{33}{104}$. Thus $\epsilon_3 = -\frac{15}{104}$, and

(4.79) $$X^3 = \left\{ \left(\tfrac{1}{4}, \tfrac{33}{104}, \tfrac{33}{104}, \tfrac{3}{26} \right) \right\}$$

contains the single imputation $x^* = \left(\frac{1}{4}, \frac{33}{104}, \frac{33}{104}, \frac{3}{26} \right)$. No further variation of x is possible: we have reduced unfairness as much as is possible. Thus x^* is the solution of the CFG; and Jed, Ned, Ted and Zed's fare shares of the time saved are $\frac{13}{4}$ hours for Jed, $\frac{33}{8}$ hours each for Ned and Ted, and $\frac{3}{2}$ hours for Zed. Ned must still arrive at 9:00, but now leaves at 11:52:30, when Ted arrives; Ted now leaves at 1:45, when Jed arrives; and Jed now leaves at 3:30, when Zed arrives for the last turn of duty. Ned and Ted benefit most, because their advertised hours have the greatest overlap with those of the others.

Whether this is the conclusion that Jed, Ned, Ted and Zed would actually reach, however, is an open question. Would it be fairer to give each dealer $\frac{13}{4}$ hours to begin with, so that Zed can just stay at home, and transfer 7 minutes and 30 seconds of the 15 minutes that Zed doesn't need to each of Ned and Ted? Well, what do you think?

Our procedure for successively eliminating unfair imputations, until only a single imputation remains, can be generalized to apply to arbitrary n-player characteristic function games. We construct by recursion a nested sequence

$$(4.80) \qquad X = X^0 \supset X^1 \supset X^2 \ldots X^\kappa = X^*$$

of sets of imputations, a nested sequence

$$(4.81) \qquad \Sigma^0 \supset \Sigma^1 \supset \Sigma^2 \ldots \Sigma^\kappa$$

of sets of coalitions, a decreasing sequence

$$(4.82) \qquad \epsilon_1 > \epsilon_2 \ldots \epsilon_\kappa$$

of real numbers and a sequence

$$(4.83) \qquad \phi_0, \phi_1, \ldots, \phi_{\kappa-1}$$

of functions such that the domain of ϕ_k is X^k. The recursion is defined for $k \geq 1$ by

$$(4.84a) \qquad \phi_k(x) = \max_{S \in \Sigma^k} e(S, x)$$

$$(4.84b) \qquad \epsilon_k = \min_{x \in X^{k-1}} \phi_{k-1}(x)$$

$$(4.84c) \qquad X^k = \left\{ x \in X^{k-1} \mid \phi_{k-1}(x) = \epsilon_k \right\}$$

$$(4.84d) \qquad \Sigma^k = \left\{ S \in \Sigma^{k-1} \mid e(S, x) < \epsilon_k \text{ for some } x \in X^k \right\},$$

where $X^0 = X$ and Σ^0 consists of all coalitions except $\varnothing$ and N. If we perform this recursion for increasing values of the positive integers, then eventually we reach a value of k for which the set X^k contains a single imputation; for a proof of this result see, e.g., [**233**, p. 146]. If we denote the final value of k by κ, then $1 \leq \kappa \leq n-1$ (because the dimension of X^k is reduced by at least one at every step of the recursion); and if $X^\kappa = \{x^*\}$ then, according to our criterion of fairness, the imputation x^* is the fair solution of the game. The set $X^\kappa = \{x^*\}$ is called the *nucleolus*.

If $\kappa = 1$, of course, then the nucleolus coincides with least rational core, as in (4.28) and (4.50), where $n = 3$; or in (4.42), where $n = 4$. More commonly, however, we have $\kappa > 1$, as in (4.64), where $\kappa = 2$ and $n = 3$; or in (4.79), where $\kappa = 3$ and $n = 4$. Technically, the task of calculating the nucleolus is an exercise in repeated linear programming,[10] which—at least for games of moderate size—can safely be delegated to a computer. For a discussion of this and other matters related to the nucleolus see, e.g., Owen [**173**] or Wang [**233**].

A final remark is in order. Technically, x^* is not the nucleolus; rather, it is the (only) imputation that the nucleolus contains. To observe the distinction is often a nuisance, however, and we shall not hesitate to breach linguistic etiquette by referring to both x^* and $\{x^*\}$ as the nucleolus whenever it is convenient to do so (as, e.g., in §4.7).

4.6. Team long-jumping. An improper game

To encourage team spirit among schoolboys who compete as individuals, a high school has instituted a long-jumping competition for teams of two or three. The rules of the competition stipulate that each team can make up to 12 jumps, of which six count officially towards the final outcome—the best three jumps of each individual in a 2-man team, or the best two jumps of each individual in a 3-man team. Furthermore, a local business has agreed to pay a dollar per foot for each official foot jumped in excess of 15 feet. So unless—which is unlikely—nobody jumps more than 15 feet, winning the competition is the same thing as maximizing dollar uptake. But the existence of prize money brings the added problem of distributing the money fairly among the team. We will use the resultant CFG to clarify the difference between what we have called the rational ϵ-core and what is known in the literature as the ϵ-core.

Let's suppose that young Jed (Player 1), young Ned (Player 2) and young Ted (Player 3) competed as a team in this year's competition. They agreed to take four jumps each, and their best three jumps are recorded (in feet) in Table 4.1; obviously, Jed had a bad day. The dollar equivalents of these jumps are recorded in Table 4.2. We see at once that Jed, Ned and Ted have earned themselves the

---

[10]For a thoroughly modern approach to this topic, see Castillo et al. [**35**].

Table 4.1. Team's best jumps (in feet)

	J	N	T
First jump	15	20	19
Second jump	15	18	19
Third jump	15	18	18

Table 4.2. Monetary values (dollars)

	J	N	T
First jump	0	5	4
Second jump	0	3	4
Third jump	0	3	3

grand sum of $0 + 0 + 5 + 3 + 4 + 4 = 16$ dollars. How should they divvy it up?

Let the benefit of cooperation, $\bar{\nu}(S)$, to coalition S be the prize money that the players in S would have earned if they had competed as a team. Then, of course, $\bar{\nu}(\{i\}) = 0$, $i = 1, 2, 3$, because one-man teams are not allowed; and, from Table 4.2, we have $\bar{\nu}(\{1,2\}) = 11 = \bar{\nu}(\{1,3\})$, $\bar{\nu}(\{2,3\}) = 22$ and $\bar{\nu}(N) = 16$. Thus, from (4.11), the characteristic function is defined by

$$(4.85) \qquad \nu(\{1,2\}) = \tfrac{11}{16} = \nu(\{1,3\}), \quad \nu(\{1,3\}) = \tfrac{11}{8}$$

(and, of course, $\nu(\{i\}) = 0$, $i = 1, 2, 3$, $\nu(N) = 1$). We see at once that there is a coalition of less than all the players who could obtain more than the grand coalition. Thus, if it were possible for Ned and Ted to dissolve the grand coalition and become a 2-man team, then it would certainly be in their interests to do so. Unfortunately, however, they have already declared themselves a 3-man team. The rules of competition enforce the grand coalition.

We now digress from our long-jump competition to introduce some more general terminology. If

$$(4.86) \qquad \nu(S \cup T) \geq \nu(S) + \nu(T)$$

for any disjoint coalitions S and T ($S \cap T = \varnothing$), then the characteristic function ν is said to be *superadditive*, and the associated game is said to be *proper*. If, on the other hand, there exist S and T such that (4.86) is violated, then the game is said to be *improper*. Thus all of the games we studied in §§4.1-4.5 were proper games, whereas the 3-player game defined by (4.85) is an improper game.

Now, whenever a game is improper, one can make a case for relaxing the condition that allocations should be nonnegative. Could not Ned and Ted demand that Jed reimburse them for the potential earnings that they have lost through joining forces with Jed? Accordingly, let the vector $x = (x_1, x_2, \ldots, x_n)$ be called a *pre-imputation* if it satisfies (4.12b), but not necessarily (4.12a); that is, if

$$(4.87) \qquad x_1 + x_2 + \ldots + x_n = 1,$$

but there may exist some $i \in N$ for which $x_i < 0$. Thus preimputations are vectors that are group rational, but not necessarily individually rational for every player. We will denote the set of all pre-imputations by $\overline{X}$. Clearly $\overline{X} \supset X$: $\overline{X}$ is an $(n-1)$-dimensional hyperplane, which contains the simplex X.

With the concept of pre-imputation, we can define the core as

$$(4.88) \qquad C(0) = \left\{ x \in \overline{X} \mid e(S, x) \leq 0 \text{ for } all \text{ coalitions } S \right\}.$$

Because $e(\{i\}, x) \leq 0$ implies $x_i \geq 0$, this definition is equivalent to (4.19). We now define the ϵ-core by

$$(4.89) \qquad C(\epsilon) = \left\{ x \in \overline{X} \mid e(S, x) \leq \epsilon \text{ for } all \text{ } S \in \Sigma^0 \right\}.$$

Thus the rational ϵ-core is

$$(4.90) \qquad C^+(\epsilon) = C(\epsilon) \cap X;$$

and by analogy with (4.23)-(4.24), if we first define a function $\overline{\phi}_0$ from $\overline{X}$ to the real numbers by

$$(4.91) \qquad \overline{\phi}_0(x) = \max_{S \in \Sigma^0} e(S, x),$$

then we can define the ϵ-core more succinctly by

$$(4.92) \qquad C(\epsilon) = \left\{ x \in \overline{X} \mid \overline{\phi}_0(x) \leq \epsilon \right\}.$$

Obviously, $C^+(\epsilon) \subset C(\epsilon)$. The contours of $\overline{\phi}_0$ coincide with those of ϕ_0 on X; however, whereas contours of ϕ_0 that meet the boundary of X must either dead-end or follow the boundary, contours of $\overline{\phi}_0$ may continue across. For example, in Figure 4.4 the contours $\overline{\phi}_0 = 0$ and $\overline{\phi}_0 = -\frac{1}{21}$ would be triangles with a vertex in the region above the line $x_1 + x_2 = 0$ (which lies inside $\overline{X}$ but outside X because $x_3 < 0$).

By analogy with results for the rational ϵ-core, there is a least value of ϵ, namely,

$$(4.93) \qquad \overline{\epsilon}_1 = \min_{x \in \overline{X}} \overline{\phi}_0(x),$$

for which the ϵ-core exists. We refer to $C(\bar{\epsilon}_1)$ as the *least core*. Because the minimum over a larger set cannot exceed the minimum over a smaller set, a comparison of (4.93) with (4.25) reveals that

$$(4.94) \qquad\qquad\qquad \bar{\epsilon}_1 \;\leq\; \epsilon_1.$$

Furthermore, if the game has a core then $\bar{\epsilon}_1 = \epsilon_1$ (and $C(\bar{\epsilon}_1) = X^1$, i.e., the least core coincides with the least rational core).

If the game is both coreless and improper, however, then the possibility[11] arises that $\bar{\epsilon}_1 < \epsilon_1$. We can illustrate this circumstance by returning now to our long-jumping game, which is both improper and coreless, because an improper 3-player CFG is always coreless (Exercise 4.13). From (4.85) and (4.92), we have

$$(4.95) \quad C(\epsilon) \;=\; \left\{ x \in \overline{X} \mid -\epsilon \leq x_1 \leq -\tfrac{3}{8} + \epsilon, \quad -\epsilon \leq x_2 \leq \tfrac{5}{16} + \epsilon \right.$$
$$\left. \tfrac{11}{16} - \epsilon \leq x_1 + x_2 \leq 1 + \epsilon \right\}.$$

By the method of §4.3, we readily find that the least value of ϵ for which $C(\epsilon) \neq \varnothing$ is $\bar{\epsilon}_1 = \tfrac{1}{4}$, and that $C(1/4)$ contains only the pre-imputation $\left(-\tfrac{1}{8}, \tfrac{9}{16}, \tfrac{9}{16}\right)$. Thus, if the least core were a fair solution, then not only should Ned and Ted take all the prize money and divide it between themselves equally, but also Jed should pay them a dollar apiece to atone for jumping so badly.

On the other hand, the least value of ϵ for which $C^{+}(\epsilon) \neq \varnothing$ is $\epsilon_1 = \tfrac{3}{8}$; whence (Exercise 4.14)

$$(4.96) \qquad X^1 \;=\; \left\{ x \in X \mid x_1 = 0, \; \tfrac{5}{16} \leq x_2 \leq \tfrac{11}{16} \right\},$$

and the nucleolus is

$$(4.97) \qquad\qquad X^2 \;=\; \{x^*\} \;=\; \left\{ \left(0, \tfrac{1}{2}, \tfrac{1}{2}\right) \right\}.$$

The fair solution is for Ned and Ted to take all the prize money and divide it equally.

4.7. The Shapley value

An alternative solution concept for cooperative games emerges when we revert from the core to the reasonable set and try to imagine the order in which the grand coalition might actually form. From §4.1, if Player i is the j-th player to join the grand coalition of an n-player

---

[11] But not the inevitability; see Exercise 4.15.

characteristic function game, and if $T - \{i\}$ denotes the $j - 1$ players who joined prior to i, then $\nu(T) - \nu(T - \{i\})$ is a fair allocation for i. As we remarked in §4.1, however, we do not know the identity of T: half of all possible coalitions (including $\varnothing$) contain i (Exercise 4.16), and T could be any of them. So all we know about T is that $T \in \Pi^i$, where Π^i is defined by (4.14). If we are going to reduce the reasonable set to a single fair imputation, then mustn't we know more about T than just that?

It is possible, however, that there is no more about T that can be known, at least in advance—and any solution concept, if it is going to be at all useful, must certainly be known in advance of its application. Suppose, for example, that the n players have decided to meet in the town hall at 8:00 p.m. to bargain over fair allocations, and that their order of arrival is regarded as the order in which the grand coalition formed. Then who can say what the order will be in advance? Although all will aim to be there at 8:00, some will be unexpectedly early and others unavoidably late, in ways that cannot be predicted with certainty. In other words, the order of arrival is a random variable. Therefore, if Player i is j-th to arrive, then j is a random variable, and the j-person coalition of which he becomes the last member is also a random variable. Let us denote it by Y_i. Then, as already agreed, $\nu(Y_i) - \nu(Y_i - \{i\})$ is a fair allocation for i; but because Y_i is a random variable, which can take many values, so also is $\nu(Y_i) - \nu(Y_i - \{i\})$. From all these values, how do we obtain a single number that we can regard as a fair allocation for Player i? Perhaps the best thing to do is to take the expected value, denoted by E. Then, denoting Player i's fair allocation by x_i^S, we have

$$(4.98) \qquad x_i^S = \mathrm{E}\big[\nu(Y_i) - \nu(Y_i - \{i\})\big]$$
$$= \sum_{T \in \Pi^i} \{\nu(T) - \nu(T - \{i\})\} \cdot \mathrm{Prob}(Y_i = T),$$

where $\mathrm{Prob}(Y_i = T)$ is the probability that the first j players to arrive are T and the summation is over all coalitions containing i.

If we assume that the $n!$ possible orders of arrival are all equally likely, then the fair imputation

$$(4.99) \qquad x^S = \big(x_1^S, x_2^S, \ldots, x_n^S\big)$$

Table 4.3. Possible orders of formation of grand coalition of 4 players.

1234	1342	2134	3142	2314	4213	2341	3421
1243	1423	2143	4123	3214	3412	2431	4231
1324	1432	3124	4132	2413	4312	3241	4321

Table 4.4. 3-player Shapley values and nucleoli compared to egalitarian imputation.

Game	x^S	x^*	x^E
Car pool: $d = 9$	$\frac{1}{126}(46, 43, 37)$	$\frac{1}{126}(50, 44, 32)$	$\frac{1}{126}(42, 42, 42)$
Car pool: $d = 1$	$\frac{1}{60}(28, 22, 10)$	$\frac{1}{60}(33, 21, 6)$	$\frac{1}{60}(20, 20, 20)$
Log hauling	$\frac{1}{60}(17, 20, 23)$	$\frac{1}{60}(14, 20, 26)$	$\frac{1}{60}(20, 20, 20)$
Long-jumping	$\frac{1}{96}(10, 43, 43)$	$\frac{1}{96}(0, 48, 48)$	$\frac{1}{96}(32, 32, 32)$

is known as the *Shapley value*. Suppose, for example, that $n = 4$, so that the 24 possible orders of arrival are as shown in Table 4.3, and that $i = 1$. Then, because $Y_1 = \{1\}$ in the first two columns of the table and all orders of arrival are equally likely, $\text{Prob}(Y_1 = \{1\}) = \frac{6}{24} = \frac{1}{4}$. Likewise, because $Y_1 = \{1, 2, 3, 4\}$ in the last two columns of the table, we have $\text{Prob}(Y_1 = \{1, 2, 3, 4\}) = \frac{6}{24} = \frac{1}{4}$ also. The first two entries in column 3 yield $\text{Prob}(Y_1 = \{1, 2\}) = \frac{2}{24} = \frac{1}{12}$; and similarly, $\text{Prob}(Y_1 = T) = \frac{1}{12}$ for all 2- and 3-player coalitions in $\Pi^1 = \left\{\{1\}, \{1, 2\}, \{1, 3\}, \{1, 4\}, \{1, 2, 3\}, \{1, 2, 4\}, \{1, 3, 4\}, \{1, 2, 3, 4\}\right\}$. So, from (4.98) and using $\nu(\{i\}) = 0 = \nu(\varnothing)$, we have

$$(4.100) \quad x_1^S = \tfrac{1}{12}\big\{\nu(\{1, 2, 3\}) + \nu(\{1, 2, 4\}) + \nu(\{1, 3, 4\})$$
$$+ \nu(\{1, 2\}) + \nu(\{1, 3\}) + \nu(\{1, 4\}) - \nu(\{2, 3\})$$
$$- \nu(\{2, 4\}) - \nu(\{3, 4\})\big\} + \tfrac{1}{4}\big\{1 - \nu(\{2, 3, 4\})\big\}.$$

The other three components of the Shapley value for a 4-player CFG now follow easily (Exercise 4.18), and similar expressions for the Shapley allocations in a 2-player or 3-player CFG are also readily found (Exercise 4.17). Note that because (4.98) satisfies (4.15) from Exercise 4.19, x^S always belongs to the reasonable set.

We are now in a position to calculate x^S for all games considered so far. Shapley values for 3-player games are presented in Table 4.4 and those for 4-player games in Table 4.5, the results being taken from

Table 4.5. 4-player Shapley values and nucleoli compared to egalitarian imputation.

Game	x^S	x^*	x_i^E
Car pool: $d = 2$	$\frac{1}{144}(43, 35, 25, 41)$	$\frac{1}{144}(45, 33, 21, 45)$	$\frac{36}{144}$
Antique minding	$\frac{1}{312}(80, 92, 92, 48)$	$\frac{1}{312}(78, 99, 99, 36)$	$\frac{78}{312}$

Exercises 4.17-4.18. For each game, the Shapley value is compared with both the nucleolus and the egalitarian imputation

$$(4.101) \qquad x^E = \left(\tfrac{1}{n}, \tfrac{1}{n}, \dots, \tfrac{1}{n}\right),$$

which distributes the benefits of cooperation uniformly among the players. We see at a glance that the Shapley value is, on the whole, a much more egalitarian imputation than the nucleolus, i.e., most allocations are at least as close to $\frac{1}{n}$ under the Shapley value as under the nucleolus. (Exceptions are provided by Player 2's allocation in the 3-person car pool with $d = 1$ and Player 1's allocation in the antique-minding game, but in each of these cases the Shapley and nucleolus allocations are both close to egalitarian.) Note, in particular, that the Shapley value would pay young Jed \$1.67 for his short long jumps in §4.6, whereas the nucleolus wouldn't pay him anything.

Intuitively, the Shapley value is more egalitarian than the nucleolus because the nucleolus gives priority to the most dissatisfied coalitions, whereas the Shapley value grants all coalitions equal status. The nucleolus derives from core-minded thinking; whereas the Shapley value derives from reasonable-set thinking, and so it is hardly surprising that the two solutions almost never coincide (except, of course, for 2-player games—see Exercise 4.17). If the Shapley value belongs to the core (as in the car-pool games), then it does so more by accident than by design.[12] That the core need not contain the Shapley value is immediate, because the Shapley value always exists, whereas the core does not. But even a non-empty core may not contain the Shapley value—witness the antique-minding game, where the Shapley value fails to satisfy the third inequality in (4.66b) when $\epsilon = 0$. By contrast, if the core exists then it always contains the nucleolus; and in any event, the nucleolus always belongs to the least rational

---

[12]Except that the core of a convex game must contain its Shapley value; see [**205**].

core (though not necessarily the least core—witness the long-jumping game). On the other hand, we shouldn't make too much of the difference between core-minded thinking and reasonable-set thinking, because the nucleolus always belongs to the reasonable set—and so does the core, when the core exists (Exercise 4.26).

Although the formulae derived in Exercises 4.17-4.18 are adequate for our examples, an explicit expression for the Shapley value of an n-player game is easily obtained. First we define $\#(S)$ to be the number of elements in set S, so that, e.g., $\#(N) = n$, $\#\left(\Pi^i\right) = 2^{n-1}$, $\#\left(\Sigma^0\right) = 2^n - 2$, $\#(X^\kappa) = 1$, $\#(\varnothing) = 0$, etc. In terms of this notation, if Player i is the j-th player to join the grand coalition, and if T denotes the first j players to join the grand coalition, then $j = \#(T)$; $\#(T) - 1$ players have joined the grand coalition before Player i; and $n - j = n - \#(T)$ will join after him. Moreover, a permutation of the order of either the first $\#(T) - 1$ players or the last $n - \#(T)$ players cannot alter the fact that T is the coalition that Player i completes. There are $(\#(T) - 1)!$ permutations of the first kind, and $(n - \#(T))!$ permutations of the second kind. Accordingly, $(\#(T)-1)! \cdot 1 \cdot (n-\#(T))!$ is the number of orders in which the players can join the grand coalition in such a way that Player i completes coalition T. But there are precisely $n!$ orders of any kind in which the players can join the grand coalition; and so, if all orders are equally likely, we have

$$(4.102) \qquad \mathrm{Prob}(Y_i = T) \;=\; \tfrac{1}{n!}\left(\#(T) - 1\right)!\,(n - \#(T))!$$

Thus, from (4.98), Player i's allocation under the Shapley value is

$$(4.103) \quad x_i^S \;=\; \tfrac{1}{n!} \sum_{T \in \Pi^i} (\#(T) - 1)!\,(n - \#(T))!\,\{\nu(T) - \nu(T - \{i\})\},$$

where Π^i is defined by (4.14). Despite the ease with which we obtained this result, however, the Shapley value of an n-player game is not in general an easy vector to calculate, because each component involves a summation over 2^{n-1} coalitions.

Which is the fairer imputation—the nucleolus or the Shapley value? Well, what do you think?[13]

---

[13]For an interesting empirical analysis of opinions on these and other solution concepts for 3-player games in three different nations, see [**208**]. Also, see the discussion of this question in Chapter 7 (p. 309).

4.8. Simple games. The Shapley-Shubik index

If 0 and 1 are the only values assigned by a characteristic function ν—as, for example, in (4.46)—then the CFG defined by ν is said to be *simple*. In other words, a CFG is simple if either $\nu(S) = 0$ or $\nu(S) = 1$ for every coalition S (not just $\varnothing$ and N).

Simple games arise most readily in the context of voting. Suppose, for example, that the mathematics department of a small liberal arts college is recruiting for a new professor. Candidates are interviewed one at a time, beginning with the one whose credentials on paper look most impressive. Immediately after each interview, the existing faculty meet to vote on the candidate; and if the vote is sufficiently favorable, then the candidate is offered a position—and no more candidates are interviewed unless, and until, he turns it down.

Now, in this department there has always been tension between pure and applied mathematicians, with a temptation for one group to exploit the other in the conflict over hiring. To guard against this temptation, it is enshrined in the department's constitution that a candidate is nominated if, and only if, he secures the vote of at least 50% of both the pure and the applied mathematicians on the existing faculty. Thus every new professor, whether pure or applied, has at least some measure of broad support throughout the existing faculty.

In this recruitment process, the purpose of cooperation among the faculty is to ensure that a candidate is hired. Every coalition of faculty is either strong enough to ensure the recruitment of its particular candidate, in which case it reaps the entire benefit of cooperation; or too weak to ensure that its candidate is hired, in which case it reaps no benefit at all. We can therefore partition the set of all coalitions into the set of winning coalitions—denoted by W—and the set of losing coalitions, which comprises all others. It is then natural to define the characteristic function, ν, by

$$(4.104) \qquad \nu(S) \;=\; \begin{cases} 0 & \text{if } S \notin W \\ 1 & \text{if } S \in W. \end{cases}$$

Thus the recruitment process yields a simple game. Indeed the characteristic function of any simple game has the form of (4.104), where W is the set of coalitions to which ν assigns the value 1.

To see that the Shapley value of a simple game can be interpreted as an index of power, let us denote by P^i the set of coalitions in which Player i's vote is crucial for victory; i.e., the set of winning coalitions that would become losing if Player i were removed. In symbols:

$$(4.105) \qquad P^i = \left\{ S \in \Pi^i \mid S \in W, \; S - \{i\} \notin W \right\}.$$

If $S \in W$ and $S - \{i\} \in W$, then $\nu(S) = 1 = \nu(S - \{i\})$; whereas if $S \notin W$ and $S - \{i\} \notin W$, then $\nu(S) = 0 = \nu(S - \{i\})$. Thus $\nu(S) - \nu(S - \{i\}) = 0$ unless $S \in W$ and $S - \{i\} \notin W$, or $S \in P^i$; in which case, $\nu(S) - \nu(S - \{i\}) = 1 - 0 = 1$. It follows from (4.98) that

$$(4.106) \qquad x_i^S = \sum_{T \in P^i} \mathrm{Prob}(Y_i = T),$$

where Y_i is the coalition that Player i was last to join. The right-hand side of (4.106) is the probability that Player i's vote is crucial (to some coalition), and it is therefore a measure of Player i's voting power. In particular, if all possible orders of coalition formation are equally likely, then from (4.102) we have

$$(4.107) \qquad x_i^S = \frac{1}{n!} \sum_{T \in P^i} (\#(T) - 1)! \, (n - \#(T))!$$

We refer to (4.107) as the Shapley-Shubik index.

Suppose, for example, that there are five members of faculty: three pure mathematicians—Players 1 to 3, say—and two applied (Players 4 and 5). Then at least two pure mathematicians and one applied mathematician must vote for a candidate to secure his nomination, so that the set of winning coalitions W contains $\{1, 2, 4\}$, $\{1, 2, 5\}$, $\{1, 3, 4\}$, $\{1, 3, 5\}$, $\{2, 3, 4\}$, $\{2, 3, 5\}$, $\{1, 2, 3, 4\}$, $\{1, 2, 3, 5\}$, $\{1, 2, 4, 5\}$, $\{1, 3, 4, 5\}$, $\{2, 3, 4, 5\}$ and $N = \{1, 2, 3, 4, 5\}$. Thus $P^1 = \big\{\{1, 2, 4\}, \{1, 2, 5\}, \{1, 3, 4\}, \{1, 3, 5\}, \{1, 2, 4, 5\}, \{1, 3, 4, 5\}\big\}$ and $P^4 = \big\{\{1, 2, 4\}, \{1, 3, 4\}, \{2, 3, 4\}, \{1, 2, 3, 4\}\big\}$ from (4.105), with similar expressions for P^2, P^3 and P^5; see Exercise 4.21. It follows from (4.107) that the Shapley-Shubik index is $x^S = \left(\frac{7}{30}, \frac{7}{30}, \frac{7}{30}, \frac{3}{20}, \frac{3}{20} \right)$. A pure mathematician has $\frac{7}{30}$ of the voting power, whereas an applied mathematician has only $\frac{3}{20}$. Collectively, applied mathematicians are 40% of the faculty but have only 30% of the voting power. Of course, this seems unfair; and so we had better hope—for their sake—that an applied mathematician is hired. Or had we? See Exercise 4.22. Then try Exercises 4.23-4.24.

4.9. Commentary

In Chapter 4 we have introduced characteristic function games (§4.1) and a variety of CFG solution concepts, namely, imputations and the reasonable set (§4.1); excess, the core, rational ϵ-core and least rational core (§4.2); the nucleolus (§4.5); superadditivity, proper games, pre-imputations and the ϵ-core (§4.6); and the Shapley value (§4.7). We have applied these ideas to a number of bargaining problems, e.g., sharing the costs of a car pool among three (§§4.1-4.2) or four (§4.3) individuals, and sharing duties in an antique dealership (§4.5); and we have illustrated that a realistic CFG can be either coreless (§4.4) or improper (§4.6). For all of these games we have compared the nucleolus and Shapley value with the egalitarian imputation (§4.7). For applications of CFGs to water resources, see [**142**, pp. 127-130]. For a comparison of the nucleolus and Shapley value in the assessment of airport landing fees, see §XI.4 of [**173**].

In §4.8 we briefly considered simple games. We adapted the Shapley value for use as an index of power, and we applied it to power sharing in an academic department. This Shapley-Shubik index is not the only index of power, however, a popular alternative being the Banzhaf-Coleman index; see Exercise 4.25 and Chapter X of [**173**], where both indices are applied to voting in U. S. presidential elections. For more on simple games, see [**221**].

Exercises 4

1. Show from (4.15) that the reasonable set is the shaded hexagon in Figure 4.3 for §4.1's car pool with $d = 9$, and the partially shaded hexagon in Figure 4.5 for the car pool with $d = 1$.

2. Find the rational ϵ-core of the 3-person car pool with $d = 1$.

3. (a) Find the least rational core for the 3-person car pool when $d = 9$, i.e., deduce (4.28) from (4.27).

    (b) Find the least rational core for the 3-person car pool when
$d = 6$.

4. Find the least rational core for the 4-person car pool when $d = 1$.

5. Find the least rational core for the 4-person car pool when $d = 8$.

6. Find the least rational core for the 4-person car pool when $d = 18$.

7. **(a)** For a 2-player CFG, verify that the set of imputations X
is a 1-dimensional line segment imbedded in 2-dimensional
space. Sketch a 1-dimensional representation of X, and
verify that $C^+(0) = X$ (thus $\epsilon_0 = 0$). What is the characteristic function? What is the least rational core?

    (b) What is the least rational core for the game defined by
(4.46)?

8. Calculate the rational ϵ-core of the log-hauling game, i.e., verify
(4.49). Find the least rational core.

9. **(a)** Show that the game defined by (4.51) is convex only if
$\max(b+c, c+a, a+b) \le 1$, and hence that the game defined
by (4.10) is never convex.

    (b) Show that the game defined by (4.51) has a core if and only
if $a + b + c \le 2$, and hence that the game defined by (4.10)
always has a core whereas that in §4.4 is coreless.

10. Jed, Ned and Ted are antique dealers who conduct their businesses in separate but adjoining rooms of a common premises.
Jed's advertised hours are from 12:00 noon until 4:00 p.m., Ned's
hours are from 9:00 a.m. until 3:00 p.m. and Ted's from 1:00
p.m. until 5:00 p.m. Because the dealers have other jobs and the
store is never so busy that one individual could not take care of
everyone's customers, it is in the dealers' interests to pool their
time in minding the store; there is no need for two people between noon and 1:00 p.m. or between 3:00 p.m. and 4:00 p.m, or
for three people between 1:00 p.m. and 3:00 p.m. Thus Jed can
arrive later than noon or leave earlier than 4:00 p.m., Ted can
arrive later than 1:00 p.m. and Ned can leave earlier than 3:00
p.m. But how much later or earlier? What is a fair allocation of
store-minding duty for each of the dealers?

11. Verify (4.65)-(4.71).

12. Consider the 4-player CFG whose characteristic function is defined by $\nu(S) = \frac{1}{2}$ if $\#(S) = 2$ or $\#(S) = 3$, except for $\nu(\{1,3\}) =$

$\frac{1}{4}$ and $\nu(\{2,4\}) = 0$; of course, $\nu(\varnothing) = 0$, $\nu(N) = 1$ and $\nu(\{i\}) = 0$ for $i \in N$. Show that the least rational core is the line segment $X^1 = \left\{ x \mid x = \frac{1}{8}(1 + 3t, 3 - 3t, 1 + 3t, 3 - 3t), \ 0 \le t \le 1 \right\}$, but that its mid-point $\left(\frac{5}{16}, \frac{3}{16}, \frac{5}{16}, \frac{3}{16} \right)$ is not the nucleolus.[14]

13. Show that every improper 3-player CFG is coreless.

14. Verify (4.96)-(4.97).

15. How should young Jed, young Ned and young Ted have split the proceeds from the long-jump competition described in §4.6 if Jed's best two jumps had been 18 and 16 feet, instead of 15 (but Ned and Ted's jumps had been the same)?

16. Show that, in any n-player CFG, each player belongs to precisely 2^{n-1} coalitions.

17. (a) Find the Shapley value of an arbitrary 2-player CFG.
    (b) Find the Shapley value of an arbitrary 3-player CFG.
    (c) Hence verify Table 4.4.

18. (a) Verify (4.100), and hence obtain a complete expression for the Shapley value of an arbitrary 4-player CFG.
    (b) Verify Table 4.5.

19. Prove that (4.98) satisfies (4.15).

20. A characteristic function game is said to be *symmetric* if every coalition's bargaining strength depends only on the number of players in it, i.e., if there exists a function f from the set of real numbers to itself such that $\nu(T) = f(\#(T))$ for all coalitions T. Prove that if a symmetric CFG has a core, then its core must contain the egalitarian imputation x^E defined by (4.101).

21. (a) Find P^1, P^2, P^3, P^4 and P^5 for the voting game in §4.8.
    (b) Hence obtain its Shapley-Shubik index.

22. In §4.8's example of a mathematics department, there are currently three pure and two applied mathematicians. Which would increase an applied mathematician's voting power more (in terms of hiring the seventh member of the faculty)—hiring an applied mathematician or hiring a pure mathematician? Elucidate.

23. In any simple game, Player i is a *dummy* if $P^i = \varnothing$ and a *dictator* if $P^i = \Pi^i$.

---

[14]This exercise is from [**128**, p. 335].

(a) Can there be more than one dummy? Either prove that this is impossible, or produce an example of a simple game with more than one dummy.

(b) Prove that there cannot be more than one dictator. Produce an example of a game with a dictator.

24. Nassau County in New York State is run by a board of six supervisors. When this board votes on an issue, each supervisor casts—either for or against the motion—a number of votes that until December, 1937 was roughly proportional to the size of his electoral district. At that time, votes were allotted to each district—at roughly one per 10,000 of population—according to the second column of the following table:

District	December 1937	January 1938
Glen Cove	1	1
Hempstead #1	9	7
Hempstead #2	9	7
Long Beach	1	1
North Hempstead	6	6
Oyster Bay	3	3

(Grofman and Scarrow, [**79**], pp. 178-179). Because the supervisors had a total of 29 votes among them, 15 votes were required in favor to carry a motion by simple majority. In January, 1938, however (without any change in the population), the numbers of votes allotted to two of the districts were reduced according to the third column of the table, thus reducing the total number of votes from 29 to 25. Nevertheless, the number of votes required to carry a motion was not reduced to 13; it remained instead at 15. In terms of game theory—why?

25. An alternative to the Shapley-Shubik index as a power index for a simple game is the Banzhaf-Coleman index, i.e, the imputation $x^B = \left(x_1^B, x_2^B, \ldots, x_n^B\right)$ defined by

$$x_i^B = \frac{\#(P^i)}{\#(P^1) + \#(P^2) + \ldots \#(P^n)}, \quad i = 1, \ldots, n$$

where $\#$ is defined on p. 164 and P^i is defined by (4.105).

(a) Why is it reasonable to interpret x^B as an index of power?

   (b) In §4.8's Mathematics Department, does an applied mathematician have more or less power according to Banzhaf and
Coleman than according to Shapley and Shubik?

   (c) Which do you consider more suitable as an index of power,
x^S or x^B?

26. Prove that the core of a characteristic function game (if it exists)
must be a subset of the reasonable set.

27. Consider the 3-player characteristic function game defined by
(4.51)-(4.52). Let x^* denote its nucleolus, and let x^T and x^E be
defined, respectively, by (4.53) and (4.101) with $n = 3$.

   (a) Show that (4.52) implies superadditivity.

   (b) Show that $m \leq \frac{1}{3}$ implies $x^* = x^E$.

   (c) Show that (4.54) implies $x^* = x^T$.

   (d) Show that 4 is the least value of d for which $x^* = x^T$ in the
3-player car pool defined by (4.10).

Chapter 5

Cooperation and the Prisoner's Dilemma

To team—or not to team? That is the question. Chapters 3 and 4 have already shown us that circumstances abound in which players do better by cooperating than by competing. Indeed if ν denotes a normalized characteristic function, then Player i has an incentive to cooperate with coalition $S - \{i\}$ whenever $\nu(S) > 0$. But an incentive to cooperate does not imply cooperation, for we have also seen in §3.4 that if one of two players is committed to cooperation, then it may be rational for the other to cheat and play noncooperatively. Then how, in such circumstances, is cooperation enforced?

To reduce this question to its barest essentials, we focus on the symmetric, 2-player, noncooperative game with two pure strategies C (strategy 1) and D (strategy 2) whose payoff matrix

$$(5.1) \qquad \begin{bmatrix} R & S \\ T & P \end{bmatrix}$$

satisfies

$$(5.2) \qquad R > \tfrac{1}{2}\max(2P, S + T).$$

The combined payoffs associated with the four possible strategy combinations CC, CD, DC and DD are, respectively, $2R$, $S + T$, $T + S$ and $2P$. From (5.2), the best combined payoff $2R$ can be achieved only if both individuals select C. We therefore say that C is a cooperative strategy and that D (for defect) is a noncooperative strategy. We will refer to this game as the cooperator's dilemma [152].[1]

If it is also true that

$$(5.3) \qquad\qquad T > R, \quad P > S$$

then the cooperator's dilemma reduces to the "prisoner's dilemma," with which we are familiar from the exercises.[2] We discovered in Exercise 1.25 that DD is the only Nash-equilibrium strategy combination (pure or mixed); in Exercise 2.7 that D is a strong ESS; and in Exercise 3.9 that CC is the Nash bargaining solution. We can consolidate these findings by saying that, because $T > R$ and $P > S$, D is the unique best reply to both C and D; in other words, D is a (strongly) dominant strategy. Thus, by symmetry, it is rational for each player to select strategy D, whence each obtains payoff P. If each were to cooperate by selecting C, however, then each would obtain a higher payoff, namely, R. We have discovered the paradox of the prisoner's dilemma: although mutual cooperation would yield a higher reward, mutual defection is rational (but only because there exists no mechanism for enforcing cooperation).

To see why this game is called the prisoner's dilemma, imagine that two prisoners in solitary confinement are suspected of some heinous crime—for which, however, there is no hard evidence. A confession is therefore needed, and the police attempt to persuade each prisoner to implicate the other. Each prisoner is told that her sentence will depend on whether she remains silent or implicates the other prisoner, according to the payoff matrix in Table 5.1. Let us suppose that long, medium, short and very short sentences are, respectively, d years, c years, b years and a years, so that $d > c > b > a$; and that a long sentence is at least twice as long as a short sentence, or $d > 2b$. Then because the prisoners would like to spend as little time in jail as possible, we have $R = -b, S = -d, T = -a$ and

---

[1] Not every problem of cooperation can be reduced to its barest essentials this way: CD and DC will yield the best combined payoff if cooperation requires players to take complementary actions [46]. But the cooperator's dilemma has broad applicability.

[2] This game descends from Merrill Flood, Melvin Dresher and A. W. Tucker; see [217]. In defining the prisoner's dilemma, some authors require in addition to (5.2) that $2P < S + T$; see, e.g., [24, 183, 187]. The additional requirement is satisfied by two of our principal examples (§§5.1-5.2), but not necessarily by the third (§5.8).

Table 5.1. Payoff matrix for the classic prisoner's dilemma

	REMAIN SILENT	IMPLICATE
REMAIN SILENT	Short	Long
IMPLICATE	Very short	Medium

$P = -c$, so that (5.3) is satisfied; and $b < \frac{1}{2}(a + d)$, so that (5.2) is also satisfied. Although the prisoners would both prefer a short sentence, they will settle for a medium sentence, because neither can be sure that the other will cooperate: if one player were to remain silent, in the hope that the other would also remain silent, then the second player could cheat—by implicating the first player—and thus obtain the shortest sentence of all. Therefore, unable to guarantee the other's silence, each player will implicate the other.

In this chapter, the prisoner's dilemma and related games will enable us to investigate rationales for cooperation. For the sake of simplicity, we will consider only games restricted to pure strategies. We begin with some concrete examples of the prisoner's dilemma. We then study ways of escape from its paradox.

5.1. A laboratory prisoner's dilemma

We can obtain the prisoner's dilemma with fewest parameters if we set $S = 0, T = P + R$ in (5.1). Then the game has payoff matrix

$$(5.4) \qquad \begin{bmatrix} R & 0 \\ R + P & P \end{bmatrix}, \quad R > P > 0.$$

This matrix suggests an experiment in which the prisoner's dilemma could be played repeatedly by laboratory animals. The apparatus for this experiment consists of a partitioned cage, each half of which contains a food chute, a decision lever with spring attachment, and an "exhaust" column for rejected food items. In Figure 5.1, XY represents the partition, which is transparent. The food chutes, which are themselves partitioned (along FK), slope gradually downward from EG or HL to the top of the partition XY. One side of each chute (extending from EF or KL) supplies food items of value R, and the other side (extending from FG or HK) supplies food items of value P, where $P < R$. The sides of the chutes slope slightly upward from

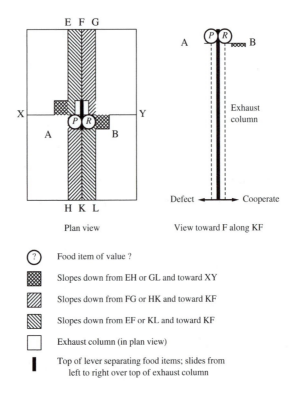

Plan view   View toward F along KF

⊙ (?)	Food item of value ?
▦	Slopes down from EH or GL and toward XY
▨	Slopes down from FG or HK and toward KF
▧	Slopes down from EF or KL and toward KF
▯	Exhaust column (in plan view)
▮	Top of lever separating food items; slides from left to right over top of exhaust column

Figure 5.1. Sketch of apparatus for laboratory prisoner's dilemma.

the line FK—just enough to ensure that items of food, having rolled down the chute, come to rest against the partition (see the plan view in Figure 5.1). The lower end of the partition that divides a chute in half is also the top of a lever, which in its neutral position prevents a food item from falling into the exhaust column beneath; see the side view in Figure 5.1, where the level AB is presumed beyond an animal's reach from the floor of the cage. By sliding one of the levers to the left or right (or operating some other mechanism to the same effect), an animal in either half of the cage can push one of the food items over the side of the chute. The other item will fall into the exhaust column and through the floor of the cage (and the animal will be unable to intercept it). If the animal pushes the lever to the left, then she chooses strategy D: the item of value P falls into her own half of the cage (and the item of value R is lost). If she pushes the lever to the right, however, then she chooses strategy C: the item of value R slides onto the cross-hatched area in Figure 5.1, which slopes downward from EH or GL and toward XY. Thus the item falls into the other animal's half of the cage (and the item of value P is lost).

For each play of the game, a pair of food items is supplied simultaneously to each half of the cage. Each animal has two choices. She can either defect, in which case she retains the item of value P; or she can cooperate, in which case she gives the item of value R to the other animal. Thus the payoff to each for mutual cooperation is R, whereas the payoff to each for mutual defection is P; and if one animal defects while the other cooperates, then the former's payoff is $P + R$ and the latter's zero. In other words, the game has payoff matrix (5.4). Of course, if an animal's daily appetite is a, and if the game is played n times per day, then it is necessary to ensure that $a \geq nR$ (whence $a/n > P > 0$).

The suggested apparatus should not be overly difficult to construct. Obviously, one must prevent the animals from climbing up the partition and obtaining food without using the levers. But this could be achieved, even for large mammals, by modifying the apparatus so that all food items must pass through a sufficiently long pipe attached at A (or the corresponding point in the other half of the cage) before falling to the floor of the cage. To ensure that neither animal is rewarded before the other, it would be necessary to cap the food pipes with trap doors that opened only when the second animal had made her choice. It would also be necessary to make the exhaust columns opaque on any side facing the other half of the cage; then, even if the animals operated their levers at different times, no information about the first mover's choice would be transferred to the second mover in the meantime, and the animals' choices could be regarded as simultaneous.[3]

The animals could be trained to play the game via two preliminary manœuvres. During one of these manœuvres, an animal would play solitaire in a cage whose partition had been removed. Thus payoffs would be either P or R, because the animal could retrieve items of value R from the other half of the cage. During the other manœuvre, an animal would play solitaire with the partition restored. Thus the payoffs would be either 0 or P. After some combination of these manœuvres, the animals should be ready to play.

Even two trained animals would enable one to observe a sequence of plays between fixed opponents, and the results should be interesting in their own right. Nevertheless, one would wish to train several animals, say $N + 1$, where N is significantly larger than 1—not only to collect data from up to $\frac{1}{2}N(N + 1)$ pairwise interactions, but also to observe animals in a repeated prisoner's dilemma against randomly chosen opponents, as suggested in Exercise 5.24.[4]

5.2. A game of foraging among oviposition sites

For a further example of the prisoner's dilemma, let the players be a pair of insects foraging over a patch of N oviposition sites. They forage randomly, and their searches are independent, so that all sites are equally likely to be visited next. Each site has the potential to support one or two eggs, and each insect begins to forage with a plentiful supply (N eggs or more). If one egg is laid at a site, then the probability that it will survive to maturity is r_1. If two eggs are laid, then the corresponding survival probability for each is r_2 (and if three or more were laid, then the survival probability for each would

---

[3]Even with this precaution, there remains the possibility that an animal will fail to push either lever. To allow for this, let us add a third strategy E (for eschew). Then payoff matrix (5.4) is replaced by

$$\begin{bmatrix} R & 0 & 0 \\ R + P & P & P \\ R & 0 & 0 \end{bmatrix}.$$

Clearly, D is still the dominant strategy and C the best cooperative strategy; and so this possibility does not detract from the value of the experiment (although it complicates the mathematical description of it).

[4]One would also require several trained animals as an insurance against mortality; and, of course, the larger the sample size with respect to any given context, the more likely it is that behavior observed in that context is representative.

be zero). Thus the expected number of offspring from a site is r_1 or $2r_2$, according to whether one or two eggs are laid, and we will assume that $r_1 > 2r_2$.

We assume that the insects arrive at the patch simultaneously, and we measure time discretely from the moment of their arrival. Let the duration of the game be n units of (discrete) time, or periods. In each period, let λ be the probability that an insect survives and remains on the patch, hence $1 - \lambda$ the probability that she leaves the patch or dies; and if an insect has survived on the patch, let ϵ be the probability that she finds a site, and $1 - \epsilon$ the probability that she finds no site (hence zero the probability that she finds more than one site). We will assume that an insect oviposits only during periods she survives on the patch, and that she never oviposits more than once per visit to a site. Henceforward, we will use survival to mean surviving and remaining on the patch.

In each period, if an insect finds a site where less than two eggs have been laid, then she can behave either cooperatively or noncooperatively. A cooperative insect will oviposit only if a site is empty, but a noncooperative insect will lay a second egg at sites where an egg has been laid by the other insect; we assume that insects recognize their eggs. The rationality behind such noncooperation is provided by the inequality $r_1 > 2r_2$: a second egg at the other insect's site yields a payoff of r_2, which is positive, whereas a second egg at the insect's own site yields a payoff of $2r_2 - r_1$, which is negative. If an insect behaves cooperatively during every period, then we shall say that she plays strategy C; whereas, if the insect behaves noncooperatively during every period, then we shall say that she plays strategy D. No other strategies for oviposition will be considered. (In particular, as stated at the outset, we do not consider mixed strategies.)

For the sake of simplicity, we now further assume that $n = 1$, so that the length of the game is a single period. The corresponding "tree" of events is shown upside down in Figure 5.2, with the "root" of the tree on top. Each branch of this tree corresponds to a *conditional* event, i.e., an event that can happen only if the preceding (conditional) event has happened; and the number on the branch is the corresponding *conditional* probability, i.e., the probability that

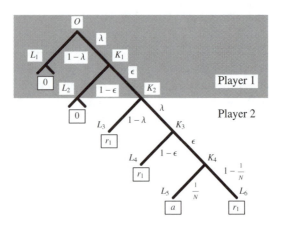

Figure 5.2. The event tree for the single-period foraging game, showing payoffs to Player 1.

the event happens, given that the event corresponding to the previous branch has already happened. Each path through the tree from the root to a "leaf" represents a sequence of conditional events; and the number in a rectangle at the end of this path is the corresponding payoff to Player 1. To obtain the probability of this payoff, we simply multiply together all conditional probabilities along the path. Note that branches in the shaded region represent events that can happen to Player 1, whereas the remaining branches represent events in the life of Player 2.

Let us now begin at the root. The first right-hand branch, from O to the vertex at K_1, corresponds to the event that Player 1 survives the period, which happens with probability λ. The next right-hand branch, from K_1 to the vertex at K_2, corresponds to the event that Player 1 locates a site—conditional, of course, upon survival. Conditional upon surviving and locating a site, Player 1 is assured of a non-zero payoff; however, the size of that payoff depends on Player 2. The left-hand branch from K_2 to the leaf at L_3 corresponds to the event that Player 2 does not survive the period; the payoff to Player 1 is then r_1. The right-hand branch from K_2 to the vertex at K_3 plus the left-hand branch from K_3 to the leaf at L_4 represent the event that Player 2 survives the period but fails to find a site (conditional, of course, upon Player 1 surviving and locating a site); in which case, the payoff to Player 1 is still r_1. Continuing in this manner, the right-hand path from K_3 through K_4 to the leaf at L_6 corresponds to the event that Player 2 locates one of the $N-1$ sites that Player 1 did not locate—with (conditional) probability $(N-1)/N$, because all sites are equally likely to be found. Again, the payoff to Player 1 is r_1. If both players locate the same site, however —which corresponds to the path from O to the leaf at L_5, then the payoff to Player 1 depends upon the players' strategies. Let us denote it by $a(u_1, v_1)$, where u_1 is Player 1's strategy, and v_1 is Player 2's. Finally, the left-hand branches from O and K_1 represent the events that Player 1 either does not survive, or survives without finding a site; in either case, the payoff is zero. (Note that, strictly, the tree in Figure 5.2 is incomplete; L_1 and L_2 are vertices, not leaves, and to complete the tree we would have to append at each vertex the unshaded part of the tree. Because the payoff to Player 1 would still be zero, however, there is nothing to be gained from doing so.)

We can now compute the expected value of Player 1's payoff in terms of a. All we need do is to multiply each payoff by the product of all conditional probabilities on branches leading to that payoff, then add. Thus Player 1's reward is

$$(5.5) \quad 0 \cdot (1-\lambda) + 0 \cdot \lambda \cdot (1-\epsilon) + r_1 \cdot \lambda\epsilon(1-\lambda)$$
$$+ r_1 \cdot \lambda^2\epsilon(1-\epsilon) + a \cdot \tfrac{\lambda^2\epsilon^2}{N} + r_1 \cdot \lambda^2\epsilon^2 \left(1 - \tfrac{1}{N}\right)$$
$$= \epsilon\lambda \left(\left\{1 - \tfrac{\epsilon\lambda}{N}\right\} r_1 + \tfrac{\epsilon\lambda a}{N} \right).$$

If our insects locate the same site, and if both are C-strategists— that is, if $u_1 = C = v_1$—then the first to arrive will oviposit, whereas the second will not; in other words, the first to arrive will obtain r_1, whereas the second to arrive will obtain zero. Let us assume that, if they do locate the same site, then they are equally likely to locate it first. Then Player 1's (conditional) expected payoff is

$$(5.6a) \qquad a(C, C) \;=\; \tfrac{1}{2} \cdot r_1 + \tfrac{1}{2} \cdot 0 \;=\; \tfrac{1}{2} r_1.$$

If both are D-strategists—that is, if $u_1 = D = v_1$—then both will oviposit, regardless of who is first to arrive, and each will obtain payoff r_2. Thus

(5.6b) $a(D, D) \; = \; r_2.$

If, on the other hand, Player 1 is a C-strategist but Player 2 a D-strategist, then Player 1 will oviposit only if she is the first to arrive; in which case, she will obtain only r_2, because Player 2 will oviposit behind her. Thus

(5.6c) $a(C, D) \; = \; \frac{1}{2} \cdot r_2 + \frac{1}{2} \cdot 0 \; = \; \frac{1}{2} r_2.$

Finally, if Player 1 is a D-strategist but Player 2 is a C-strategist, then Player 1 will oviposit even if Player 2 arrives first, in which case Player 1 will obtain r_2; whereas if Player 1 arrives first, then Player 2 will fail to oviposit, and so Player 1's payoff will be r_1. Accordingly,

(5.6d) $a(D, C) \; = \; \frac{1}{2} \cdot r_1 + \frac{1}{2} \cdot r_2 \; = \; \frac{1}{2}(r_1 + r_2).$

Substitution from (5.6) into (5.5) yields the payoff matrix $\left[\begin{smallmatrix} R & S \\ T & P \end{smallmatrix} \right] =$

(5.7) $\epsilon\lambda \begin{bmatrix} \frac{\epsilon\lambda}{N} \cdot \frac{r_1}{2} + \left(1 - \frac{\epsilon\lambda}{N}\right) r_1 & \frac{\epsilon\lambda}{N} \cdot \frac{r_2}{2} + \left(1 - \frac{\epsilon\lambda}{N}\right) r_1 \\ \frac{\epsilon\lambda}{N} \cdot \frac{r_1 + r_2}{2} + \left(1 - \frac{\epsilon\lambda}{N}\right) r_1 & \frac{\epsilon\lambda}{N} \cdot r_2 + \left(1 - \frac{\epsilon\lambda}{N}\right) r_1 \end{bmatrix}.$

Because $r_1 > 2r_2$, we have $T > R > P > S$ and $2R > S + T$, so that (5.2)-(5.3) are satisfied. In other words, the game is a prisoner's dilemma.

Now might be a good time to attempt Exercise 5.25.

5.3. Tit for tat: champion reciprocative strategy

Having seen two examples of the prisoner's dilemma, we now investigate possible escapes from its paradox. An important idea in this regard is that of *reciprocity*: one good turn deserves another—and one bad turn deserves another, too. To be more precise, reciprocity in the sense of "reciprocal altruism" [**9, 223, 224**] means that one good turn *now* deserves another *later*, and similarly for bad turns. Thus reciprocity is an inherently dynamic concept: it is impossible to reciprocate if the game is played only once, and we shall therefore assume that it is played repeatedly. Indeed it is convenient to define a brand new game, of which a single play consists of all the plays of the prisoner's dilemma that an animal—whether bird or mammal as in §5.1, insect as in §5.2 or human as in §5.8—makes within some specified interval of time. We call this brand new game the *iterated* *prisoner's dilemma*, or IPD; and whenever a prisoner's dilemma is imbedded in an iterated prisoner's dilemma, we shall refer to each play of the prisoner's dilemma as a *move* of the IPD.

An animal who cooperates on the first move of an IPD might well cooperate at all subsequent times. But whereas her strategy in the prisoner's dilemma would be C, as defined above, her strategy in the IPD would be

$$(5.8) \qquad ALLC \;=\; (C, C, C, C, \dots, C),$$

for "at all times cooperate." More generally, in the iterated prisoner's dilemma, pure strategy u—and we consider only pure strategies—consists of a sequence $u_1, u_2, \dots, u_n$ of prisoner's-dilemma strategies, where prisoner's dilemma strategy u_k is used on move k of the IPD and n is the number of moves; or, regarding the sequence as a vector,

$$(5.9) \qquad u \;=\; (u_1, u_2, \dots, u_n).$$

Although C and D are the only values that u_k can take, there would be 2^n values that u could take even if strategies had to be unconditional; (5.8) is one such value, and

$$(5.10) \qquad ALLD \;=\; (D, D, D, D, \dots, D),$$

for "at all times defect," is another. But IPD strategies need not be unconditional: they can be contingent upon opponents' strategies—a prerequisite for reciprocity. The paragon of such a strategy is "tit for tat," or TFT, which cooperates on the first move and subsequently plays whatever prisoner's-dilemma strategy an opponent used on the previous move. Thus, if the prisoner's-dilemma strategies adopted by a player's opponent are denoted by $v_1, v_2, \dots$, then

$$(5.11) \qquad TFT \;=\; (C, v_1, v_2, \dots, v_{n-1})$$

—if, that is, n is finite. We shall now suppose, however, that n may be infinite (though with vanishingly small probability).

Let the number of moves in an IPD be denoted by the (integer-valued) random variable M, so that $\text{Prob}(M \geq 1) = 1$. Conditional upon there being a k-th move in the IPD, let $\phi(u_k, v_k)$ be that move's payoff to prisoner's-dilemma strategy u_k against prisoner's dilemma strategy v_k; thus, from (5.1), $\phi(C, C) = R, \phi(C, D) = S, \phi(D, C) = T$ and $\phi(D, D) = P$. Then the actual payoff to prisoner's-dilemma strategy u_k against prisoner's dilemma strategy v_k from move k of the IPD is the random variable

$$(5.12) \qquad F_k(u_k, v_k) = \begin{cases} \phi(u_k, v_k) & \text{if } M \geq k \\ 0 & \text{if } M < k; \end{cases}$$

and the reward from move k of the game is

$$(5.13) \quad \mathrm{E}\,[F_k(u_k, v_k)] = \phi(u_k, v_k) \cdot \mathrm{Prob}(M \geq k) + 0 \cdot \mathrm{Prob}(M < k)$$
$$= \phi(u_k, v_k)\,\mathrm{Prob}(M \geq k),$$

where E denotes expected value. Thus, if $f(u, v)$ is the reward to strategy u against strategy v from (all moves of) the IPD, then

$$(5.14) \quad f(u, v) = \sum_{k=1}^{\infty} \mathrm{E}\,[F_k(u_k, v_k)] = \sum_{k=1}^{\infty} \phi(u_k, v_k)\,\mathrm{Prob}(M \geq k).$$

Let us now suppose that there are only three (pure) strategies, namely, $ALLC, ALLD$ and TFT, where $ALLC$ is strategy 1, $ALLD$ is strategy 2 and TFT is strategy 3. Then a 3×3 payoff matrix A for the IPD can be computed directly from (5.14), with $a_{12} = f(ALLC, ALLD), a_{23} = f(ALLD, TFT)$, and so on. For example,

$$(5.15) \quad a_{23} = \phi(D, C)\,\mathrm{Prob}(M \geq 1) + \sum_{k=2}^{\infty} \phi(D, D)\,\mathrm{Prob}(M \geq k)$$
$$= T + P \sum_{k=2}^{\infty} \mathrm{Prob}(M \geq k).$$

But (Exercise 5.1)

$$(5.16) \qquad \mathrm{E}[M] = \sum_{k=1}^{\infty} k \cdot \mathrm{Prob}(M = k) \equiv \sum_{k=1}^{\infty} \mathrm{Prob}(M \geq k)$$

(assuming, of course, that both series converge). Hence, if we define

$$(5.17) \qquad\qquad\qquad \mu = \mathrm{E}[M],$$

so that $\mu \geq 1$, then

$$(5.18) \qquad\qquad \sum_{k=2}^{\infty} \mathrm{Prob}(M \geq k) = \mu - 1.$$

Thus, from (5.15), $a_{23} = T + P(\mu - 1)$. Similarly, $a_{32} = S + P(\mu - 1)$. Furthermore, it is clear that the expected payoff to $ALLC$ against $ALLD$ is just $\phi(C, D)$ times the expected number of moves, or $a_{12} =$

$S\mu$; and that $a_{11} = a_{13} = a_{31} = a_{33} = R\mu$, because TFT always cooperates with $ALLC$. Thus the payoff matrix is

$$(5.19) \qquad A = \begin{bmatrix} R\mu & S\mu & R\mu \\ T\mu & P\mu & T + P(\mu - 1) \\ R\mu & S + P(\mu - 1) & R\mu \end{bmatrix}.$$

We see at once that each entry of the third row is at least as great as the corresponding entry of the first row and in one case strictly greater, provided only that $\mu > 1$; in other words, strategy 1 is (weakly) dominated by strategy 3. Let us therefore remove strategy 1 from the game (although see Exercise 5.27), and consider instead the IPD with only two (pure) strategies, $ALLD$ and TFT, with $ALLD$ re-defined as strategy 1 and TFT as strategy 2. The payoff matrix for this reduced IPD is

$$(5.20) \qquad A = \begin{bmatrix} P\mu & T + P(\mu - 1) \\ S + P(\mu - 1) & R\mu \end{bmatrix}.$$

Let us define

$$(5.21) \qquad \mu_c = \frac{T-P}{R-P}.$$

If $\mu < \mu_c$, then each entry of the first row in (5.20) exceeds the corresponding entry of the second row, so that $ALLD$ is a dominant strategy, and hence also a strong ESS (Exercise 2.21). If $\mu > \mu_c$, however, then both $a_{11} > a_{21}$ and $a_{22} > a_{12}$; whence, from (2.37), both $ALLD$ and TFT are evolutionarily stable strategies. Thus initial conditions will determine which strategy emerges as the winning strategy in a large population of animals, some of whom adopt $ALLD$, the remainder of whom adopt TFT. According to the dynamics of §2.6, for example, TFT will emerge as victorious if (on substituting from (5.20) into (2.44)) its initial frequency exceeds the critical value

$$(5.22) \qquad \gamma = \frac{P-S}{P-S+(R-P)(\mu-\mu_c)}.$$

The greater the amount by which μ exceeds μ_c, the easier it is to satisfy this condition. Thus maintenance of cooperation via reciprocity—specifically, via TFT—requires two things: first, that the average number of interactions be sufficiently high, or (which is essentially the same thing) that the probability of further interaction be sufficiently high; and second, that the initial proportion of TFT strategists be sufficiently large.

Until now, we have allowed M to be any (integer-valued) random variable with $\mu > 1$. Further progress is difficult, however, unless we specify the distribution of M. Accordingly, we assume henceforth that there is constant probability w of further interaction, which implies (Exercise 5.2) that M has a geometric distribution defined by

$$(5.23) \qquad \text{Prob}(M \geq k) \;=\; w^{k-1}, \; k \geq 1.$$

Then

$$(5.24) \qquad \mu = \tfrac{1}{1-w},$$

so that condition $\mu > \mu_c$ for TFT to be evolutionarily stable becomes

$$(5.25) \qquad w > \tfrac{T-R}{T-P},$$

on using (5.21).

Robert Axelrod [6, pp. 40-54] used the payoff matrix (5.1) with

$$(5.26) \qquad P = 1, \; R = 3, \; S = 0, \; T = 5$$

for a computer tournament. The game that was played in the second round of this tournament, which had 63 contestants, was the IPD. The strategies were TFT (submitted by Anatol Rapoport), $RNDM$ (defined in Exercise 5.10) and 61 other strategies submitted by contestants in six countries and from a variety of academic disciplines. The result? TFT was easily the most successful strategy.

Axelrod chose w so that the median number of moves in a game would be 200;[5] in other words, so that $\text{Prob}(M \leq 200) = \tfrac{1}{2}$, or $\text{Prob}(M \geq 201) = \tfrac{1}{2}$. So, from (5.23), Axelrod's value of w is defined by $w^{200} = \tfrac{1}{2}$. We will refer to the IPD defined by (5.26) and $w = (0.5)^{0.005} = 0.99654$ as Axelrod's prototype, and from time to time we will use his payoffs for illustration. Note, however, that they are purely arbitrary.

5.4. Other reciprocative strategies

Tit for tat is a *nice, forgiving* and *provocable* strategy based on reciprocity. It is nice, because it always begins by cooperating. It is forgiving, because if an opponent—after numerous defections—suddenly begins to cooperate, then TFT will cooperate on the following move.

---

[5] In the first round of the tournament, which had fewer contestants, the number of moves was fixed at 200 [6, pp. 30-40].

And TFT is provocable, because it always responds to a defection with a defection. By contrast, ALLD is a *nasty* strategy, that is, a strategy that always defects on the first move. For that matter, ALLD always defects, so it is the meanest strategy imaginable.

TFT is not, however, the only example of a nice, forgiving and provocable strategy based on reciprocity. A more forgiving nice strategy is $TF2T$, or tit for two tats, which always cooperates on the first two moves, but thereafter plays *TFT*; in other words,

$$(5.27) \qquad TF2T \;=\; (C, C, v_2, v_3, \dots),$$

where v denotes the other player's strategy. In a sense, $TF2T$ is one degree more forgiving than *TFT*. A homogeneous *TFT* population is indistinguishable from a homogeneous $TF2T$ population—or indeed any mixture of the two strategies—because both populations always cooperate; but the strategies are distinguishable in the presence of any nasty strategies, because $TF2T$ will forgive an initial defection, whereas *TFT* will punish it on the second move. In particular, if *TFT* were to play against $STCO = (D, C, C, ...)$, for "slow to cooperate," then mutual cooperation would be established only at the third move; whereas if $TF2T$ were to play against $STCO$, then mutual cooperation would be established at the second move. The question therefore arises: is *TFT* too mean? See Exercise 5.3.

In a sense, $STCO$ is the least exploitative of all the nasty strategies, and $ALLD$ is the most exploitative. In between lies $STFT$, for "suspicious tit for tat," which always defects on the first move, but thereafter plays *TFT*; i.e.,

$$(5.28) \qquad STFT \;=\; (D, v_1, v_2, \dots).$$

A homogeneous $ALLD$ population is indistinguishable from a homogeneous $STFT$ population (or any mixture of the two strategies), because both populations always defect; but the strategies are distinguishable in the presence of any nice strategies, because $STFT$ will reciprocate an initial cooperation, whereas $ALLD$ will exploit it. To see how *TFT* fares against these nasty strategies, let us compute the payoff matrix for the IPD with $ALLD$ (strategy 1), TFT (strategy 2) and $STFT$ (strategy 3). Of course, a_{11}, a_{12}, a_{21} and a_{22} are still defined by (5.20); moreover, because (5.10) and (5.28) imply that $STFT$ and

$ALLD$ always suffer mutual defection, we have $a_{13} = a_{31} = a_{33} = P\mu$. If, on the other hand, $u = TFT$ and $v = STFT$, then from (5.11) and (5.28) we have $\phi(u_1, v_1) = \phi(C, D) = S$, $\phi(u_2, v_2) = \phi(D, C) = T$, $\phi(u_3, v_3) = \phi(C, D) = S$, $\phi(u_4, v_4) = \phi(D, C) = T$, and so on; in other words, for all $j \geq 1$,

$$(5.29) \qquad \phi(u_{2j-1}, v_{2j-1}) = S, \quad \phi(u_{2j}, v_{2j}) = T.$$

TFT and $STFT$ are caught in an endless war of reprisal. From (5.14), (5.23) and (5.29), $a_{23} = f(TFT, STFT) = \sum_{k=1}^{\infty} \phi(u_k, v_k) w^{k-1} = \sum_{j=1}^{\infty} Sw^{2j-2} + \sum_{j=1}^{\infty} Tw^{2j-1} = (S + wT) \sum_{j=1}^{\infty} (w^2)^{j-1} = \frac{S+wT}{1-w^2} = \lambda_0 \mu(S + wT)$, where μ is defined by (5.24),

$$(5.30) \qquad \lambda_0 = \frac{1}{1+w}$$

and we have set $\beta = w^2$ in the standard formula

$$(5.31) \qquad \sum_{j=1}^{\infty} \beta^{j-1} = \frac{1}{1-\beta}, \qquad 0 \leq \beta < 1$$

for the sum of a geometric series. Similarly (Exercise 5.4), $a_{32} = \lambda_0 \mu(T + wS)$. Hence the payoff matrix is

$$(5.32) \qquad A = \begin{bmatrix} P\mu & T + Pw\mu & P\mu \\ S + Pw\mu & R\mu & \lambda_0(S + wT)\mu \\ P\mu & \lambda_0(T + wS)\mu & P\mu \end{bmatrix}.$$

Applying (2.37), we discover that neither $ALLD$ nor $STFT$ is evolutionarily stable. $ALLD$ is collectively stable, because $a_{11} \geq a_{j1}$ for $j = 2, 3$; but it is not evolutionarily stable, because neither $a_{11} > a_{31}$ nor $a_{13} > a_{33}$. $STFT$ is collectively stable if $a_{33} \geq a_{23}$, or

$$(5.33) \qquad w \leq \frac{P - S}{T - P};$$

but even then it is not evolutionarily stable, because neither $a_{33} > a_{13}$ nor $a_{31} > a_{11}$. On the other hand, (5.25) implies $a_{22} > a_{12}$; and if also $a_{22} > a_{32}$, that is, if $w > \frac{T-R}{R-S}$, then TFT is evolutionarily stable. In other words, TFT is the sole ESS of this IPD if

$$(5.34) \qquad w > \max\left(\frac{T-R}{T-P}, \frac{T-R}{R-S}\right).$$

When (5.34) is satisfied, a TFT population can be invaded by neither a small army of $STFT$-strategists nor a small army of $ALLD$-strategists—nor any combination of the two, because TFT is a strong ESS of the game defined by (5.32).[6]

Of course, (5.34) does not imply that TFT can invade an $ALLD$ population, because $ALLD$ is collectively stable (and evolutionarily stable without $STFT$). But if (5.33) does not hold, i.e., if $w > \frac{P-S}{T-P}$, then $STFT$ has no such resistance to TFT. In particular, if

$$(5.35) \qquad S + T \ \geq \ P + R,$$

then

$$(5.36) \qquad w \ > \ \frac{T - R}{R - S}$$

implies (5.34), which in turn implies $w > \frac{P-S}{T-P}$; so that, not only is a TFT population immune to invasion by $STFT$ (in the IPD defined by (5.32), but also a single TFT-strategist is enough to conquer an entire $STFT$ population. Note that (5.35) is satisfied with strict inequality by Axelrod's payoffs (5.26), and with equality by both the laboratory game of §5.1 and the foraging game of §5.2. Nevertheless, not every prisoner's dilemma satisfies (5.35); see §5.8 for an exception.

We have seen that if the probability of further interaction is sufficiently high, specifically, if (5.34) is satisfied, then TFT is uninvadable by the pair of exploitative strategies, $STFT$ and $ALLD$. But could a more forgiving strategy do just as well? The obvious way to answer this question is to compute the payoff matrix for the IPD with $ALLD$ (strategy 1), $STFT$ (strategy 2) and $TF2T$ (strategy 3). Because $STFT$ and $ALLD$ always suffer mutual defection, we now have $a_{11} = a_{12} = a_{21} = a_{22} = P\mu$; and because $TF2T$ always cooperates with itself, we have $a_{33} = R\mu$. From (5.10) and (5.27), if $u = ALLD$ and $v = TF2T$, then $\phi(u_1, v_1) = \phi(D, C) = T, \phi(u_2, v_2) = \phi(D, C) = T$ and $\phi(u_k, v_k) = \phi(D, D) = P$ for $k \geq 3$. Thus $a_{13} = f(ALLD, TF2T) = \sum_{k=1}^{\infty} \phi(u_k, v_k)w^{k-1} = T + wT + \sum_{k=3}^{\infty} Pw^{k-1} = T + wT + w^2P\sum_{j=1}^{\infty} w^{j-1} = (1+w)T + w^2P\mu$, on using (5.14), a change of summation index ($j = k-2$) and (5.31) with

---

[6]Axelrod [**6**] has proved that TFT is collectively stable (but not evolutionarily stable) against any deviant strategy (not just $ALLD$ or $STFT$) when w is greater than or equal to the right-hand side of (5.34).

$\beta = w$, μ being defined by (5.24). Continuing in this manner, we find (Exercise 5.4) that the payoff matrix is

(5.37) $A = \begin{bmatrix} P\mu & P\mu & (1+w)T + w^2 P\mu \\ P\mu & P\mu & T + Rw\mu \\ (1+w)S + w^2 P\mu & S + Rw\mu & R\mu \end{bmatrix}$.

We see at once that $a_{33} - a_{23} = R - T < 0$, so that $TF2T$ is invadable by $STFT$; whereas TFT is not invadable. Even if $STFT$ were absent, we see from (5.37) that $TF2T$ can withstand $ALLD$ only if $R\mu > (1+w)T + w^2 P\mu$ or

(5.38) $$w > \sqrt{\tfrac{T-R}{T-P}};$$

whereas (5.25) implies that TFT can withstand $ALLD$ if only $w > \frac{T-R}{T-P}$, which is more readily satisfied. With Axelrod's payoffs (5.26), for example, this means that TFT, but not $TF2T$, can withstand ALLD whenever $0.5 < w < 0.7071$. The conclusion is clear: $TF2T$ is too forgiving to persist as an orthodox strategy if infiltrated by $STFT$, and less resistant to invasion (in the sense of requiring higher w) than TFT when infiltrated by $ALLD$.

Although $STFT$ can invade $TF2T$, it does not necessarily eliminate it. To see this, let us model a population's long-term dynamics in terms of strategies according to (2.52), so that if $x_k(n)$ is the proportion adopting strategy k in generation n, then

(5.39) $$x_1(n) + x_2(n) + x_3(n) = 1,$$

and proportions evolve according to

(5.40a) $\overline{W(n)}\{x_1(n+1) - x_1(n)\} = x_1(n)x_2(n)\{W_1(n) - W_2(n)\}$
$\qquad\qquad + x_1(n)x_3(n)\{W_1(n) - W_3(n)\}$

(5.40b) $\overline{W(n)}\{x_2(n+1) - x_2(n)\} = x_2(n)x_1(n)\{W_2(n) - W_1(n)\}$
$\qquad\qquad + x_2(n)x_3(n)\{W_2(n) - W_3(n)\}$

(5.40c) $\overline{W(n)}\{x_3(n+1) - x_3(n)\} = x_3(n)x_1(n)\{W_3(n) - W_1(n)\}$
$\qquad\qquad + x_3(n)x_2(n)\{W_3(n) - W_2(n)\}$

where

(5.41) $$W_k(n) = a_{k1}x_1(n) + a_{k2}x_2(n) + a_{k3}x_3(n)$$

is the reward to strategy k in generation n and

$$(5.42) \qquad \overline{W(n)} = x_1(n)W_1(n) + x_2(n)W_2(n) + x_3(n)W_3(n)$$

is the average reward to the entire population (and is clearly positive). In view of (5.39), we can determine how the population evolves by following the point with coordinates $(x_1(n), x_2(n))$ in the triangle defined by the inequalities

$$(5.43) \qquad 0 \le x_1 \le 1, \ 0 \le x_2 \le 1, \ 0 \le x_1 + x_2 \le 1.$$

As in Figure 4.2, where (x_1, x_2, x_3) was an imputation, the upper boundary $x_1 + x_2 = 1$ of triangle (5.43) corresponds to $x_3 = 0$ and x_3 increases toward the southwest, with $x_3 = 1$ at the point $(0,0)$.

We obtain $W_1 - W_2 = \mu w x_3\{T - R - w(T - P)\}$, $W_1 - W_3 = \mu x_1(1 - w^2)(P - S) + \mu x_2\{P - S - w(R - S)\} + \mu x_3\{T - R - w^2(T - P)\}$ and $W_2 - W_3 = \mu x_1(1 - w^2)(P - S) + \mu x_2\{P - S - w(R - S)\} + \mu x_3(T - R)(1 - w)$ from (5.37) and (5.41). For the sake of simplicity, let us now choose Axelrod's payoffs (5.26). Then, on substituting into (5.40) and using (5.39), we have

$$(5.44) \qquad \overline{W}(n)\{x_k(n + 1) - x_k(n)\} = \phi_k(x_1(n), x_2(n))$$

for $k = 1, 2, 3$, where we define

$$(5.45a) \qquad \phi_1(x_1, x_2) = \mu x_1(1 - x_1 - x_2)(a + cx_1 - dx_2)$$

$$(5.45b) \qquad \phi_2(x_1, x_2) = \mu x_2(1 - x_1 - x_2)(b + cx_1 - dx_2)$$

$$(5.45c) \qquad \phi_3(x_1, x_2) = \mu(1 - x_1 - x_2)(x_1 + x_2)(dx_2 - cx_1)$$
$$- \mu(1 - x_1 - x_2)(ax_1 + bx_2)$$

and

$$(5.46) \qquad a = 2(1 - 2w^2), \ b = 2(1 - w), \ c = 3w^2 - 1, \ d = 1 + w.$$

Let us at least assume that $TF2T$ can withstand $ALLD$. Then, from (5.38), $w > 1/\sqrt{2}$ and $a < 0 < c$, $d > b > 0$.

It is convenient here to define parallel lines L_1, L_2 by

$$(5.47) \qquad L_1: \ cx_1 - dx_2 + a = 0, \qquad L_2: \ cx_1 - dx_2 + b = 0.$$

These lines are sketched in Figure 5.3(a), together with both branches of the hyperbola H defined by

$$(5.48) \qquad H: \ (cx_1 - dx_2)(x_1 + x_2) + ax_1 + bx_2 = 0.$$

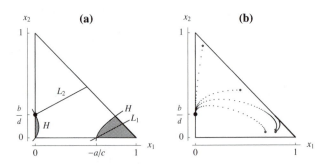

Figure 5.3. Convergence to equilibrium in the triangle (5.43) for the IPD with strategies $ALLD$ (frequency x_1), $STFT$ (frequency x_2) and $TF2T$ (frequency $1 - x_1 - x_2$). The figure is drawn for $w = 4/5$. **(a)** The domain of attraction of the equilibrium at $(0, b/d)$ contains the whole of the unshaded region and the shaded region near the origin of coordinates. **(b)** A sample of trajectories (with open circles for initial points). For example, $(x_1(n), x_2(n))$ converges to the locally stable equilibrium point $(0, 2/9)$ from $(x_1(0), x_2(0)) = (0.8, 0.07)$ but to the metastable equilibrium point $(0.828, 0.172)$ from $(x_1(0), x_2(0)) = (0.8, 0.05)$. Both of these initial points lie well within the shaded region to the southeast.

This hyperbola partitions triangle (5.43) into the shaded region of Figure 5.3(a), where $\phi_3 < 0$, and the unshaded region, where $\phi_3 > 0$ (Exercise 5.5). Because ϕ_1 is negative to the left of L_1 and positive to the right of L_1, from (5.44)-(5.45) we have $x_1(n+1) < x_1(n)$ when the point $(x_1(n), x_2(n))$ lies to the left of L_1, but $x_1(n+1) > x_1(n)$ when $(x_1(n), x_2(n))$ lies to the right of L_1. Similarly, $x_2(n+1) < x_2(n)$ when $(x_1(n), x_2(n))$ lies above L_2, whereas $x_2(n+1) > x_2(n)$ when $(x_1(n), x_2(n))$ lies below L_2; and $x_3(n+1) < x_3(n)$ or $x_3(n+1) > x_3(n)$ according to whether $(x_1(n), x_2(n))$ lies in the shaded or unshaded part of the triangle (5.43). Therefore, if $(x_1(0), x_2(0))$ lies either to the left of H or between its two branches, then $(x_1(n), x_2(n))$ converges to $(0, b/d)$ as $n \to \infty$; whereas, if $(x_1(0), x_2(0))$ lies sufficiently far to the right of H, then $(x_1(n), x_2(n))$ converges to the line $x_1 + x_2 = 1$ as $n \to \infty$. See Figure 5.3(b) for illustrations.

We say that (ξ_1, ξ_2) is an *equilibrium point* if $(x_1(n_0), x_2(n_0)) = (\xi_1, \xi_2)$ implies $(x_1(n), x_2(n)) = (\xi_1, \xi_2)$ for all $n \geq n_0$. Hence, from

(5.44), (ξ_1, ξ_2) is an equilibrium point if, and only if, $\phi_1(\xi_1, \xi_2) = 0 = \phi_2(\xi_1, \xi_2)$. Thus $(0, b/d)$, where H meets L_2 on the x_2-axis, is an equilibrium point (marked by a dot in Figure 5.3). There is also an equilibrium point, namely, $(-a/c, 0)$ on the x_1-axis. All other equilibrium points lie on the line $x_1 + x_2 = 1$. Indeed every point on the line segment that joins $(0, 1)$ to $(1, 0)$ is an equilibrium point. But those between $(0, 1)$ and $\left(\frac{d-b}{a-b+c+d}, \frac{a+c}{a-b+c+d} \right)$, where H intersects $x_1 + x_2 = 1$, are *unstable*, because the slightest increase in x_3 from 0 will displace $(x_1(n), x_2(n))$ into the unshaded region, where it begins a relentless march towards $(0, b/d)$. Similarly, $(-a/c, 0)$ is unstable, because the slightest leftward displacement of $(x_1(0), x_2(0))$ will again send $(x_1(n), x_2(n))$ to $(b/d, 0)$, and the slightest rightward displacement will send it towards the line $x_1 + x_2 = 1$. But $(b/d, 0)$ is a *locally stable* equilibrium point, because it attracts $(x_1(n), x_2(n))$ from any point $(x_1(0), x_2(0))$ in its vicinity—indeed, from any point to the left of H or between its two branches. By contrast, equilibria on $x_1 + x_2 = 1$ between $\left(\frac{d-b}{a-b+c+d}, \frac{a+c}{a-b+c+d} \right)$ and $(1, 0)$ are merely *metastable*, because a slight displacement of $(x_1(0), x_2(0))$ from (ξ_1, ξ_2) will neither return $(x_1(n), x_2(n))$ to (ξ_1, ξ_2), nor send it far away; rather, $(x_1(n), x_2(n))$ will shift to a neighbouring equilibrium on the line $x_1 + x_2 = 1$. All of these statements are readily verified by considering the signs of ϕ_1, ϕ_2 and ϕ_3 in the various regions of the triangle (5.43).

We can now confirm that, even if $STFT$ invades $TF2T$, it does not necessarily eliminate it. Observe that, in Figure 5.3, $(0, 0)$ corresponds to a homogeneous $TF2T$ population, $(1, 0)$ to a homogeneous $ALLD$ population and $(0, 1)$ to a homogeneous $STFT$ population. Infiltration of a $TF2T$ population by $STFT$ or $ALLD$ corresponds to displacing $(x_1(0), x_2(0))$ slightly from the origin, which will send $(x_1(n), x_2(n))$ to $(0, b/d)$ as $n \to \infty$. Thus $TF2T$ is not eliminated. Rather, $ALLD$ is eliminated, the proportion of $STFT$ increases to b/d, and the proportion of $TF2T$ decreases from 1 to $1 - b/d = (3w - 1)/(1 + w)$. It can be shown more generally (i.e., when payoffs other than Axelrod's are used) that, provided $w > (P - S)/(R - S)$, the final composition of the population will be a mixture of $STFT$

and $TF2T$ in which the proportion of $TF2T$ is

$$(5.49) \qquad x_3(\infty) = 1 - x_2(\infty) = \frac{w(R-S)-(P-S)}{(1-w)(T-R)+w(R-S)-(P-S)};$$

see Exercise 5.6.

Let us now restore TFT. In (5.32) it was strategy 2; whereas $ALLD$ was strategy 1, and $STFT$ was strategy 3. In (5.37), where TFT was absent, we promoted $STFT$ to strategy 2 and introduced $TF2T$ as strategy 3. With all four strategies together, it is convenient to demote $ALLD$ from 1 to 4 and relabel as follows:

$$(5.50) \qquad \begin{array}{llll} TFT: & \text{strategy 1} & STFT: & \text{strategy 3} \\ TF2T: & \text{strategy 2} & ALLD: & \text{strategy 4.} \end{array}$$

The advantage is that nice and nasty strategies are now adjacent, and from Exercise 5.7 the payoff matrix is

$$(5.51) \qquad A = \begin{bmatrix} R\mu & R\mu & \lambda_0\mu(S+wT) & S+P\mu w \\ R\mu & R\mu & S+Rw\mu & S(1+w)+w^2P\mu \\ \lambda_0\mu(T+wS) & T+Rw\mu & P\mu & P\mu \\ T+P\mu w & T(1+w)+w^2P\mu & P\mu & P\mu \end{bmatrix}$$

with μ as usual defined by (5.24) and λ_0 by (5.30). On applying (2.37), we see that $ALLD$ is collectively stable (for any w), and that TFT is collectively stable if (5.34) is satisfied. But neither strategy is evolutionarily stable: $ALLD$ is incapable of eliminating the other nasty strategy, namely, $STFT$; and TFT is incapable of eliminating the other nice strategy, namely, $TF2T$.

Indeed no strategy—that is, no pure strategy, because we have expressly forbidden mixed strategies—can resist invasion by an arbitrary mixture of infiltrators, even if w is sufficiently large.[7] Broadly speaking, the reason for this is that with any collectively stable strategy we can associate a second strategy that, although in principle quite distinct, is distinguishable in practice only in the presence of a third strategy; and if the second strategy does better than the first against the third, then the frequency of the second strategy can increase. To see this, let us first consider the IPD with strategies TFT,

---

[7]A formal proof is given by Boyd and Lorberbaum [**23**], who show that no (pure) strategy in the IPD can resist invasion if $w > \min(\{T-R\}/\{T-P\}, \{P-S\}/\{R-S\})$, hence if $w > 1/3$ with Axelrod's payoffs. But the result may have little practical significance; see §§5.9-5.10.

$TF2T$ and $STFT$, whose payoff matrix is

$$(5.52) \qquad A = \begin{bmatrix} R\mu & R\mu & \lambda_0\mu(S+wT) \\ R\mu & R\mu & S+Rw\mu \\ \lambda_0\mu(T+wS) & T+Rw\mu & P\mu \end{bmatrix},$$

obtained from (5.51) by deleting the final row and column; and let us assume that (5.34) is satisfied. Then TFT is collectively stable, and $w > \frac{T-R}{R-S}$ implies $a_{23} > a_{13}$, so that $TF2T$ is a second strategy, distinct from TFT, that does better against a third strategy, namely, $STFT$ (without which $TF2T$ would be indistinguishable from TFT). The long-term dynamics can again be described by (5.44); but because the strategies are different now, we have

$$(5.53a) \qquad \phi_1(x_1, x_2) = \mu x_1(1 - x_1 - x_2)(a - cx_1 - dx_2)$$

$$(5.53b) \qquad \phi_2(x_1, x_2) = \mu x_2(1 - x_1 - x_2)(b - cx_1 - dx_2)$$

$$(5.53c) \qquad \phi_3(x_1, x_2) = \mu(1 - x_1 - x_2)(x_1 + x_2)(cx_1 + dx_2)$$
$$- \mu(1 - x_1 - x_2)(ax_1 + bx_2)$$

in place of (5.45)-(5.46), where

$$(5.54) \quad \begin{array}{ll} a = \lambda_0\{(T-P)w - (P-S)\}, & b = w(R-S) - P + S \\ c = S+T-P-R, & d = c + (2R-S-T)w \end{array}$$

and (5.34) implies $d > b > a > c$ (Exercise 5.8).

We can again deduce the long-term dynamics by following the point with coordinates (x_1, x_2) in the triangle (5.43). Let us first suppose that (5.35) is satisfied, i.e., $S + T \geq P + R$, so that $d > b > a > c \geq 0$. If we re-define the parallel lines L_1 and L_2 by

$$(5.55) \qquad L_1: \ cx_1 + dx_2 = a, \qquad L_2: \ cx_1 + dx_2 = b$$

and the hyperbola H by

$$(5.56) \qquad H: \ (x_1 + x_2)(cx_1 + dx_2) = ax_1 + bx_2,$$

then in Figure 5.4(a) we have $\phi_1 > 0$ below L_1, $\phi_2 > 0$ below L_2 and $\phi_3 > 0$ above the upper branch of H, which intersects L_2 at the point $\left(0, \frac{b}{d}\right)$ but lies entirely above L_1. (The lower branch of H passes through the origin, but otherwise lies outside the triangle.)

The point $\left(0, \frac{b}{d}\right)$, where $x_3 = \frac{d-b}{d} > 0$, is again the only locally stable equilibrium. On $x_1 + x_2 = 1$ there exists, however, a whole line segment of metastable equilibria—and (x_1, x_2) may be attracted

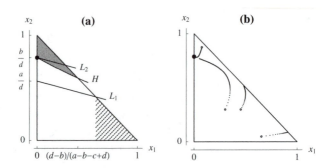

Figure 5.4. Convergence to equilibrium in the triangle (5.43) for the IPD with strategies TFT (frequency x_1), $TF2T$ (frequency x_2) and $STFT$ (frequency $1 - x_1 - x_2$). The figure is drawn for $T = 5, R = 3.5, P = 1, S = 0$ and $w = 0.725$, so that (5.57) reduces to $x_1(0) > 0.585$. **(a)** The point (x_1, x_2) is attracted to the line $x_1 + x_2 = 1$ in the hatched region, but to the equilibrium at $(0, b/d)$ in the shaded region. **(b)** A sample of trajectories (with open circles for initial points). For example, $(x_1(n), x_2(n))$ converges to the locally stable equilibrium point $(0, 0.788)$ from $(x_1(0), x_2(0)) = (0.3, 0.3)$ but to the metastable equilibrium point $(0.475, 0.525)$ from $(x_1(0), x_2(0)) = (0.45, 0.3)$.

to one of these, rather than to $\left(0, \frac{b}{d}\right)$. To see this, let us first suppose that $(x_1(0), x_2(0))$ lies in the smaller, hatched triangle with a vertex at $(1, 0)$. Then because $\phi_1 > 0, \phi_2 > 0, \phi_3 < 0$, (5.44) implies that the point $(x_1(n), x_2(n))$ will always move rightwards, upwards and toward the line $x_1 + x_2 = 1$. Thus $x_3(\infty) = 0$, and $x_1(\infty) + x_2(\infty) = 1$; furthermore, $x_2(\infty) > x_2(0)$, so that $TF2T$ invades. On the other hand, $STFT$ is eliminated; and $x_1(\infty) > x_1(0)$, so that the frequency of TFT is higher than initially. Indeed because L_1 intersects the line $x_1 + x_2 = 1$ where $x_1 = \frac{d-a}{d-c}$, we see that TFT is bound to increase in frequency whenever

(5.57) $$x_1(0) > \frac{(T-R)(1-w)+(2R-S-T)w^2}{(2R-S-T)w(1+w)}.$$

It is interesting to note that (5.57) approaches $x_1(0) > \frac{1}{2}$ as $w \to 1$. Bendor and Swistak [**14**, p. 3600] have shown that the "minimal stabilizing frequency" above which TFT can resist all possible invasions must exceed $\frac{1}{2}$ and approaches $\frac{1}{2}$ as $w \to 1$. (It is possible for the

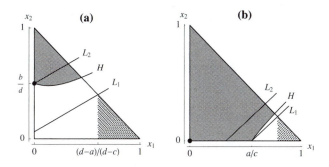

Figure 5.5. Domains of attraction in the triangle (5.43) for the IPD with strategies TFT (frequency x_1), $TF2T$ (frequency x_2) and $STFT$ (frequency $1 - x_1 - x_2$) when $P + R > S + T$. In the hatched region, (x_1, x_2) is attracted to the line $x_1 + x_2 = 1$. In the shaded region, (x_1, x_2) is attracted to the equilibrium point marked by a dot. Both figures are drawn for $T = 5, R = 3.5, P = 2$ and $S = 0$ **(a)** Here $w = 0.7$, implying $w > (P - S)/(T - P)$ and $d > b > a > 0 > c$. **(b)** Here $w = 0.525$, implying $(P - S)/(R - S) > w > (P + R - S - T)/(2R - S - T)$ and $d > 0 > b > a > c$.

right-hand side of (5.57) to be less than $\frac{1}{2}$; but [**14**] is not contradicted because, e.g., $ALLD$ is not a potential mutant in our model.)

If $(x_1(0), x_2(0))$ lies in the unshaded region of Figure 5.4(a), then $(x_1(n), x_2(n))$ may converge either to the line $x_1 + x_2 = 1$ or to the point $\left(0, \frac{b}{d}\right)$, as illustrated by Figure 5.4(b). If, however, $(x_1(0), x_2(0))$ lies in the shaded region of Figure 5.4(a), then convergence to $\left(0, \frac{b}{d}\right)$ and elimination of TFT are assured.

A similar analysis applies to the case where $P + R > S + T$. Then $\frac{P-S}{T-P} > \frac{P-S}{R-S} > \frac{T-R}{T-P} > \frac{T-R}{R-S}$ (Exercise 5.9), so that (5.34) implies $w > (T - R)/(T - P)$. Nevertheless, w can be either larger or smaller than each of $\frac{P-S}{T-P}, \frac{P-S}{R-S}$ and $\frac{P+R-S-T}{2R-S-T}$; and the third of these, though smaller than $(P - S)/(R - S)$, can be larger or smaller than $(T - R)/(T - P)$. So various cases are possible, and Figure 5.5 shows two. In Figure 5.5(a), where $\left(0, \frac{b}{d}\right)$ is the only stable equilibrium, we have $w > (P - S)/(T - P)$; whereas in Figure 5.5(b), where $(0, 0)$ is the only stable equilibrium, $\frac{P-S}{R-S} > w > \frac{P+R-S-T}{2R-S-T}$. The smaller the value of w, the larger the shaded region in which (x_1, x_2) is guaranteed to be attracted to $x_1 = 0$. Nevertheless, the hatched

Table 5.2. Numerical solution of (2.52) with TFT as strategy 1, $TF2T$ as strategy 2, $STFT$ as strategy 3 and $ALLD$ as strategy 4. Initially, the proportion of TFT is 0.7 and the other strategies are equally represented.

n	$x_1(n)$	$x_2(n)$	$x_3(n)$	$x_4(n)$
0	0.700	0.100	0.100	0.100
1	0.791	0.119	0.064	0.026
2	0.827	0.129	0.038	0.006
3	0.843	0.134	0.022	0.001
4	0.851	0.137	0.012	0.000
5	0.855	0.138	0.007	0.000
6	0.857	0.139	0.004	0.000
7	0.858	0.140	0.002	0.000
8	0.859	0.140	0.001	0.000
9	0.859	0.140	0.001	0.000
10	0.859	0.140	0.000	0.000
11	0.860	0.140	0.000	0.000

region, where TFT is bound to increase in frequency, is still finite; and if the whole population were using TFT just before infiltration by $STFT$ and $TF2T$ at time $n = 0$, then $(x_1(0), x_2(0))$ would have to lie close to $(1,0)$ and hence within the hatched region. Of course, a fresh infiltration of $STFT$ would displace (x_1, x_2) from the line $x_1 + x_2 = 1$, slightly toward the southwest. The population would subsequently evolve to a new metastable equilibrium, still on the line $x_1 + x_2 = 1$ but nearer to $(0,1)$. In the course of time, repeated infiltrations by $STFT$ could nudge (x_1, x_2) all the way along the line $x_1 + x_2 = 1$, slowly decreasing the fraction of TFT, so that (x_1, x_2) would eventually enter the shaded region, to be attracted by $x_1 = 0$. At that stage, TFT would have been eliminated (and also $TF2T$ in Figure 5.5(b)). But whether this would actually happen is far beyond the scope of our dynamic model, namely, (5.44).

The dynamics are more difficult, both to analyze and to visualize, when $ALLD$ is present also. But it is easy to solve (2.52) numerically, and Table 5.2 shows a sample calculation for Axelrod's prototype when $x_1(0) = 0.7$ and the other three strategies are equally represented initially. Notice that the two nasty strategies are summarily dispatched; and because the two nice strategies are incapable of eliminating one another, the final composition of the population is 86% TFT and 14% $TF2T$.

Two remarks are now in order. First, as we have said before, insofar as no strategy can resist invasion by every conceivable combination of deviant strategies, the IPD has no evolutionarily stable strategy. But this does not mean that TFT cannot persist as an orthodox IPD strategy, for the simple reason that strategies capable of displacing it may be either absent or too rare. As we have remarked already in §2.7 (p. 91), whether a strategy is evolutionarily stable will always depend upon the scope of the strategy set. By suitably enlarging it, most of the strategies that have ever been claimed as ESSes for one game or another could almost certainly be destabilized. For example, if in Crossroads we allowed a third strategy, Z, which stands for "instantly convert your car into a motor cycle, so that you can proceed at once regardless of what the other motorist does, and zoom," then clearly Z would be a dominant strategy, and would replace §2.3's ESS as the solution of the game. But not every motorist drives Jane Bond's car.[8] Likewise, $STFT$ might not emerge as a deviant strategy sufficiently often to sustain the process for eliminating TFT that we described in the last paragraph but one.

Second, Axelrod [**6**, pp. 48-52] has marshalled impressive empirical support for the persistence of TFT as an orthodox strategy by conducting an "ecological" simulation of the IPD in a population of computer programs submitted to his tournament; see §5.3. At first, all programs were represented; subsequently, he allowed the composition of the population to evolve, essentially according to the dynamics of §2.6, and found that the frequency of TFT was always higher than that of any other strategy and increased steadily.[9] Axelrod's results have since been rationalized by Bendor and Swistak [**14, 15**], who show that there is a sense in which TFT and other nice, retaliatory strategies are "maximally robust."

In the light of these remarks, we should interpret (5.34) as a necessary condition for TFT to be uninvadable. If this condition is

---

[8] Jane Bond is female Agent 007.

[9] TFT was also the champion of the computer tournament itself; but this was a round-robin tournament in which every strategy played every other strategy once.

satisfied, then TFT may persist as an orthodox strategy, simply because a mixture of deviant strategies that could invade it may not be represented. If (5.34) is not satisfied, however, then there is little hope that TFT will persist, because deviant strategies that can eliminate it are so simple as virtually to be guaranteed to arise.

Nevertheless, there are circumstances in which (5.34) is inappropriate even as a necessary condition, because—as we shall demonstrate in §5.6—it is based on two tacit assumptions, namely, that the population is very large; and that opponents, though randomly selected at the beginning of the game, are retained for its duration. But even if the players are all drawn at random from a large population, the sub-population with which they interact may not be large; and opponents may be drawn at random from this (sub-)population throughout the game—which would happen, for example, if §5.1's laboratory game were played repeatedly by randomly chosen pairs of trained animals. Let us therefore relax both assumptions.

5.5. Dynamic versus static interaction

Given TFT's success in Axelrod's computer tournament and the remarks at the end of the previous section, we would like to obtain some theoretical insights into the resilience of this strategy in a finite population. In this section, we shall consider two different modes of interaction: first, a mode in which players select their opponents at random, but retain them for the duration of the game; and second, a mode in which opponents are drawn randomly throughout the game.

Consider therefore a population of $N + 1$ individuals, of whom N_k play strategy k in the IPD. Thus, with m strategies, we have

$$(5.58) \qquad N_1 + N_2 + \ldots + N_m = N + 1.$$

Because opponents are not necessarily fixed, it no longer makes sense to construct a payoff matrix for the entire game; instead, we construct a payoff matrix for each move, and use it to derive expressions for the expected payoffs to the various strategies against all possible opponents. Let $\phi(k)$ denote this $m \times m$ matrix, i.e., for $1 \leq j, l \leq m$, let $\phi_{jl}(k)$ be a j-strategist's payoff from interaction k if the interaction is with an l-strategist; and, for $j = 1, 2, ..., m$, let W_j denote a j-strategist's expected payoff from the game. Then, in place of (5.13), the reward from move k of the game—conditional upon encountering an l-strategist—is

$$\phi_{jl}(k)\,\mathrm{Prob}(M \geq k), \tag{5.59}$$

where M as usual is the number of interactions.

In the case where opponents are drawn at random throughout, let π_{jl} denote the probability that a j-strategist's next interaction is with an l-strategist (assumed the same on every move). Then the (unconditional) reward from move k of the game is

$$\sum_{l=1}^{m} \pi_{jl}\,\phi_{jl}(k)\,\mathrm{Prob}(M \geq k), \tag{5.60}$$

because the opponent uses strategy l at move k with probability π_{jl}; and the reward from the entire game is

$$W_j = \sum_{k=1}^{\infty} \left\{ \sum_{l=1}^{m} \pi_{jl}\phi_{jl}(k) \right\} \mathrm{Prob}(M \geq k), \quad 1 \leq j \leq m. \tag{5.61}$$

On the other hand, in the case where players interact with fixed opponents, for some value of l a j-strategist's opponent is known to be an l-strategist from the first move onwards; whence (5.59) implies that the reward from the entire game, assessed at the first move, is

$$\sum_{k=1}^{\infty} \phi_{jl}(k)\,\mathrm{Prob}(M \geq k), \quad 1 \leq j,l \leq m. \tag{5.62}$$

Before the first move the identity of l is unknown, however; and so to obtain the reward we must average this expression over the distribution of possible opponents, which yields

$$W_j = \sum_{l=1}^{m} \pi_{jl} \left\{ \sum_{k=1}^{\infty} \phi_{jl}(k)\,\mathrm{Prob}(M \geq k) \right\}, \quad 1 \leq j \leq m. \tag{5.63}$$

Clearly, the same expression for W_j results in either case.

Let us now define $x_j = N_j/N$ for $j = 1,\dots,m$ and $\alpha = 1/N$, so that (5.58) implies

$$x_1 + x_2 + \dots + x_m = 1 + \alpha. \tag{5.64}$$

Then, because the probability that a j-strategist interacts with an l-strategist is N_l/N if $l \neq j$ but $(N_j - 1)/N$ if $l = j$, the probabilities of interaction are defined by

$$\pi_{jl} = x_l - \alpha\delta_{jl}, \quad 1 \leq j,l \leq m \tag{5.65}$$

where we define $\delta_{jl} = 0$ if $j \neq l$ but $\delta_{jj} = 1$; and, from (5.23),

$$(5.66) \qquad W_j = \sum_{k=1}^{\infty} w^{k-1} \sum_{l=1}^{m} (x_l - \alpha\delta_{jl})\phi_{jl}(k), \quad 1 \leq j,l \leq m.$$

Further progress requires an explicit expression for the $m \times m$ matrix $\phi(k)$. Let us therefore choose $m = 4$ in (5.66) and define the strategies as in (5.50). Then, because TFT or $TF2T$ always cooperate with one another, the payoff to TFT or $TF2T$ against TFT or $TF2T$ at interaction k is independent of k:

$$(5.67a) \qquad \phi_{11}(k) = \phi_{12}(k) = \phi_{21}(k) = \phi_{22}(k) = R, \quad 1 \leq k < \infty.$$

Similarly, because $STFT$ or $ALLD$ always defect against one another,

$$(5.67b) \qquad \phi_{33}(k) = \phi_{34}(k) = \phi_{43}(k) = \phi_{33}(k) = P, \quad 1 \leq k < \infty.$$

For all other values of j and l, however, $\phi_{jl}(k)$ depends on k.

Let σ_k denote the probability that an individual's opponent at move k has been encountered before, let γ_k denote the probability that the opponent has previously been encountered more than once, and let ϵ_k denote the probability that the opponent has previously been encountered an even number of times (hence $1 - \epsilon_k$ the probability that the opponent has been encountered an odd number of times). Then, because TFT cooperates with $STFT$ on odd-numbered encounters (when the opponent defects) but defects against $STFT$ on even-numbered ones (when the opponent cooperates), we have

$$(5.68a) \qquad \phi_{13}(k) = \epsilon_k S + (1 - \epsilon_k)T, \quad 1 \leq k < \infty.$$

Similarly,

$$(5.68b) \qquad \phi_{31}(k) = \epsilon_k T + (1 - \epsilon_k)S, \quad 1 \leq k < \infty.$$

Because TFT cooperates with $ALLD$ on a first encounter but thereafter defects, we have

$$(5.68c) \qquad \phi_{14}(k) = \sigma_k P + (1 - \sigma_k)S, \quad 1 \leq k < \infty.$$

Similarly,

$$(5.68d) \qquad \phi_{23}(k) = \sigma_k R + (1 - \sigma_k)S, \quad 1 \leq k < \infty$$

$$(5.68e) \qquad \phi_{32}(k) = \sigma_k R + (1 - \sigma_k)T, \quad 1 \leq k < \infty$$

$$(5.68f) \qquad \phi_{41}(k) = \sigma_k P + (1 - \sigma_k)T, \quad 1 \leq k < \infty.$$

Again, because $TF2T$ cooperates against $ALLD$ on a first or second encounter, but thereafter defects, we have

(5.68g) $$\phi_{24}(k) = \gamma_k P + (1 - \gamma_k)S, \quad 1 \le k < \infty$$

(5.68h) $$\phi_{42}(k) = \gamma_k P + (1 - \gamma_k)T, \quad 1 \le k < \infty.$$

The matrix $\phi(k)$ is now defined, and from (5.64)-(5.68) we obtain:

(5.69a) $$W_1 = \sum_{k=1}^{\infty} w^{k-1} \{(x_1 - \alpha + x_2)R + x_3(\epsilon_k S + \{1 - \epsilon_k\}T)$$
$$+ x_4(\sigma_k P + \{1 - \sigma_k\}S)\}$$

(5.69b) $$W_2 = \sum_{k=1}^{\infty} w^{k-1} \{(x_1 + x_2 - \alpha)R + x_3(\sigma_k R + \{1 - \sigma_k\}S)$$
$$+ x_4(\gamma_k P + \{1 - \gamma_k\}S)\}$$

(5.69c) $$W_3 = \sum_{k=1}^{\infty} w^{k-1} \{x_1(\epsilon_k T + \{1 - \epsilon_k\}S)$$
$$+ x_2(\sigma_k R + \{1 - \sigma_k\}T) + (x_3 - \alpha + x_4)P\}$$

(5.69d) $$W_4 = \sum_{k=1}^{\infty} w^{k-1} \{x_1(\sigma_k P + \{1 - \sigma_k\}T)$$
$$+ x_2(\gamma_k P + \{1 - \gamma_k\}T) + (x_3 + x_4 - \alpha)P\}.$$

If players keep the same opponent throughout the game, then

(5.70a) $$\sigma_1 = 0; \quad \sigma_k = 1 \text{ if } k \ge 2$$

(5.70b) $$\gamma_1 = 0 = \gamma_2; \quad \gamma_k = 1 \text{ if } k \ge 3$$

(5.70c) $$\epsilon_k = \tfrac{1}{2}\left\{1 - (-1)^k\right\}, \quad k \ge 1.$$

We shall say in this case that the interaction is *static*. Explicit expressions for $W_1, ..., W_4$ follow readily from (5.68)-(5.70). On using (5.31) and the related formula

(5.71) $$\sum_{j=1}^{\infty} j\beta^{j-1} = \frac{1}{(1 - \beta)^2}, \quad 0 \le \beta < 1$$

for the derivative of a geometric series, we obtain (Exercise 5.11):

(5.72a) $$W_1 = \mu(x_1 + x_2 - \alpha)R + \lambda_0 \mu x_3(S + wT)$$
$$+ x_4(S + \mu wP)$$

(5.72b) $$W_2 = \mu(x_1 + x_2 - \alpha)R + x_3(S + \mu wR)$$
$$+ x_4\{S(1 + w) + \mu w^2 P\}$$

(5.72c) $\quad W_3 \;=\; \lambda_0 \mu x_1 (T + wS) + x_2 (T + \mu w R)$
$$+ \mu(x_3 + x_4 - \alpha)P$$

(5.72d) $\quad W_4 \;=\; x_1 (T + \mu w P) + x_2 \{T(1 + w) + \mu w^2 P\}$
$$+ \mu(x_3 + x_4 - \alpha)P$$

where μ is defined by (5.24) and λ_0 by (5.30). Note that the difference in payoffs between the two nice strategies TFT and $TF2T$ is

(5.73) $\quad W_1 - W_2 = \lambda_0 \mu x_3 w \{T - R - w(R - S)\} + x_4 w(P - S),$

which vanishes if $x_3 = 0 = x_4$. Because the coefficient of x_4 is positive when $x_4 \neq 0$, TFT always does better than $TF2T$ against $ALLD$. Similarly, $TF2T$ does better than TFT against $STFT$ when $w > \frac{T-R}{R-S}$, because the coefficient of x_3 is then negative.

If, on the other hand, opponents are drawn at random throughout the game, then we shall call the interaction *dynamic*. Because the probability per move of meeting any specific opponent is $\alpha = \frac{1}{N}$, the probability that the opponent at move k has not been encountered during the previous $k - 1$ moves is $(1 - \alpha)^{k-1}$, whence

(5.74a) $\qquad\qquad \sigma_k = 1 - (1 - \alpha)^{k-1}, \;\; k \geq 1.$

Similarly, the probability of precisely one encounter with the opponent during the previous $k - 1$ moves is $(k - 1)\alpha(1 - \alpha)^{k-2}$, whence

(5.74b) $\qquad \gamma_k = 1 - (1 - \alpha)^{k-1} - (k - 1)\alpha(1 - \alpha)^{k-2}, \;\; k \geq 1.$

Because zero is an even number, $\epsilon_1 = 1$. Moreover, the number of previous encounters with an opponent is even at move k if either it was even at move $k - 1$ and the opponent was then different, or it was odd at move $k - 1$ and the opponent was then the same. Thus $\epsilon_k = (1 - \alpha)\epsilon_{k-1} + \alpha(1 - \epsilon_{k-1})$ for $k \geq 2$. The solution of this recurrence equation subject to $\epsilon_1 = 1$ (Exercise 5.11) is

(5.74c) $\qquad\qquad \epsilon_k = \frac{1}{2} \left\{ 1 + (1 - 2\alpha)^{k-1} \right\}, \;\; k \geq 1.$

Explicit expressions for $W_1, ..., W_4$ now follow readily from (5.68), (5.69) and (5.74). On using (5.31) and (5.71), we obtain

(5.75a) $\quad W_1 \;=\; \mu(x_1 + x_2 - \alpha)R + \frac{1}{2}x_3\{\mu(S + T) + \lambda_2(S - T)\}$
$$+ \; x_4\{\mu P + \lambda_1(S - P)\}$$

(5.75b) $W_2 = \mu(1 - x_4)R + \lambda_1 x_3(S - R)$
$$+ \ x_4\big\{\mu P + \tfrac{\lambda_1^2(S-P)}{\lambda_2}\big\}$$

(5.75c) $W_3 = \tfrac{1}{2}x_1\{\mu(S + T) + \lambda_2(T - S)\} + x_2\{\mu R + \lambda_1(T - R)\}$
$$+ \ \mu(x_3 + x_4 - \alpha)P$$

(5.75d) $W_4 = x_1\{\mu P + \lambda_1(T - P)\} + x_2\big\{\mu P + \tfrac{\lambda_1^2(T-P)}{\lambda_2}\big\}$
$$+ \ \mu(x_3 + x_4 - \alpha)P,$$

where we define

$$(5.76) \qquad \lambda_1 \ = \ \tfrac{1}{1-w+\alpha w}, \quad \lambda_2 \ = \ \tfrac{1}{1-w+2\alpha w}$$

(so that $\mu > \lambda_1 > \lambda_2 > \lambda_0$). Note that if either (2.47) or (2.52) with $m = 4$ is used for long-term dynamics, then either (2.45) or (2.50) must be replaced by (5.75) under dynamic interaction and by (5.72) under static interaction. Note also that (5.75) reduces to (5.72) when $N = 1$: we cannot distinguish between static and dynamic interaction when there are only two individuals (Exercise 5.12).

We will apply this model in the following two sections. First, however, you should attempt Exercises 5.13 and 5.22.

5.6. Stability of a nice population: static case

In this section, we consider the stability of a nice population under static interaction. Given our remarks at the end of §5.4, our goal will be to seek necessary conditions for population stability—and if a strategy is to be stable against all possible deviation, then it must at least be stable against pure infiltration by a single nasty player.

We begin by considering the stability of TFT against $ALLD$. Thus $N_2 = N_3 = 0$ (hence $x_2 = 0 = x_3$), and (5.72) implies

$$(5.77) \qquad W_1 = \mu(x_1 - \alpha)R + x_4(S + \mu wP)$$
$$W_4 = x_1(T + \mu wP) + \mu(x_4 - \alpha)P.$$

For a TFT population to be stable against pure infiltration by $ALLD$, it must at least be true that W_1 exceeds W_4 when $N_1 = N$ and $N_4 = 1$. Accordingly, we set $x_1 = 1$ and $x_4 = \alpha$ in (5.77) to obtain

$$(5.78) \quad W_1 - W_4 = \mu(R - P) - (T - P) - \alpha\{\mu(R - P) + P - S\}.$$

Straightforward algebraic manipulation now shows that this expression is positive when

$$(5.79) \qquad R - P + (1 - w)(P - S) + N\{T - R - w(T - P)\} < 0,$$

which requires both

$$(5.80) \qquad\qquad\qquad w > \frac{T-R}{T-P}$$

for a negative coefficient of N in (5.79) and

$$(5.81) \qquad\qquad\qquad N > \frac{R-P+(1-w)(P-S)}{w(T-P)-(T-R)}.$$

Next we consider the stability of TFT against $STFT$. Now $N_2 = 0 = N_4$ (hence $x_2 = 0 = x_4$), and (5.72) implies

$$(5.82) \qquad\qquad W_1 = \mu(x_1 - \alpha)R + \lambda_0\mu x_3(S + wT)$$
$$W_3 = \lambda_0\mu x_1(T + wS) + \mu(x_3 - \alpha)P,$$

where λ_0 is defined by (5.30). It follows from (5.82) that

$$(5.83) \quad W_1 - W_3 = \mu\{R + \alpha P - \lambda_0(1 + \alpha)(T + wS)\}$$
$$+ \mu(S + T - P - R)x_3.$$

For a TFT population to be stable against pure infiltration by $STFT$, it must at least be true that W_1 exceeds W_3 when $N_1 = N$ and $N_3 = 1$. Accordingly, we set $x_1 = 1$ and $x_3 = \alpha$ in (5.83). Then

$$(5.84) \quad W_1 - W_3 = \lambda_0\mu \{w(R - S) - (T - R)$$
$$- \alpha((1 - w)(T - R) + 2R - S - T)\}$$

is positive for

$$(5.85) \qquad w > \frac{T-R}{R-S}, \quad N > \frac{(1-w)(T-R)+2R-S-T}{w(R-S)-(T-R)}.$$

For sufficiently large N, (5.80)-(5.81) and (5.85) reduce to (5.34). Thus §5.3 corresponds to static interaction in a large population.

A similar analysis can be applied to the other nice strategy, $TF2T$. For $TF2T$ to be stable against pure infiltration by $ALLD$, we require (Exercise 5.15)

$$(5.86) \qquad w > \sqrt{\frac{T-R}{T-P}}, \quad N > \frac{R-P+(1-w^2)(P-S)}{w^2(T-P)-(T-R)},$$

which agrees with (5.38) when N is sufficiently large. But no value of w is large enough to make $TF2T$ stable against $STFT$. To see this, we set $x_1 = 0 = x_4$ in (5.72), obtaining $W_2 = \mu(x_2 - \alpha)R + x_3(S + \mu wR)$

and $W_3 = x_2(T + \mu w R) + \mu(x_3 - \alpha)P$. For $x_2 = 1$ and $x_3 = \alpha$, $W_2 - W_3 = R - T - \alpha(R - S)$ is always negative. Thus the frequency of $STFT$ will always increase; $TF2T$ is too nice a strategy to resist invasion. On the other hand, $TF2T$ need not be eliminated: for $x_2 = \alpha$ and $x_3 = 1$, $W_2 - W_3 = \mu(R-P) - R + S - \alpha\{T - R + \mu(R-P)\} > 0$ if $w > \frac{P-S}{R-S}$ and $N > \frac{(1-w)(T-R)+R-P}{w(R-S)-(P-S)}$. Assuming both conditions hold, let us define x_{es} by $N\{(1-w)(T-R) + w(R-S) - P + S\}x_{es} = \{w(R-S) - P + S\}N - (1-w)(R-S)$. Then, because $x_2 + x_3 = 1 + \alpha$, $W_2 - W_3$ is negative if $x_2 > x_{es}$ but positive if $x_2 < x_{es}$, so that the population will reach equilibrium at a mixture of $TF2T$ and $STFT$ in which the proportion of $TF2T$ is $x_{es}/(1+\alpha)$. In the limit as $N \to \infty$, of course, x_{es} approaches (5.49).

If $R - S > T - P$, then there is a range of values of w, specifically, $(T-R)/(T-P) < w < \sqrt{(T-R)/(T-P)}$, for which TFT is stable against pure infiltration by $STFT$ or $ALLD$, whereas $TF2T$ is not. Of course, $T-P$ may exceed $R-S$; but if also $(T-R)(T+R-S-P) < 2(R-S)^2$, then there is still a range, namely, $(T-R)/(R-S) < w < \sqrt{2(T-R)/(T+R-S-P)}$, for which TFT is stable against pure infiltration, whereas $TF2T$ is not. These conditions are satisfied, for example, by Axelrod's prototype.

We more or less knew all this in §5.4, and it could be argued that our finite-N analysis has done little more than recover the results of that section. Things are very different, however, when opponents are drawn at random throughout the game—or, as we have chosen to say, when the interaction is dynamic.

5.7. Stability of a nice population: dynamic case

We now consider the stability of a nice population under dynamic interaction. Some mathematical preliminaries will facilitate analysis, both here and in Exercises 5.20-5.21. Accordingly, let us define quadratic polynomials $\zeta_1, \zeta_2, \zeta_3$ and $\Delta_1, \Delta_2, \Delta_3$ by

$$(5.87) \qquad \zeta_j(N) = (1-w)(T-R)N^2 + (R - S - wa_j)N \\ + w(a_j - R + S)$$

$$(5.88) \qquad \Delta_j(w) = (R - S - wa_j)^2 \\ - 4w(1-w)(T-R)(a_j - R + S)$$

for $j = 1, 2, 3$ and

(5.89) $a_1 = 2R - P - S$, $a_2 = 3R - 2S - T$, $a_3 = \frac{5R - 3S - T - P}{2}$

so that (5.2)-(5.3) imply

(5.90) $a_j > R - S$, $\quad j = 1, 2, 3$.

Then because $\Delta_j(0) > 0$, $\Delta_j((R - S)/a_j) < 0$ and $\Delta_j(1) > 0$, the smaller and larger roots of $\Delta_j(w) = 0$ satisfy $0 < w < (R - S)/a_j$ and $(R - S)/a_j < w < 1$, respectively. Let ξ_j denote the larger root:

(5.91) $\xi_j = \frac{2(a_j - R + S)\sqrt{(T - R)(T - S)} + a_j(R - S) + 2(T - R)(a_j - R + S)}{a_j^2 + 4(T - R)(a_j - R + S)}$

for $j = 1, 2, 3$. Whenever $\xi_j < w < 1$, $\Delta_j(w)$ is positive; and so the equation $\zeta_j(N) = 0$ has two real roots, both of which are positive, by (5.90). Let the greater of these roots be denoted by $U_j(w)$. Then, for $j = 1, 2, 3$, we have established that $\zeta_j(N) < 0$ if

(5.92a) $\xi_j < w < 1$

(5.92b) $\frac{w(a_j - R + S)}{(1 - w)(T - R)U_j(w)} < N < U_j(w)$,

where

(5.93) $U_j(w) = \frac{wa_j - R + S + \sqrt{\Delta_j(w)}}{2(1 - w)(T - R)}$.

The lower bound in (5.92b) is the smaller root of $\zeta_j(N) = 0$. After these preliminaries, we can now proceed.

First we consider the stability of TFT against pure infiltration by $ALLD$: $N_2 = 0 = N_3$, so that (5.75) implies $W_1 = \mu(x_1 - \alpha)R + x_4\{\mu P + \lambda_1(S - P)\}$ and $W_4 = x_1\{\mu P + \lambda_1(T - P)\} + \mu(x_4 - \alpha)P$. Proceeding as in §5.7, we require for stability that $W_1 > W_4$ when $N_1 = N$ and $N_4 = 1$. On setting $x_1 = 1$ and $x_4 = \alpha$ in our expressions for W_1 and W_4, the above condition reduces (Exercise 5.17) to

(5.94) $\zeta_1(N) < 0$.

Hence necessary conditions for the stability of TFT are

(5.95a) $w > \xi_1$

(5.95b) $\frac{w(R - P)}{(1 - w)(T - R)U_1(w)} < N < U_1(w)$,

where ξ_1 is defined by (5.91) and U_1 by (5.93). For example, in Axelrod's prototype we have $\xi_1 = 0.869$ and $U_1(w) = 285.5$; the lower bound on N in (5.95b) is then $N > 1.009$, and satisfied if only $N \geq 2$. Thus we require $3 \leq N + 1 \leq 286$.

Generally, the lower bound in (5.92b) is rather trivially satisfied by $N \geq 2$ unless w is very close to ξ_j; in which case, the range of values admitted by (5.92b) is too small to be of interest. For example, with Axelrod's payoffs $P = 1, R = 3, S = 0$ and $T = 5$ but $w = 0.87$, (5.95b) requires $2.38 < N < 2.81$, which is not satisfied by any integer value of N. Thus our interest lies principally in the upper bound, U_j.

Next we consider the stability of TFT against pure infiltration by $STFT$. With $x_2 = 0 = x_4$, (5.75) implies that $W_1 = \mu(x_1 - \alpha)R + \frac{1}{2}x_3\{\mu(T + S) - \lambda_2(T - S)\}$ and $W_3 = \frac{1}{2}x_1\{\mu(T + S) + \lambda_2(T - S)\} + \mu(x_3 - \alpha)P$. We require for stability that $W_1 > W_3$ when $N_1 = N$ and $N_3 = 1$; in other words (Exercise 5.17), that

$$(5.96) \qquad \qquad \zeta_2(N) < 0.$$

Thus TFT is stable against pure infiltration by $STFT$ if

$$(5.97a) \qquad \qquad w > \xi_2$$

$$(5.97b) \qquad \qquad \frac{w(2R - S - T)}{(1 - w)(T - R)U_2(w)} < N < U_2(w),$$

where ξ_2 is defined by (5.91) and U_2 by (5.93). For example, in Axelrod's prototype we have $\xi_2 = 0.930$ and $U_2(w) = 141.5$; the lower bound on N in (5.97b) is then $N > 1.018$, and so resistance to $STFT$ requires $3 \leq N + 1 \leq 142$.

A similar analysis applies to the other nice strategy, $TF2T$. In a population of $TF2T$ and $ALLD$, $N_1 = 0 = N_3, W_2 = \mu(x_2 - \alpha)R + x_4\{\mu P - \lambda_1^2(P - S)/\lambda_2\}$ and $W_4 = x_2\{\mu P + \lambda_1^2(T - P)/\lambda_2\} + \mu(x_4 - \alpha)P$. For $TF2T$ to be stable, we require $W_2 > W_4$ when $x_2 = 1$ and $x_4 = \alpha$ or $(1 - w)^2(T - R)N^3 + (1 - w)\{2(T - R)w + (1 - w)(R - S)\}N^2 + w\{2(1 - w)(R - S) - w(R - P)\}N + w^2(R - P) < 0$, which requires in particular that $w > \frac{2(R - S)}{3R - 2S - P}$. As we would expect, the condition is more difficult to satisfy than (5.94)—yielding, for example, $3 \leq N + 1 \leq 16$ in Axelrod's prototype.

If $TF2T$ is still the orthodox strategy but $STFT$ replaces $ALLD$ as the deviant strategy, so that $N_2 = 0 = N_4$, then (5.75) implies $W_2 = \mu R - \lambda_1 x_3(R - S)$, $W_3 = x_2\{\mu R + \lambda_1(T - R)\} + \mu(x_3 - \alpha)P$; and $x_2 = 1, x_3 = \alpha$ yields $W_2 - W_3 = -\lambda_1\{T - R + \alpha(R - S)\} < 0$. Thus $STFT$ increases in number; regardless of whether interaction is static or dynamic, $TF2T$ is too nice a strategy to resist invasion by $STFT$. On the other hand, $TF2T$ need not be eliminated: for $x_2 = \alpha$ and $x_3 = 1$, $W_2 - W_3 = \lambda_0(R-P) - \lambda_1(R-S) - \alpha\{\lambda_0(R-P) + \lambda_1(T-R)\}$, which is positive if

$$(5.98) \quad \psi(N) = (1-w)(P-S)N^2$$
$$+ \{T - P - w(T + R - 2P)\}N + w(R - P)$$

is negative, requiring in particular that $w > \frac{T-P}{T+R-2P}$. Routine manipulations (Exercise 5.18) now establish that $\psi(N) < 0$ whenever

$$(5.99) \qquad w > \xi_4, \qquad \frac{w(R-P)}{(1-w)(P-S)U_4(w)} < N < U_4(w),$$

where

$$(5.100) \qquad \xi_4 = \frac{(T-P)(T+R-2P) + 2(R-P)(P-S) + 2(R-P)\sqrt{(P-S)(T-S)}}{(T+R-2P)^2 + 4(P-S)(R-P)}$$

satisfies $\frac{T-P}{T+R-2P} < \xi_4 < 1$ and $U_4(w)$ is the larger root of $\psi(N) = 0$ (whose existence is guaranteed by $w > \xi_4$). Assuming that (5.99) is satisfied—which, for example, requires $3 \leq N + 1 \leq 571$ in Axelrod's prototype—let us define $x_{\mathrm{ed}} = \frac{w(2R - P - S) - R + S - N(1-w)(P-S)}{w(R-P) + N(1-w)(S+T-P-R)}$. Then, because $x_2 + x_3 = 1 + \alpha$, $W_2 - W_3$ is negative if $x_2 > x_{\mathrm{ed}}$ but positive if $x_2 < x_{\mathrm{ed}}$, so that the population will reach equilibrium at a mixture of $TF2T$ and $STFT$ in which the proportion of $TF2T$ is $x_{\mathrm{ed}}/(1+\alpha)$.

Three conclusions can be drawn from this analysis. First, we have shown that TFT cannot be stable under dynamic interaction unless

$$(5.101\mathrm{a}) \qquad\qquad w > \max(\xi_1, \xi_2)$$

$$(5.101\mathrm{b}) \qquad\qquad N < \min(U_1(w), U_2(w)),$$

where ξ_1, ξ_2, U_1 and U_2 are defined by (5.91) and (5.93). Routine algebra (Exercise 5.19) shows that (5.2)-(5.3) imply $\xi_1 > \frac{T-R}{T-P}$, $\xi_2 > \frac{T-R}{R-S}$. So, comparing (5.101) with (5.34), we see that the probability of further encounters must be higher under dynamic interaction than under static interaction for TFT to be a stable orthodox strategy. Note that if $S = 0$, $T = P + R$ (as in §5.1), then $\xi_1 = \xi_2$, $U_1 = U_2$ and (5.101) simplifies greatly; see Exercise 5.24. Second, we see that (5.94), (5.96) and the corresponding inequalities for $T2FT$ are all false as $N \to \infty$. So, when interaction is dynamic, reciprocity cannot maintain cooperation if the population is too large, essentially because nasty strategies can then profit from too many first encounters and too few second or higher encounters—in other words, too few opportunities for punishment. And third, $TF2T$ is again too forgiving to be a stable orthodox strategy, but it can coexist with $STFT$ if the population is small enough.

Subsequently, it will be convenient to refer to animals or other organisms—e.g., bacteria [**9**, p. 1392]—as highly mobile if their pattern of interaction for an IPD is dynamic; and as sessile if instead it is static. These labels need not apply to the organism's entire life history. For example, impala live in very loose social groups and on the whole are better described as highly mobile; but pairs are sessile during an IPD consisting of repeated bouts of "allogrooming" to remove parasites from areas that partners cannot reach [**53**, pp. 91-94].

5.8. Mutualism: common ends or enemies

We have seen that it is possible to sustain cooperation via reciprocity in the iterated prisoner's dilemma—provided, of course, that a player's probability of further interaction is high enough. But how is cooperation initiated? Recall that in the simplest IPD we considered, namely, the game in which $ALLD$ and TFT are the only strategies, both strategies are evolutionarily stable if $\mu > \mu_c$, where μ_c is defined by (5.21); and that TFT will become the orthodox strategy only if $x_2(0)$, the initial fraction of TFT strategists in the game with matrix (5.20), exceeds the value γ defined by (5.22). But what if $x_2(0) < \gamma$? How is cooperation initiated then? One possibility is that sometimes the game with payoff matrix (5.1) is a prisoner's dilemma, and sometimes C is a dominant strategy, according to whether the environment in which the game is played is lenient or harsh. If so, then cooperation could be initiated by selfish behavior during a harsh phase, and sustained throughout a lenient phase by reciprocity. Then, as it were, the environment would play the role of a common enemy—the ultimate enforcer of cooperation. How many times have you heard it said that the only way to accomplish wholesale cooperation among humans would be to invite aliens from outer space to invade our planet?

To illustrate the effect of a common enemy, we turn to another kind of game—wild animals—for a further example of the prisoner's dilemma. Suppose that an army of rangers in jeeps is employed to protect an endangered species. These rangers patrol at random, and are advised to confront poachers—who are quite ruthless—only in pairs, the accepted custom being that any poacher is intercepted by the nearest two rangers. Let us suppose that a poacher has been spotted, and that the nearest two rangers have been identified (because, say, they are all in radio contact). If one of these rangers confronts the poacher, then we shall say that she selects strategy C, or cooperates; whereas if she desists from confrontation, then we shall say that she selects strategy D, or defects. The poacher, however, will not be granted the status of a player; rather, she is the rangers' common enemy. This does not mean that poachers have no strategic possibilities; but because our interest is the behavior of the rangers, we disregard the poachers' strategies (and instead incorporate their behavior into the parameters of our model, namely, ϵ, q and Q defined below).

There are two kinds of encounter between ranger and poacher, deliberate or accidental. The first arises when the ranger elects to confront the poacher, in which case an encounter between ranger and poacher is certain. The second arises when the ranger desists from confrontation, but is nevertheless unlucky enough to find herself on the poacher's path; in which case, we suppose that an encounter between ranger and poacher occurs with probability $\epsilon(< 1)$, regardless of whether the other ranger cooperates or defects (think of a defector as stationary, and a cooperator or poacher as always moving). When a poacher encounters a ranger, let the probability that the poacher inflicts injury on the ranger be Q or q, according to whether the ranger is isolated or reinforced by her colleague; and assume that if a poacher encounters a defecting ranger accidentally, then the encounter will be isolated only if the other ranger is also defecting (because if she is cooperating, then her jeep will be right on the poacher's heels when the poacher meets the other ranger). Then, if the payoff to a ranger is the probability of avoiding injury, the interaction between rangers has payoff matrix

$$(5.102) \qquad A = \begin{bmatrix} R & S \\ T & P \end{bmatrix} = \begin{bmatrix} 1-q & 1-Q \\ 1-\epsilon q & 1-\epsilon Q \end{bmatrix};$$

and it is straightforward to show (Exercise 5.23) that the game with this matrix is a prisoner's dilemma whenever

$$(5.103) \qquad\qquad\qquad q < \epsilon Q$$

(implying in particular that $Q > q$). Note that if there are $N + 1$ rangers, then the iterated game is an example of the IPD we analyzed in the previous section. Note also that (5.102)-(5.103) imply $P + R > S + T$, so that (5.35) is violated.

Now, in constructing our payoff matrix, we have assumed that the only difference between cooperating and defecting lies in the probability of accidental discovery by a poacher—the rows of (5.102) would be identical if ϵ were equal to 1. We are therefore assuming that, conditional upon an encounter, a lone ranger who attacks a poacher has no advantage over a lone ranger who is suprised by a poacher on the run. If there is any truth at all to the old adage that attack is the best form of defense, however, then this assumption is false; rather, if the probability of injury while confronting alone is Q, then the probability of injury while defecting alone should be, not Q, but Z, where $Z > Q$. Similarly, if the probability of injury while confronting in pairs is q, then the probability of injury to a defector whose opponent is cooperating should be, not q, but z, where $z > q$—partly because the defector does not have the advantage of attack, but also because the cooperator may not be so hot on the heels of the poacher as to guarantee effective reinforcement (especially if she is mad at her colleague for defecting). We also expect $Z > z$. To allow (5.102) as a special case of the analysis that follows, however, we will weaken the inequalities $Z > Q, z > q$ to $Z \geq Q, z \geq q$. Furthermore, we will assume that poachers are becoming more and more ruthless as time goes by in inflicting injuries on rangers they surprise, so that, if $\delta \geq 0$ is a measure of ruthlessness, then z and Z are increasing functions of δ. Let $\delta = 0$ correspond to (5.102). Then the payoff matrix for the game at ruthlessness δ becomes

$$(5.104) \qquad A(\delta) \;=\; \begin{bmatrix} R & S \\ T(\delta) & P(\delta) \end{bmatrix} \;=\; \begin{bmatrix} 1 - q & 1 - Q \\ 1 - \epsilon z(\delta) & 1 - \epsilon Z(\delta) \end{bmatrix},$$

where

$$(5.105) \qquad z(0) = q, \; Z(0) = Q$$

and

$$(5.106) \qquad z(\delta) < Z(\delta), \; z'(\delta) > 0, \; Z'(\delta) > 0, \quad 0 < \delta < \infty.$$

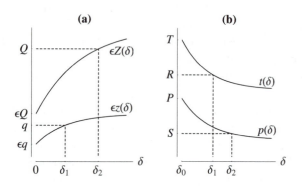

Figure 5.6. Variation with δ of defector's payoffs in games against a common enemy.

Clearly, confrontation is the only rational strategy if poachers are infinitely ruthless against defectors. Therefore C must be dominant as $\delta \to \infty$ or

(5.107) $$\epsilon z(\infty) > q, \quad \epsilon Z(\infty) > Q.$$

Together, (5.105)-(5.107) imply the existence of unique δ_1 and δ_2 such that

(5.108) $$\epsilon z(\delta_1) = q, \quad \epsilon Z(\delta_2) = Q.$$

It need not be true that $\delta_1 < \delta_2$. Nevertheless, we shall assume this; see Figure 5.6a.

It is clear from inspection of (5.102) that the rangers' game at ruthlessness δ is a prisoner's dilemma only if $0 \le \delta \le \delta_1$. For $\delta_1 < \delta < \delta_2$, we have $R > T(\delta)$ and $S < P(\delta)$, so that both C and D are evolutionarily stable, and the final composition of the ranger population is determined by the initial fraction of cooperators; if most of the rangers are cooperating initially, then all will be cooperating eventually, and vice versa. For $\delta_2 < \delta < \infty$, however, we have $R > T(\delta)$ and $S > P(\delta)$, so that C is a dominant strategy; what is best for the group is then also best for the individual. In summary, if the poachers are sufficiently ruthless (but not infinitely ruthless) then cooperation is the only rational strategy; whereas if the poachers are not ruthless enough, and if the initial proportion of cooperators is too small, then the rangers remain locked in a prisoner's dilemma.

If we think of the poachers as a harsh environment for the rangers, then what we have shown is that cooperation can emerge under adverse conditions purely as a response to changes in environmental parameters, and without any need for reciprocity. More generally, we can conceive of games with payoff matrix

$$(5.109a) \qquad A(\delta) = \begin{bmatrix} R & S \\ t(\delta) & p(\delta) \end{bmatrix}$$

satisfying

$$(5.109b) \qquad t(\delta_0) = T > R > \tfrac{1}{2}(S+T), \ p(\delta_0) = P > S, \ R > S$$

$$(5.109c) \qquad t(\infty) < R, \ p(\infty) < S,$$

and

$$(5.109d) \qquad R > p(\delta), \ t(\delta) > p(\delta), \ p'(\delta) < 0, \ t'(\delta) < 0$$

for $0 \leq \delta_0 < \delta < \infty$, so that there exist unique δ_1 and δ_2 satisfying

$$(5.109e) \qquad t(\delta_1) = R, \quad p(\delta_2) = S;$$

see Figure 5.6b, which is drawn for the case where $\delta_1 < \delta_2$. Here δ is a parameter that measures adversity; and δ_0 is a base value, at which adversity is so slight that the players are still locked in a prisoner's dilemma. Obviously, the ranger game is a special case of (5.109), in which $R = 1 - q, S = 1 - Q, t(\delta) = 1 - \epsilon z(\delta), p(\delta) = 1 - \epsilon Z(\delta)$ and $\delta_0 = 0$, but other such games can be constructed.[10] Indeed it can be shown that if the oviposition game in §5.2 is extended to n periods, and if the insects are unable to recognize their eggs, and if half the number of sites exceeds the probability per period of finding a site times the average number of periods an insect survives on the patch— a perfectly natural assumption if we wish sites to be abundant when the insects start to forage—then the game is an example of (5.109) in the limit as $n \to \infty$, with δ equal to the ratio of survival probabilities, r_1/r_2, and $\tfrac{5}{2} = \delta_0 < \delta_1 < \delta_2 < \tfrac{11}{4}$ [138]. What this means is that a very modest increase in environmental adversity—here, a decrease in the ratio of the survival probability of a paired egg to that of a solitary egg from $\tfrac{2}{5} = 0.4$ to $\tfrac{4}{11} \approx 0.36$—can be all it takes to convert

---

[10]See, e.g., [34, 165], especially in the context of [149, p. 274].

a game in which only defection pays $(\delta < \delta_1)$ to one in which only cooperation pays $(\delta > \delta_2)$.

Note, finally, that there is no difference in principle between cooperation against a common enemy and cooperation toward a common end (for much the same reason that minimizing a function is the same thing as maximizing its negative). In either case, cooperation is an incidental consequence of ordinary selfish behavior; and so this category of cooperation is known to biologists as "byproduct mutualism" [**30, 152**], or simply mutualism [**154**]. What distinguishes reciprocity from mutualism is the presence or absence of scorekeeping. For reciprocators, benefits are conferred or costs extracted by specific individuals, and it is necessary to keeps tabs on their past behavior. For mutualists, by contrast, benefits are conferred or costs extracted by the common environment—both players and non-players—with which all interact; even if scorekeeping is possible it is unnecessary, because the risk is too high that anyone who tries to exploit others for short-term gain will only penalize herself. In other words, although there is no direct feedback between individuals—only indirect feedback from the environment—benefits exceed costs over the time scale on which rewards are measured, and so there is no incentive to cheat.

For further examples of cooperation via mutualism, see §§6.5-6.6.

5.9. Much ado about scorekeeping

Many instances of cooperation among animals have been interpreted as either mutualism or reciprocity. Examples of the first kind include cooperative hunting in lions [**184**, pp. 276-277] and sentinel duty in various species [**13**]. Examples of the second kind include predator inspection by guppies or sticklebacks [**53**, pp. 63-70], allogrooming by impala [**53**, pp. 92-94] and blood sharing by bats [**53**, pp. 113-114]. But the evidence is inconclusive, and so debate has been spirited. A key issue is whether putative reciprocators are sessile (have fixed partners) for the purposes of an IPD; because even if (5.101) holds, cooperation among highly mobile organisms is arguably more likely to be mutualism than reciprocity. To see why, and also demonstrate that mutualistic $ALLC$ and reciprocative TFT may be distinguishable even in the absence of noncooperative $ALLD$, we embroil those three strategies in an iterated cooperator's dilemma under dynamic interaction in this section, and under static interaction in §5.10. As in §5.3, $ALLC$, $ALLD$ and TFT are strategies 1, 2 and 3, respectively.

Now, the entire analysis of §5.7 was predicated on the ability of animals to distinguish one another: without it, reciprocity cannot be stable under dynamic interaction (Exercise 5.22). But there may be costs associated with recognizing previous partners and remembering whether they cooperated or defected; in other words, with keeping score. We include these costs here by supposing that TFT-strategists store a memory of each partner at cost c_0; and that, at move k, they compare the current partner with each of the $k-1$ previous partners, at a cost of c_1 per comparison. Thus, on using (5.24) and (5.71), the total scorekeeping cost is $\sum_{k=1}^{\infty}\{c_0 + (k-1)c_1\}w^{k-1} = \mu\bar{c}$, where $\bar{c} = c_0 + \mu c_1 w$ is the average cost of keeping score per move.

Next we calculate a reward matrix A for the entire iterated co-operator's dilemma, in which a_{ij} is the reward to an individual playing strategy i against N individuals playing strategy j. For example, a TFT-strategist among N $ALLD$-strategists obtains $P - \bar{c}$ at move k if she has previously met her partner, i.e., with probability σ_k, where σ_k is defined by (5.74a); and $S - \bar{c}$ at move k if her partner is new, i.e., with probability $1 - \sigma_k$. Thus $a_{32} = \sum_{k=1}^{\infty}\{(P - \bar{c})\sigma_k + (S - \bar{c})(1 - \sigma_k)\}w^{k-1} = \mu(P - \bar{c}) + \lambda_1(S - P)$, where $\mu = \frac{1}{1-w}$ and λ_1 is defined by (5.76). We find in this way that

$$(5.110) \quad A = \begin{bmatrix} \mu R & \mu S & \mu R \\ \mu T & \mu P & \mu P + \lambda_1(T - P) \\ \mu(R - \bar{c}) & \mu(P - \bar{c}) + \lambda_1(S - P) & \mu(R - \bar{c}) \end{bmatrix},$$

agreeing with (5.19) when $\bar{c} = 0$ and $N = 1$.

Now, if $\bar{c}$ were zero, then $ALLC$ would be (weakly) dominated by TFT (because $\mu > \lambda_1$) and we could remove $ALLC$ from the game as in §5.3; moreover, TFT would be an ESS for $w > \frac{T-R}{T-R+\alpha(R-P)}$, agreeing with (5.25) for $N = 1$. Under dynamic interaction, however, there must surely be significant costs associated with keeping score; and if $\bar{c}$ is positive, no matter how small, then TFT is no longer even collectively stable (because $a_{13} > a_{33}$). Rather, $ALLC$ is an ESS if $R > T$, and otherwise only $ALLD$ is an ESS. Thus our model predicts that only mutualism can sustain cooperation among highly mobile organisms, although reciprocity could work for sessile organisms if the cost of scorekeeping were zero. But mustn't there always be some cost associated with scorekeeping, because of the necessity to recall opponents' previous moves? Can TFT never resist $ALLC$? We will return to this point in the following section.

Meanwhile, we conclude this short section by noting that, for discrete games in a finite population, there is a subtle difference between the standard conditions for strategy j to be an ESS—namely, $a_{jj} > a_{ij}$ for all $i \neq j$, where a_{ij} is the reward to an i-strategist among N other individuals, all of whom are j-strategists—and the stability conditions applied in §5.7. By comparing a_{jj} to a_{ij}, as here, we imagine that a potential mutant who is currently a j-strategist compares her current reward to the one she would instead obtain if she switched to strategy i (and actually switches if that reward is greater). By contrast, in §5.7 we imagined that an individual who is already an i-strategist compares her reward to the one she would obtain if instead she were one of the other N individuals (and switches back if that reward is greater). In other words, in applying the standard conditions, we ask whether a mutant can enter a finite population; whereas in §5.7 we asked whether an infiltrator can be expelled. An answer to the second question yields more stringent conditions for cooperation than an answer to the first; but in practice the difference is often negligible, the standard conditions are usually easier to apply, and we will use them exclusively in §§6.5-6.6.

5.10. The comedy of errors

Everything so far assumes that players do not make errors in executing their strategies. But if they did, what would be their effect? There is no simple answer: errors can destabilize TFT, but they can also favor reciprocity by making an ESS out of a different reciprocative strategy that is otherwise only collectively stable.

To broach the issues involved, we revisit the game of the previous section—but under static interaction, so that scorekeeping costs are small. For the sake of simplicity, we assume initially that they are precisely zero. Then, in the absence of errors, the payoff matrix A would reduce to (5.19). But we assume instead that each player makes a mistake—that is, defects when she intended to cooperate or vice versa—with the same small probability ϵ on every move. Thus, at their initial encounter, two TFT-strategists achieve their intended outcome CC of mutual cooperation with probability $(1 - \epsilon)^2$. With probability $2\epsilon(1 - \epsilon)$, one player makes a mistake, so that the actual outcome is CD or DC; and with probability ϵ^2 they both make an error, so that the outcome is DD. If ϵ is sufficiently small, however, then we can safely neglect terms that are second-order in ϵ, which we denote collectively by $O(\epsilon^2)$. So, with negligible error, the probabilities of outcomes CC, CD, DC and DD for the initial encounter between two TFT-strategists are, respectively, $1 - 2\epsilon$, ϵ, ϵ and 0.

If we label CC, CD, DC and DD as outcomes 1, 2, 3 and 4, respectively, then their probabilities at subsequent encounters can be found with the help of an update matrix U, in which u_{ij} is defined to be the probability of outcome j at move $k + 1$, conditional upon outcome i at move k. For TFT versus TFT, it is readily shown that

$$(5.111a) \qquad U = \begin{bmatrix} (1-\epsilon)^2 & (1-\epsilon)\epsilon & \epsilon(1-\epsilon) & \epsilon^2 \\ \epsilon(1-\epsilon) & \epsilon^2 & (1-\epsilon)^2 & (1-\epsilon)\epsilon \\ (1-\epsilon)\epsilon & (1-\epsilon)^2 & \epsilon^2 & \epsilon(1-\epsilon) \\ \epsilon^2 & \epsilon(1-\epsilon) & (1-\epsilon)\epsilon & (1-\epsilon)^2 \end{bmatrix}$$

$$(5.111b) \qquad = \begin{bmatrix} 1-2\epsilon & \epsilon & \epsilon & 0 \\ \epsilon & 0 & 1-2\epsilon & \epsilon \\ \epsilon & 1-2\epsilon & 0 & \epsilon \\ 0 & \epsilon & \epsilon & 1-2\epsilon \end{bmatrix} + O(\epsilon^2).$$

For example, $u_{32} = (1 - \epsilon)^2 = 1 - 2\epsilon + O(\epsilon^2)$ because DC is followed by the intended outcome CD only if neither player errs; if Player 1 or Player 2, respectively, errs, then the outcome instead is DD or CC. Hence, on defining $\Delta_\pm(k) = \frac{1}{2}(1 - 2k\epsilon)\{1 \pm (-1)^k\}$, we obtain

$$(5.112) \qquad U^k = \begin{bmatrix} 1-2k\epsilon & k\epsilon & k\epsilon & 0 \\ k\epsilon & \Delta_+(k) & \Delta_-(k) & k\epsilon \\ k\epsilon & \Delta_-(k) & \Delta_+(k) & k\epsilon \\ 0 & k\epsilon & k\epsilon & 1-2k\epsilon \end{bmatrix} + O(\epsilon^2)$$

(Exercise 5.28). But if $x_i(k)$ denotes the probability of outcome i at move k, for $i = 1, \dots, 4$, then $x(k) = x(0) U^k$. So, given that $x(0) = (1, 0, 0, 0)$, the probabilities of CC, CD or DC and DD at move k are $x_1(k) = 1 - 2k\epsilon$, $x_2(k) = k\epsilon = x_3(k)$ and $x_4(k) = 0$, respectively, on neglecting terms of order ϵ^2; and so

$$(5.113) \quad a_{33} = \sum_{k=1}^{\infty} \{Rx_1(k) + Sx_2(k) + Tx_3(k) + Px_4(k)\}w^{k-1}$$

$$= \sum_{k=1}^{\infty} \{R - (2R - S - T)k\epsilon\}w^{k-1}$$

$$= R\mu - (2R - S - T)\mu^2\epsilon$$

to first order in ϵ, on using (5.31) and (5.71). A similar calculation for $ALLC$ versus TFT (Exercise 5.29) shows that, again to first order in ϵ, $x(1) = (1 - 2\epsilon, \epsilon, \epsilon, 0)$ and $x(k) = (1 - 3\epsilon, 2\epsilon, \epsilon, 0)$ for $k \geq 2$. So $a_{13} = (1 - 2\epsilon)R + \epsilon S + \epsilon T + \sum_{k=2}^{\infty}\{(1 - 3\epsilon)R + 2\epsilon S + \epsilon T\}w^{k-1}$, or

$$(5.114) \qquad a_{13} = R\mu - \{2R - S - T + w(R - S)\}\mu\epsilon$$

after simplification. The corresponding calculation for $ALLD$ versus TFT (Exercise 5.29) yields, to the same order in ϵ,

$$(5.115) \quad a_{23} = T + \mu wP - \{2T - R - P + (3P - 2T - S)w\mu\}\epsilon.$$

From (5.113-(5.115), although $a_{33} = a_{13}$ in the absence of errors, for $\epsilon > 0$ (no matter how small) we have $a_{33} > a_{13}$ if $w < \frac{T-R}{R-S}$. The terms of order ϵ in (5.115) do not affect the lower bound on w, which is still (5.25). So TFT is an ESS if $\frac{T-R}{T-P} < w < \frac{T-R}{R-S}$.[11] This condition is always satisfied for some range of values of w when $S + T > P + R$, e.g., for $\frac{1}{2} < w < \frac{2}{3}$ with Axelrod's payoffs (5.26). The upper bound on w arises because the probability $x_1(k)$ of mutual cooperation at move k continually decreases with k for TFT against itself but remains constant for $ALLC$ against TFT if $k \geq 2$, so that an $ALLC$ mutant has an advantage over TFT if w is sufficiently large.

Returning now to the question of whether TFT can ever resist $ALLC$, note that a small scorekeeping cost $\bar{c}$ does not affect the evolutionary stability of TFT as long as $\bar{c} < \{T - R - w(R - S)\}\mu^2 w\epsilon$. It is therefore possible for errors to offset the negative effect of the cost of scorekeeping on the stability of TFT. The important point, however, is not that TFT may be stable after all, but rather the more general point that one small effect can easily be offset by another.

---

[11]Of course, if $ALLC$ and $ALLD$ correspond to absolutely fixed behavior, then it is unrealistic to suppose that either would ever behave like the other, even by mistake. So suppose instead that only TFT makes errors. Then the only possible states between $ALLC$ and TFT are CC and CD, and (5.114) becomes $a_{13} = \sum_{k=1}^{\infty}\{(1 - \epsilon)R + \epsilon S\} = R\mu - (R - S)\mu\epsilon$. The condition for a_{33} to exceed this quantity turns out to be exactly the same as before, i.e., $w < (T - R)/(R - S)$. Similarly, (5.115) becomes $a_{23} = P\mu + (T - P)\mu\epsilon = P\mu + O(\epsilon)$ as before. So our result is unaffected.

Unfortunately for TFT, the condition that makes it immune to $ALLC$ is precisely the condition that renders it vulnerable to $STFT$; see (5.36). But TFT is merely one of many reciprocative strategies that are capable of sustaining cooperation. Another such strategy, namely, "contrite tit for tat" or $CTFT$, which cooperates unconditionally after defecting by mistake, turns out with errors to be uniquely the best reply to itself—and therefore a strong ESS—if $w > \max\left(\frac{P-S}{R-S}, \frac{T-R}{R-S}\right)$ [**22, 219**].[12] Extrapolating from TFT to $CTFT$: a small scorekeeping cost need not prevent reciprocity from sustaining cooperation among sessile organisms, because its effect could be offset by, e.g., that of errors in executing strategies.

This, however, is by no means the end of the story because, e.g., $CTFT$ is still susceptible to errors in perception of an opponent's move [**19**]. More generally, execution errors and scorekeeping costs are merely two among a panoply of possible further effects that can significantly modify the outcome of the game; see, e.g., [**53**, pp. 25-28] and §5.11. So many assumptions differ between models that no clear picture of their overall effect has yet emerged, and further discussion would take us too far afield.

5.11. Commentary

In Chapter 5, we have explored conditions for the evolution of cooperation in terms of the cooperator's dilemma. We began with two illustrations of the prisoner's dilemma, a laboratory game in §5.1 and a foraging game in §5.2. Then, in §§5.3-5.4, we analyzed the iterated prisoner's dilemma or IPD as a discrete population game, showing how the reciprocative strategy tit for tat or TFT might sustain cooperation. In §5.5, we constructed a model of the IPD in a finite population; and in the next two sections we used this model to develop necessary conditions for TFT to be a stable population strategy under both static (§5.6) and dynamic (§5.7) patterns of interaction. These three sections are based on [**141**] and are further developed in [**151**], where it is argued that the conditions for stable reciprocity

---

[12]Indeed when Wu & Axelrod [**240**] repeated Axelrod's ecological simulation (p. 199) with errors and a handful of additional strategies, including $CTFT$, the honor of most successful strategy clearly devolved from TFT to $CTFT$.

may be hard to satisfy.[13] In §5.8, we showed how cooperation could emerge without reciprocity, as mutualism. These ideas are further developed in [**54, 152**]. Finally, in §§5.9-§5.10, we explored implications of scorekeeping costs and execution errors. For further discussion of mutualism versus reciprocity, see [**39**], [**154**] and [**184**].[14]

Experiments on humans equivalent to the laboratory experiment with payoff matrix (5.4)—which reduces in essence to a choice for each animal of giving either P to itself or R $(> P)$ to the other—have been performed with sums of money in lieu of P and R [**59, 183**]. Other experiments on humans have instead used the matrix (5.1), e.g., with $S + T = 0$ [**186**]. For a review, see Chapter 7 of [**40**]. By and large, these experiments underscore the difficulties of achieving mutual cooperation in an IPD. Sustained mutual defection was the characteristic outcome; moreover, even when subjects did cooperate, it was impossible to confirm or deny reciprocity—$ALLC$ versus $ALLC$, TFT versus TFT, and TFT versus $ALLC$ all generate the same cooperative outcome in a repeated binary interaction. On the other hand, more recent experiments—critically reviewed by Romp [**195**, pp. 229-240]—indicate that humans are more cooperative in an ordinary prisoner's dilemma than game theory predicts. Why? As we have seen, there is really only one escape from the prisoner's dilemma, and that is to discover that the game being played (or, in experiments, the game that subjects perceive themselves as playing) is not really the prisoner's dilemma after all. In other words, if A is the payoff matrix, strategy 1 is cooperative, strategy 2 is noncooperative and there appears to exist a time scale over which $a_{21} > a_{11}$ then, when all relevant (including spatial) factors are accounted for, either that appearance must turn out to have been an illusion; or else there must exist a longer time scale over which $a_{11} > a_{21}$, and which is also the relevant time scale for tallying benefits and costs [**154**].

Whereas prisoner's-dilemma experiments with humans date from the 1950s, experiments with other animals did not begin until the

---

[13][**151**] considers a possible continuum of mobility between sessile and highly mobile; there is no spatial structure, and conditions become steadily less favorable to reciprocity as mobility increases. But Ferriere and Michod find that spatial structure can make conditions most favorable to reciprocity at intermediate mobility [**62**].

[14]According to [**184**, p. 275], I suggest "reciprocity is widepread"—but if anything, I suggest only the opposite! See [**138**, p. 264], [**54**, p. 203] and [**152**, p. 278].

1980s. They include experiments with rats [64], starlings [188] and blue jays [39]. On the whole, these experiments reaffirmed the difficulties of sustaining cooperation through reciprocity; in particular, Clements and Stephens [39] found that cooperation neither developed nor persisted in an IPD, but mutualism persisted in the corresponding repeated game with the signs of $T - R$ and $P - S$ in (5.1) reversed.

These experiments mimic circumstances in which cooperation has been observed in the wild, having evolved by natural selection; assuming cooperation, the experiments ask how cooperation is maintained. By contrast, the experiment suggested in §5.1 asks whether there is cooperation (but if so, then it also asks how). In the first case, cooperation is—as it were—learned by natural selection; in the second case, cooperation—if it arises—is learned by the organism's own (as opposed to its ancestors') experience. For a discussion of learning rules in the context of evolutionary stability, see [88, 101]. For learning in the context of the prisoner's dilemma, see [214]. For models of learning in general, see [66], Chapter 11 of [38] and references therein.

Although Axelrod [6] is still obligatory and Gadagkar [67] is highly recommended, the best overall introduction to the evolution of cooperation is now Dugatkin [53], which cites most of the relevant literature; among the few exceptions are [7, 8, 29, 166]. Earlier work on the IPD—e.g., [61], in which different players have different probabilities of further interaction—typically considers only pure strategies. Later work on the IPD [168, 170] allows for mixed strategies, which we in this chapter have not discussed; but neglecting them in §§5.5-5.7 is to some extent bolstered by Vickery's contention that only pure strategies can be ESSes in a finite population [227, 228]. In §5.7, dynamic interaction implies pairwise encounters, but it is possible to assume instead that each individual interacts with every other individual in the population. The rationality of cooperation has been studied in this context by Schelling [199]; and also by Boyd [24], whose main conclusion—that conditions allowing the evolution of reciprocal cooperation become extremely restrictive as group size increases—parallels our finding in §5.7 that TFT is stable only if the population does not exceed a certain critical size.

The idea that reciprocity may be found only in sessile organisms (§§5.9-5.10) implicitly recognizes that a population's spatial structure is important: sessile individuals interact only with their neighbors. Some studies that incorporate this spatial effect explicitly exclude reciprocation; for example, Nowak et al. [**169**] consider binary choices between cooperation and defection, whereas Killingback et al. [**110**] consider unilateral investments in cooperation (with zero investment equivalent to defection). In the first case, the game is a spatial version of the ordinary prisoner's dilemma; in the second case, spatial extension is added to a continuous prisoner's dilemma or CPD. Either way, any cooperation is due to benefits of clustering, not reciprocity. In other studies, however, the relevant game is a spatial IPD in either one [**62**] or two [**28, 78**] dimensions; or an iterated CPD without spatial structure, which Roberts and Sherratt [**193**] and Wahl and Nowak [**232**] use to predict that successful strategies should gradually increase investment in the welfare of others, "raising the stakes" against strategies that are initially reluctant to cooperate. But further discussion of these or related studies would distract us too far from our goals, not least because spatial games are apt to rely so heavily on stochastic numerical simulations or "cellular automata" [**238**]; whereas we have chosen to emphasize analytical models. Note, however, that analytical progress in spatial games is also possible, to varying degrees [**62, 105, 185**].

Whither TFT? We saw in §5.10 that execution errors can rob TFT of its robustness, but that contrition can restore it. Another potential restorative is generosity, i.e., a tendency to forgive others their defections; see Chapter 8 of Sigmund [**209**], and references therein. Subsequent studies of the IPD, e.g., [**112, 113, 121, 171**] have tended to focus on greater strategic sophistication. Much of this literature implicitly assumes that reciprocative strategies—though not TFT—solve the problem of cooperation in nature; but Crowley and Sargent [**48**] identify situations where TFT might be important in nature, relative to more complex cooperative strategies. Furthermore, studies typically do not account for scorekeeping costs; or temporal discounting, i.e., the tendency to value instant gratification more highly than delayed gratification. But Crowley et al. [**47**, p. 61] find that even low memory costs can greatly disfavor reciprocity; and Stephens et al. [**213, 215**] find that temporal discounting is significant, and greatly disfavors reciprocity, because its effect is equivalent to reducing the value of w in an IPD.[15] So one shouldn't forget that the implicit assumption may be false: reciprocity need not always, or even often, solve the problem of cooperation—and where it does, the strategies may not need to be sophisticated.

On the other hand, mutualism is not the only alternative to reciprocity—merely the most parsimonious [**39**]. Other categories of cooperation include "group-selected" behavior [**237**], in which an individual's success is determined (for better or worse) by that of her group; and "kin-selected" behavior [**81, 235**], whose surface we will scratch in §6.3. No two categories are mutually exclusive; e.g., a group may consist of kin (as will be implicit in §6.3). But there are at least some cases of cooperation, including the one we will discuss in §6.5, where kin- and group-selected behavior can be ruled out a priori [**154**, p. 556]. And so, for the sake of simplicity, we have chosen to focus on mutualism versus reciprocity.

Exercises 5

1. Establish (5.16).
2. (a) Show that if there is constant probability w of further interaction in the iterated prisoner's dilemma, then the number of moves has distribution (5.23).
   (b) Verify (5.24).
3. Under the conditions of §5.3:
   (a) Show that $TF2T$ can be invaded by $STCO$, whereas TFT cannot be invaded by $STCO$ if (5.36) is satisfied.
   (b) Determine the final composition of the population when $STCO$ invades $TF2T$.
4. Verify (5.32) and (5.37).
5. Verify that H defined by (5.48) is a hyperbola, and that ϕ_3 is negative in the shaded region of Figure 5.3.
6. Verify (5.49).
7. Verify (5.51).
8. Verify (5.53)-(5.54) and that (5.34) implies $d > b > a > c$.

---

[15]But not every kind of discounting is bad for cooperation; see §2.4 (p. 71).

9. Show that $P + R > S + T$ in a prisoner's dilemma implies

    (a) $\frac{P-S}{T-P} > \frac{P-S}{R-S} > \frac{T-R}{T-P} > \frac{T-R}{R-S}$

    (b) $\frac{P-S}{R-S} > \frac{P+R-S-T}{2R-S-T}$.

10. Strategy $RNDM$ means cooperating or defecting with probability $\frac{1}{2}$. Extend the IPD of §5.4 to five strategies, with $RNDM$ as strategy 5; in other words, define the fifth row and column of the 5×5 move-k payoff matrix $\phi(k)$.

11.   **(a)** Obtain (5.72).

    (b) Obtain (5.74c).

12. Verify that (5.75) reduces to (5.72) when $N = 1$.

13. With regard to the iterated prisoner's dilemma, we showed in §5.4 that $TF2T$ does better than TFT against $STFT$ whenever $w > \frac{T-R}{R-S}$, assuming of course that no other strategies are present; see the remarks after (5.52). But $TF2T$'s first two payoffs against $STFT$ are S and R, whereas TFT's first two payoffs are S and T. Thus, because $S + wT > S + wR$, $TF2T$'s advantage over TFT (against $STFT$) does not emerge until the third encounter. Under dynamic interaction, however, we would expect third encounters with the same individual to be rather infrequent. Does this mean that $TF2T$ loses this advantage over TFT when opponents are drawn at random throughout the game? Why or why not? Use (5.75).

14. Show that TFT is stable against $RNDM$ (defined in Exercise 5.10) for sufficiently large N if (5.34) is satisfied.

15. Obtain (5.86).

16. Find necessary conditions for $TF2T$ to be stable against infiltration by $RNDM$ (defined in Exercise 5.10).

17. Verify (5.94) and (5.96).

18. Verify that (5.98) implies $\psi(N) < 0$, where ψ is defined by (5.99).

19. Show that (5.2)-(5.3) imply $\xi_1 > \frac{T-R}{T-P}$ and $\xi_2 > \frac{T-R}{R-S}$, where ξ_1 and ξ_2 are defined by (5.91).

20. Let TFT be the orthodox strategy and $RNDM$ (defined in Exercise 5.10) the only deviant strategy in an iterated prisoner's dilemma. With $RNDM$ as strategy 5 (and $N_2 = N_3 = N_4 = 0$), find expressions for the expected payoffs W_1 and W_5 under dynamic interaction. Hence find conditions for TFT to be stable against $RNDM$ when opponents are drawn at random.

21. Let $TF2T$ (strategy 2) be the orthodox strategy and $RNDM$ (defined in Exercise 5.10) the only deviant strategy in an iterated prisoner's dilemma. With $RNDM$ as strategy 5 (and $N_1 = N_3 = N_4 = 0$), find expressions for the expected payoffs W_2 and W_5 under dynamic interaction. Hence find conditions for $TF2T$ to be stable against $RNDM$ when opponents are drawn at random.

22. The entire analysis of §5.7 is predicated on perfect recognition and recall; that is, we assume that players always recognize opponents they have met before, and always remember whether they cooperated or defected. Show that if recognition were absent, then dynamic interaction would make TFT unstable at all values of w and N, because ALLD would always invade.

23. Verify that, subject to (5.103), (5.102) satisfies (5.2)-(5.3).

24. How could §5.1's experiment be used to test the theory developed in §5.7? Discuss constraints on the parameters.

25. According to Pruitt [**183**], a prisoner's dilemma is decomposable if choosing C can be made equivalent to keeping payoff σ_c for oneself and giving ω_c to the other player, and if choosing D can be made equivalent to keeping payoff σ_d for oneself and giving ω_d to the other player.

   (a) Show that this requires $\sigma_c + \omega_d = S$, and find three other equations that $\sigma_c, \sigma_d, \omega_c$ and ω_d must satisfy.

   (b) What condition—other than (5.2)-(5.3)—must T, R, P and S satisfy, if the prisoner's dilemma is decomposable?

   (c) If the game is decomposable, is its decomposition unique?

   (d) Is the laboratory game of §5.1 decomposable?

   (e) Is the foraging game of §5.2 decomposable?

   (f) Is the ranger game of §5.8 decomposable?

26. (a) A further example of a decomposable prisoner's dilemma is provided by Dawkins [**49**, pp. 184-186], who considers a population of animals that are unable to remove parasites from certain parts of their bodies. In an encounter between two such animals, a cooperator grooms the other animal, thereby improving her ability to survive and reproduce; whereas a noncooperator refuses to groom the other animal (but allows herself to be groomed). Let b be the benefit (in terms of reproductive success) of being groomed

and c the cost (in the same units) of grooming another animal. Write down the payoff matrix for an interaction between two animals, together with a condition that makes it a prisoner's dilemma.

(b) Under dynamic interaction (§5.7), Dawkins considers an IPD with the following three (pure) strategies:

Grudger: Always groom a stranger, but never groom an animal who has defected against you (even once) in the past.

Cheat: Never groom (but allow yourself to be groomed).

Sucker: Always groom (regardless of previous behavior).

To which three strategies in Chapter 5 do Grudger, Cheat and Sucker correspond?

(c) Obtain necessary conditions for Grudger to be uninvadable.

(d) Show that, if these conditions are satisfied, Grudger will eliminate both Cheat and Sucker if $N_1 > \frac{N(1-w)(Nc+b)}{w(b-c)} + 1$ initially, where w is the probability of further interaction and there are N_1 Grudgers in a population of $N+1$ animals. (Necessary conditions **(c)** ensure that the right-hand side of this inequality does not exceed the population size.)

27. Consider the IPD in which TFT is subject to mixed infiltration by $ALLC$ and $ALLD$. Show that TFT—though not an ESS of the game with matrix (5.19)—will nevertheless increase in frequency under static interaction if both (5.80) and (5.81) are satisfied, provided only that its initial frequency be sufficiently large. Repeat your analysis for dynamic interaction and obtain the corresponding result.

28. **(a)** Verify (5.111).

(b) Show that $x(k) = x(0) U^k$, where U is any update matrix and $x_i(k)$ is the probability of outcome i at move k.

(c) Verify (5.112).

(d) What are the update matrices for $ALLC$ and $ALLD$ against TFT in §5.10? Verify that $U^{k+1} = U^k$ for $k \geq 2$.

29. Verify (5.113)-(5.115).

Chapter 6

More Population Games

Here we study further examples of both continuous and discrete population games. Our goal is twofold: to demonstrate that games are valuable tools in the study of animal (including human) behavior, and to illustrate the breadth of possible applications. All but one of the continuous games are *separable*; i.e., if their strategies are p-dimensional vectors, then their reward functions can be written in the form

$$(6.1) \qquad f(u,v) = \sum_{i=1}^{p} f_i(u_i, v),$$

where $v = (v_1, v_2, \ldots, v_p)$ is the strategy adopted by the population and $u = (u_1, u_2, \ldots, u_p)$ is a potential mutant strategy. Separability (which is guaranteed when $p = 1$, as in §2.4) reduces the problem of maximizing $f(u,v)$ for given v to p one-dimensional problems and thus greatly simplifies calculation of the rational reaction set

$$(6.2) \qquad R = \left\{ (u,v) \mid f(u,v) = \max_{\overline{u}} f(\overline{u}, v) \right\},$$

although the tractability of the analysis still usually increases with p; for example, in §6.8, where $p = 4$, the analysis is appreciably more complicated than in §6.4, where $p = 2$. Note, however, that separability is by no means a prerequisite for analysis, as illustrated by §6.7 for $p = 3$.

6.1. Sex allocation: a game with a weak ESS

We begin with a simple model of sex allocation in animals.[1] Suppose that an animal can determine the sex of its offspring. Then what proportion of its offspring should be male, and what proportion female? In answering this question, we will make matters simple by supposing that the animal is the female of her species, that there is a fixed amount of resources to invest in her brood and that sons and daughters are equally costly to raise, so that after mating she produces the same number of children, say C, regardless of how many are sons or daughters. According to Darwin, our animal will behave so as to transmit as many of her genes as possible to posterity. We can capture this idea most simply by supposing that our animal will behave so as to maximize the expected number of genes she transmits to the second generation, i.e., to her grandchildren. Sex ratio clearly cannot affect the size of the first generation—for if our animal's objective were simply to maximize number of children, then what would it matter if they were male or female?

Let our animal's strategy be the proportion of her offspring that is male, which we denote by u. Then she always has uC sons and $(1 - u)C$ daughters after mating. In principle, uC and $(1 - u)C$ should both be integers; in practice, they may not be integers. But that will scarcely matter if we first assume that our animal is the kind that lays thousands of eggs—a fish, perhaps. We will imagine that she plays a game against every other female in her population, and that all such females—say N in all—adopt strategy v; that is, they have vC sons and $(1 - v)C$ daughters after mating. Let σ_m denote the proportion of sons who survive to maturity and σ_f the proportion of daughters. Then the number of daughters in the next generation is $d = \sigma_f(1-u)C + N\sigma_f(1-v)C$ and the number of sons is $s = \sigma_m u C + N\sigma_m v C$. If the population is so large that our animal's choice of strategy will have negligible effect on proportions, then an excellent approximation is

$$(6.3) \qquad \frac{d}{s} = \frac{\sigma_f(1 - v)}{\sigma_m v}.$$

---

[1] The model derives from Fisher [63], but here is adapted from Charnov [36] and Maynard Smith [132].

Formally, of course, (6.3) obtains in the limit as $N \to \infty$.

Let us now assume that females always find a mate, because some males will mate more than once if $d > s$; and that males are equally likely to find a mate if $d < s$, equally likely to find a second mate if $s < d < 2s$, etc. Thus, if M is the number of times that a son mates, then M is a random variable with expected value $\mathrm{E}[M] = d/s$.[2]

Now, the number of genes that our animal transmits to the second generation is proportional to her number of surviving grandchildren, and hence to

$$(6.4) \qquad F = \sigma_f(1 - u)C \cdot C + \sigma_m uC \cdot CM,$$

because $\sigma_f(1 - u)C$ of her daughters and $\sigma_m uC$ of her sons survive to maturity; and because daughters all produce C offspring, whereas sons produce C offspring per mating.[3] The expected value of this payoff is $\mathrm{E}[F] = \sigma_f(1-u)C^2 + \sigma_m uC^2\mathrm{E}[M]$. Thus, on setting $\mathrm{E}[M] = d/s$ and using (6.3), our animal's reward is

$$(6.5) \qquad f(u,v) = C^2\sigma_f\{v + (1 - 2v)u\}/v.$$

It is now easy to calculate the rational reaction set R and show that $v = \frac{1}{2}$ is a weak ESS (Exercise 6.1). Thus, in the highly idealized circumstances described, our animal's population should evolve to produce equal numbers of sons and daughters—irrespective of the proportions of sons and daughters that survive to maturity.

6.2. Damselfly duels: a war of attrition

The science of behavioral ecology[4] thrives on paradoxes, baffling inconsistencies between intuition and evidence that engage our attention and stimulate further investigation. A paradox arises because evidence fails to support an intuition, which (assuming the evidence

---

[2]There are two ways to derive this result. The first is simply to observe that it is obvious. The second is to suppose that $(k - 1)s \leq d < ks$, where $k \geq 1$ is an integer, and observe that $M = k - 1$ if a son fails to mate a k-th time but $M = k$ if the son does mate a k-th time; then, because the probability of a k-th mating is, by assumption, $(d - (k - 1)s)/s$, and the probability of no k-th mating therefore $(ks - d)/s$, we have $\mathrm{E}[M] = (k - 1) \cdot (ks - d)/s + k \cdot (d - (k - 1)s)/s = d/s$.

[3]If sons and daughters mated, then the daughter's genes would be counted by the first term in (6.4) and the son's genes by the second; see Exercise 6.27. Here, however, we simply assume that this does not happen: the population outbreeds.

[4]See [114] for an introduction to the subject at large.

to be sound) can happen only if the intuition relies on a false assumption about behavior, albeit an implicit one. So the way to resolve the paradox is to spot the false assumption. In other words, if a paradox of animal behavior exists, then we have wrongly guessed which game best models how a real population interacts, and to resolve this paradox we must guess again—if necessary, repeatedly—until eventually we guess correctly. Assuming the validity of our solution concept for population games, i.e., assuming that observed behavior corresponds to some ESS, our task is to construct a game whose ESS corresponds to the observed behavior. Then the resolution of the paradox lies in the difference between the assumptions of this new model and the assumptions we had previously been making about the observed behavior (whether we realized it or not). Of course, a model population is only a caricature of a real population. But a paradox is only a caricature of real ignorance. So, in terms of realism, a game and a paradox are a perfect match.

In this regard, our next example will illustrate how a population game can help to resolve a paradox. The game is a simple model of nonaggressive contest behavior, in which animals vie for an indivisible resource by displaying only—unlike in the Hawk-Dove game of §1.2, where the animals sometimes fight. The cost of displaying increases with time, and the winner is the individual whose opponent stops displaying first. We'll refer to this game as the war of attrition.

A common expectation for such contests, confirmed by experimental studies on a variety of animals, is that each animal compares his own strength to that of his opponent and withdraws when he judges himself to be the probable loser.[5] We will call this expectation the assessment hypothesis. The duration of such contests is greatest when opponents are of nearly equal fighting ability, so that it is more difficult to judge who is stronger and therefore the likely winner. But a series of contests over mating territories between male damselflies, staged by Jim Marden and Jonathan Waage [126], failed to follow this logic. Although the weaker animal ultimately conceded to his opponent in more than 90% of encounters, there was no significant negative correlation between contest durations and differences in strength. Why? We explore this question in terms of game theory.

---

[5] For references see [150, 156], on which §6.2 is based.

A possible answer is that the damselflies were not assessing one another's strength. Animals who contest indivisible resources will vary in reserves of energy and other factors, but an animal's state need not be observable to his opponent, and so we will assume that an animal has information only about his own condition. Let us also assume that variation in reserves is continuous, and that an animal's state is represented by the maximum time he could possibly display before he would have to cede the resource for want of energy. Let this time be denoted by T_{max}, and let T be the time for which the animal has already displayed. Then the larger the value of $T_{max} - T$, the more likely it is that continuing to display will gain the animal the resource. Thus the animal should persist if his perception of $T_{max} - T$ is sufficiently large. We will refer to $T_{max} - T$ as the animal's current reserves, and to T_{max} as his initial reserves.[6]

Our war-of-attrition model requires a result from psychophysics. Let π (> 0) denote the intensity of the stimulus of some physical magnitude, e.g., size or time, and let $\rho(\pi)$ be an animal's subjective perception of π. Then, in general, $\rho'(\pi) > 0$ and $\rho''(\pi) < 0$, where a prime denotes differentiation; that is, perception increases with intensity of stimulus, but the greater that intensity, the greater the increment necessary for perception. Now, for all sensory modalities in humans, the ratio between a stimulus and the increment required to make a difference just noticeable is approximately constant over the usable (middle) range of intensities.[7] Therefore, to the extent that human psychophysics also applies to other animals, it is reasonable to assume that if a is the least observable intensity and π is increased steadily beyond a, then a first difference will be noticed when $\pi = a(1+b)$, a second when $\pi = a(1+b)^2$, a third when $\pi = a(1+b)^3$, and so on, where b is the relevant constant. If it is further assumed that these just noticeable differences are all equal—to c, say—and if zero on the subjective scale corresponds to a on the objective scale, then an animal's subjective perception of the stimulus $\pi = a(1 + b)^k$ is $\rho = kc$. Thus $\pi = a(1 + b)^{\rho/c}$ or

$$(6.6) \qquad \rho(\pi) \;=\; \gamma \ln{(\pi/a)}$$

---

[6] An implicit assumption is that energy is proportional to time [**156**, p. 72].
[7] See, for example, [**200**], pp. 15-16.

where $\gamma = c/\ln(1+b)$ is a constant. In psychophysics, (6.6) is usually known as Fechner's law (but occasionally as the relativity principle), and clearly satisfies $\rho'(\pi) > 0$, $\rho''(\pi) < 0$. We will assume throughout that it provides an adequate model of the relationship between sensory and physical magnitudes.[8]

Now, let H and L be physical magnitudes whose difference determines the probability that a favorable outcome will be achieved if a certain action is taken. Then an animal should take the action if he perceives H to be sufficiently large compared to L. If he perceives each magnitude separately,[9] then he should take the action if $\rho(H) - \rho(L)$ is sufficiently large. But (6.6) implies that $\rho(H) - \rho(L) = \gamma \ln(H/L)$. So the action should be taken if H/L is sufficiently large. Thus, on setting $H = T_{max}$ and $L = T$ above, an animal should persist if T_{max}/T is sufficiently large—bigger than, say, $1/w$—but otherwise give up. Accordingly, we define strategy w to mean

(6.7)
$$\begin{array}{ll} \text{STAY} & \text{if} \quad T < wT_{\max} \\ \text{GO} & \text{if} \quad T \geq wT_{\max}. \end{array}$$

We interpret w as the proportion of an animal's initial reserves that he is prepared to expend on a contest. Thus, if the initial reserves of a u-strategist and a v-strategist are denoted by X and Y, respectively, where X and Y are both drawn randomly from the distribution of T_{max}, then the u-strategist will depart after time uX, and the v-strategist will depart after time vY.

We assume that the value of the contested resource is an increasing function of current reserves, a reasonable assumption when the resource is a mating territory (although the assumption would clearly be violated in a contest for food). For simplicity, we assume that the increase is linear, i.e., the value of the resource is $\alpha(T_{max} - T)$ where $\alpha > 0$. The parameter α has the dimensions of EFRS (p. 7) per unit of time, and so we can think of α as the rate at which the victor is able to translate his remaining reserves into future offspring. Let $\beta\ (> 0)$ denote the cost per unit time of persisting, in the same units. Then,

---

[8] For further discussion see [**143**, pp. 233-234]. The extent to which human psychophysics also applies to other animals is largely unknown, because humans are the subject of most empirical work. See, e.g., [**212**].

[9] As argued by [**156**, p. 75]. If the animal actually perceives the difference, of course, then he should instead take the action if $\rho(H - L)$ is sufficiently large, and hence if $H - L$ is sufficiently large; but results are qualitatively the same [**155**].

because X and Y are (independent) random variables, the payoff to a u-strategist against a v-strategist is also a random variable, say F defined by

$$(6.8) \qquad F(X,Y) = \begin{cases} \alpha(X - vY) - \beta vY & \text{if } uX > vY \\ -\beta uX & \text{if } uX < vY \\ 0 & \text{if } u = 0 = v \end{cases}$$

because the u-strategist wins if $uX > vY$ but loses if $uX < vY$, and the cost of display is determined by the loser's persistence time for both contestants. We ignore the possibility that $uX = vY$ (other than where $u = 0 = v$) because we assume that reserves are continuously distributed, and so it occurs with probability zero. We also assume the resource is sufficiently valuable that $\alpha > \beta$; or $\theta < 1$, where

$$(6.9) \qquad \theta = \beta/\alpha.$$

Let us now assume that initial reserves T_{max} are distributed over $(0, \infty)$ with probability density function g. Then the reward to a u-strategist is $f(u,v) = \mathrm{E}[F] = \int_0^\infty \int_0^\infty F(x,y)dA$, where we have used (2.22) and E, as usual, denotes expected value. On using (6.8), we obtain

$$(6.10) \quad f(u,v) = \int_0^\infty g(x) \left\{ \int_0^{ux/v} \{\alpha x - (\alpha + \beta)vy\}g(y)\,dy \right\} dx$$
$$- \beta u \int_0^\infty g(y) \left\{ \int_0^{vy/u} xg(x)\,dx \right\} dy$$

if $(u,v) \neq (0,0)$ but $f(0,0) = 0$. For example, if T_{max} is uniformly distributed with mean μ according to

$$(6.11) \qquad g(t) = \begin{cases} \frac{1}{2\mu} & \text{if } 0 < t < 2\mu \\ 0 & \text{if } 2\mu < t < \infty \end{cases}$$

then (Exercise 6.2)

$$(6.12) \quad f(u,v) = \begin{cases} \frac{\alpha\mu u}{3v}\{2 - 3\theta v - (1-\theta)u\} & \text{if } u \leq v, v \neq 0 \\ \frac{\alpha\mu\{3(1-\{1+\theta\}v)u^2 + (\{2+\theta\}u-1)v^2\}}{3u^2} & \text{if } u > v \\ 0 & \text{if } u = 0 = v \end{cases}$$

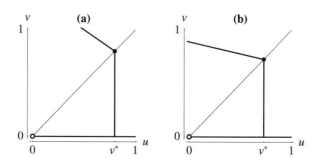

Figure 6.1. Rational reaction set for the war of attrition with uniform distribution of initial reserves when **(a)** $0 < \theta < \frac{2}{3}$ and **(b)** $\frac{2}{3} < \theta < 1$, where θ is defined by (6.9) and v^* by (6.14). R is drawn for **(a)** $\theta = 1/2$ and **(b)** $\theta = 3/4$.

and R is defined by

$$(6.13) \qquad u = \begin{cases} \dfrac{2-3\theta v}{2-2\theta} & \text{if } u \le v, v \ne 0 \\[2mm] u = \dfrac{2}{2+\theta} & \text{if } \quad u > v \\[2mm] \text{any } u \in (0,1] & \text{if } \quad v = 0. \end{cases}$$

Note that $(0,0) \notin R$. Thus the unique ESS is $v = v^*$ where

$$(6.14) \qquad\qquad v^* = \frac{2}{2+\theta},$$

as illustrated by Figure 6.1. At this ESS, each animal is prepared to expend at least two thirds of his initial reserves, although only the loser actually does so. Moreover, because each animal is prepared to persist for time $v^* T_{max}$ at the ESS, the victor is always the animal with the higher value of T_{max}—even though an opponent's reserves are assumed to be unobservable. Thus victory by the stronger animal need not imply that reserves are being assessed directly (although animals still respond to the distribution of reserves among the population). We will refer to the expectation that reserves are not being (directly) assessed as the no-assessment hypothesis.

Now, judicious approximation is the essence of modelling: any effect that is small in a real population is typically absent from a model. So we would expect a 90% win rate for stronger males in the world of real damselflies to translate into a 100% win rate for stronger males in a model world. And this is precisely what we have predicted: because both contestants are prepared to deplete their initial reserves by the same proportion, the weaker one invariably gives up first. But it is also what the assessment hypothesis predicts, at least in the absence of assessment errors, because the weaker animal withdraws as soon as he has judged himself to be the probable loser. So are the two competing hypotheses indistinguishable? We will return to this question later.

Although the uniform distribution is not the only one for which an ESS always exists (Exercise 6.3), for many distributions an ESS exists only if θ is sufficiently small. For example, suppose that T_{max} is distributed parabolically with mean μ according to

$$(6.15) \qquad g(t) = \begin{cases} \frac{3t(2\mu - t)}{4\mu^3} & \text{if } 0 < t < 2\mu \\ 0 & \text{if } 2\mu < t < \infty. \end{cases}$$

Then, on interchanging the order of integration in the second term of (6.10) and substituting from (6.15), we have

$$(6.16a) \quad f(u,v) = \int_0^{2\mu} g(x) \left\{ \int_0^{ux/v} \{\alpha x - (\alpha + \beta)vy\}g(y)\,dy \right\} dx$$

$$- \beta u \int_0^{2\mu} xg(x) \left\{ \int_{ux/v}^{2\mu} g(y)\,dy \right\} dx$$

$$= \alpha\mu u \left\{ \frac{u(5u\{(3-\theta)u - 4\} + 14v\{3 - (2-\theta)u\})}{35v^3} - \theta \right\}$$

for $u \le v, v \ne 0$; whereas, on interchanging the order of integration in the first term of (6.10) before substituting from (6.15), we have

$$(6.16b) \quad f(u,v) = \int_0^{2\mu} g(y) \left\{ \int_{vy/u}^{2\mu} \{\alpha x - (\alpha + \beta)vy\}g(x)\,dx \right\} dy$$

$$- \beta u \int_0^{2\mu} g(y) \left\{ \int_0^{vy/u} xg(x)\,dx \right\} dy =$$

$$\alpha\mu \left\{ 1 - (1+\theta)v + \frac{(14u\{(3+\theta)uv - 2v\} + 5v^2\{3 - (4+\theta)u\})v^2}{35u^4} \right\}$$

for $u \ge v$ (Exercise 6.4). Differentiating, for $u \ge v$ we have

$$(6.17) \qquad \frac{\partial f}{\partial u} = \frac{\alpha\mu v^3 (3u\{28 + 5(4+\theta)v\} - 60v - 28\{3+\theta\}u^2)}{35u^5},$$

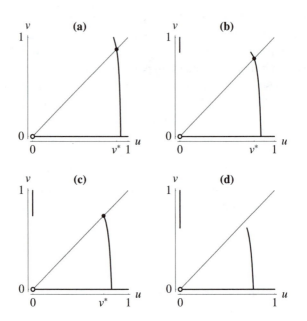

Figure 6.2. Rational reaction set for the war of attrition with parabolic distribution of initial reserves when **(a)** $\theta = \frac{1}{4}$, **(b)** $\theta = \frac{1}{2}$, **(c)** $\theta = \frac{108}{169}$ and **(d)** $\theta = \frac{7}{8}$; θ is defined by (6.9) and v^* by (6.19).

which implies that f has a local maximum where $u = \hat{u}$, defined by

$$(6.18) \qquad \hat{u} = \frac{84+15(4+\theta)v+\sqrt{\{84+15(4+\theta)v\}^2-6720(3+\theta)v}}{56(3+\theta)}.$$

But this local maximum is also a global one only if $f(\hat{u}, v) \geq f(0, v) = 0$; in which case, if $\hat{u} = v$ then v is an ESS. That is, if

$$(6.19) \qquad v^* = \frac{24}{24+13\theta}$$

then $v = v^*$ is an ESS when $f(v^*, v^*) \geq 0$ or

$$(6.20) \qquad \theta \leq \frac{108}{169}$$

(Exercise 6.5). When $\theta > \frac{108}{169}$, however, a population of v^*-strategists can be invaded by mutants who play $u = 0$, i.e., give up immediately. These results are illustrated by Figure 6.2, where R is sketched.

Although giving up immediately can invade strategy v^* if $\theta > \frac{108}{169}$, it is advantageous only when rare, and so does not eliminate v^*. For suppose that giving up immediately (strategy 1) and expending proportion v^* (strategy 2) are found at frequencies x_1 and x_2, respectively. Then the reward to strategy 1, which entails neither benefits nor costs, is $W_1 = 0$; and W_2, the reward to strategy 2, is $f(v^*, 0) = \alpha\mu$ times the probability of meeting strategy 1 plus $f(v^*, v^*)$ times the probability of meeting strategy 2. So

$$(6.21) \qquad \begin{aligned} W_1 - W_2 &= -\alpha\mu x_1 - f(v^*, v^*)x_2 \\ &= -f(v^*, v^*) - \{\alpha\mu - f(v^*, v^*)\}x_1 \end{aligned}$$

where $f(v^*, v^*) < 0$. For small x_1, $W_1 - W_2$ is positive, so that strategy 1 increases in frequency; for large x_1, $W_1 - W_2$ is negative, so that x_1 decreases. The population stabilizes where $W_1 = W_2$ or

$$(6.22) \qquad x_1 = \frac{2(169\theta - 108)}{13(48 + 61\theta)} < \frac{122}{1417}.$$

In other words, although there is no monomorphic ESS when $\theta > \frac{108}{169}$, there still exists a polymorphic ESS at which the proportion of those who give up immediately is invariably less than 9%. The polymorphism persists because negative payoffs to strategy 2 on meeting itself are balanced by large positive payoffs on rarer occasions when it meets strategy 1. Thus the alternative strategies do equally well on average, and there is no incentive to switch from one to the other.

Neither a uniform nor a parabolic distribution of initial reserves is especially realistic. Nevertheless, the results of this section are readily generalized to other distributions (Exercise 6.7), enabling one to find an acceptable fit to the data on damselfly energy reserves collected by Marden and Waage (and later Bob Rollins [**125**]). Now, we discovered earlier that, with respect to victory by the stronger contestant, these data are consistent with both the assessment and the no-assessment hypotheses. But there is also a difference. In our no-assessment model, a contest ends when the loser gives up after using a fixed proportion of its reserves, and so we predict a positive correlation between final loser reserves and contest duration; whereas, as discussed earlier, the assessment hypothesis predicts a negative correlation between strength difference and contest duration. This difference in predictions has been explored elsewhere,[10] although with inconclusive results. For us, however, the important point is that game-theoretic models are capable of yielding testable predictions.

6.3. Games among kin versus games between kin

We discovered in §6.2 that the war of attrition need have no monomorphic ESS if the cost of display is sufficiently high.[11] We assumed, however, that contestants are unrelated. Here we study how nonzero relatedness modifies the conditions for a strategy to be an ESS, with particular reference to the war of attrition. For the sake of simplicity, we assume throughout that the decision set is the unit square (as in §§6.1-6.2).

According to Darwin, animals will behave so as to transmit as many as possible of their genes to posterity. By descent, any two blood relations share a non-negligible proportion of genes for which there is variation in the population at large, and hence have a tendency to exhibit the same behavior—or to have the same strategy. But animals may also behave identically because of cultural association. Accordingly, let r be the probability that a strategy encounters itself by virtue of kinship, where kinship can be interpreted to mean either blood-relationship or similarity of character (as in common parlance); then $1 - r$ is the probability that the strategy encounters an opponent at random (still possibly itself). We will call r the *relatedness*, and assume that $r < 1$. Nothing in our analysis will depend on whether animals tend to behave identically by virtue of shared descent or shared culture—except, as remarked in §2.3 (p. 65), the time scale of the dynamic by which an ESS can be reached.

Let the population contain proportion $1 - \epsilon$ of an orthodox strategy v and proportion ϵ of a mutant strategy u; and let $f(u, v)$ denote, as usual, the reward to a u-strategist against a v-strategist. Then, because u and v are encountered with probabilities ϵ and $1 - \epsilon$, respectively, the reward to strategy s against a random opponent is

$$(6.23) \qquad w(s) = \epsilon f(s, u) + (1 - \epsilon)f(s, v),$$

---

[10] See [**150**, **156**].

[11] And variation in reserves is sufficiently low; see [**150**, **156**].

where s may be either u or v. Thus the payoff to s, allowing for s to encounter either itself with probability r or a random opponent with probability $1 - r$, is

$$(6.24) \quad W(s) = rf(s,s) + (1-r)w(s) =$$
$$rf(s,s) + (1-r)f(s,v) + \epsilon(1-r)\{f(s,u) - f(s,v)\}.$$

Again, s may be either u or v. Now, strategy v is an ESS among kin if $W(v) > W(u)$ for all $u \neq v$ when ϵ is sufficiently small. From (6.24), however, $W(v) - W(u)$ has the form $A + B\epsilon$ with

$$(6.25a) \qquad A = f(v,v) - f(u,v) + r\{f(u,v) - f(u,u)\},$$
$$(6.25b) \qquad B = (1-r)\{f(v,u) - f(v,v) - f(u,u) + f(u,v)\}.$$

So v is an ESS among kin if, for all $u \neq v$, either $A > 0$ or $A = 0$ and $B > 0$. That is, v is an ESS among kin if for all $u \neq v$, EITHER

$$(6.26a) \qquad\qquad f(v,v) > (1-r)f(u,v) + rf(u,u)$$

OR

$$(6.26b) \qquad \begin{aligned} f(v,v) &= (1-r)f(u,v) + rf(u,u) \\ f(v,u) &+ rf(u,v) > (1+r)f(u,u) \end{aligned}$$

(Exercise 6.10). If (6.26a) is satisfied for all $u \neq v$ then v is a strong ESS among kin, and otherwise v is a weak ESS among kin.

These conditions are simpler to deal with if first we define ϕ by

$$(6.27) \qquad\qquad \phi(u,v) = (1-r)f(u,v) + rf(u,u).$$

Then it follows from (6.26) that v is an ESS among kin when, for all $u \neq v$, EITHER

$$(6.28a) \qquad\qquad\qquad \phi(v,v) > \phi(u,v)$$

OR

$$(6.28b) \qquad \begin{aligned} \phi(v,v) &= \phi(u,v) \\ \phi(v,u) &> \phi(u,u) \end{aligned}$$

(Exercise 6.10). These conditions are merely (2.11) with ϕ in place of f. So v is an ESS among kin of the game with reward function f when v is an ordinary ESS of the game with reward function ϕ, which we therefore interpret as the kin-modified reward of the original game. The rational reaction set—obtained by setting $f = \phi$ in (6.2)—is found in the usual way, and if R intersects the line $u = v$ uniquely (i.e., if ϕ has a unique maximum at $u = v$) then v is a strong ESS.

Suppose, for example, that the war of attrition is played among kin and that initial reserves are uniformly distributed with mean μ so that (6.12) and (6.27) imply, for $u \leq v \neq 0$:[12]

$$(6.29) \qquad \phi(u, v) = \tfrac{1}{3}\alpha\mu \left\{ 2r + 2(1 - r)\left(\tfrac{1}{v} - \tfrac{1}{\lambda}\right)u - \tfrac{(1-r)(1-\theta)u^2}{v} \right\}$$

where θ is the cost/value ratio. Let us define $\lambda = \frac{2(1-r)}{r+(3-r)\theta}$ and

$$(6.30) \qquad\qquad v^* = \frac{2(1-r)}{2-r+(1+r)\theta}.$$

Then ϕ has a unique maximum along the line $(1 - \theta)\lambda u + v = \lambda$ if $v^* \leq v \leq \lambda$ but at $u = 0$ if $\lambda \leq v \leq 1$ (Exercise 6.11). For $u \geq v \neq 0$, on the other hand, we have

$$(6.31) \qquad \phi(u, v) = \frac{\alpha\mu(1-r)v^2}{3u}\left(2 + \theta - \tfrac{1}{u}\right)$$
$$+ \tfrac{1}{3}\alpha\mu\{3 - r - (1 + 2\theta)ru - 3(1 - r)(1 + \theta)v\},$$

which has a unique maximum where u is the only positive root of the cubic equation

$$(6.32) \qquad (1 + 2\theta)ru^3 + (2 + \theta)(1 - r)v^2u = 2(1 - r)v^2$$

(Exercise 6.11). In either case, the maximum occurs where $u = v$ if, and only if, $v = v^*$ (Figure 6.3). So v^* is the unique ESS among kin. Note that v^* decreases with r: the higher the relatedness, the lower the proportion of his initial reserves that an animal should be prepared to expend on contesting a resource. The upshot is that the average population reward at the ESS, namely,

$$(6.33) \qquad\qquad \phi(v^*, v^*) = f(v^*, v^*) = \frac{2(1-\theta+3\theta r)}{2-r+(1+r)\theta}$$

(Exercise 6.11) is an increasing function of r. But the higher the average reward to the population, the more cooperative the outcome. Thus kinship induces cooperation, as in the prisoner's dilemma among kin; see Exercise 6.15.

For a parabolic distribution of initial reserves, the effect of kinship is even more striking because it can ensure that there is always

---

[12] If $v = 0$ then ϕ has no maximum; it approaches its least upper bound $\alpha\mu(1-r/3)$ as $u \to 0$, but $\phi(0, 0) = 0$.

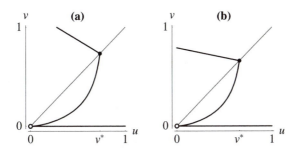

Figure 6.3. Rational reaction set for the war of attrition among kin with uniform distribution of initial reserves when (a) $3\theta + (3 - \theta)r < 2$ and (b) $3\theta + (3 - \theta)r > 2$, where θ is defined by (6.9) and v^* by (6.30). R is drawn for $r = 0.1$ and (a) $\theta = 1/2$, (b) $\theta = 3/4$. See Exercise 6.11.

a monomorphic ESS, even when θ is very close to 1 (although the required degree of relatedness, almost $\frac{1}{8}$, is perhaps rather high). From (6.15) and (6.27), we deduce after simplification that

$$(6.34a) \qquad \left.\frac{\partial\phi}{\partial u}\right|_{u=v} = \frac{\alpha\mu}{35v}\{24(1 - r) - \{24 - 11r + 13(1 + r)\theta\}v\}$$

$$(6.34b) \qquad \left.\frac{\partial^2\phi}{\partial u^2}\right|_{u=v} = -\frac{12\alpha\mu}{35v^2}(1 - r)(3 - \{1 + 2\theta\}v) < 0$$

(Exercise 6.12). So ϕ invariably has a local maximum on $u = v$ where $v = v^*$, defined by

$$(6.35) \qquad\qquad v^* = \frac{24(1-r)}{24-11r+13(1+r)\theta}.$$

This local maximum is also a global maximum if $\phi(v^*, v^*) = (v^*, v^*) \geq 0$. But (6.15) and (6.35) imply that

$$(6.36) \qquad f(v^*, v^*) = \frac{2\alpha\mu\{108-169\theta+35(1+13\theta)r\}}{35\{24-11r+13(1+r)\theta\}},$$

which is invariably positive if

$$(6.37) \qquad\qquad r \geq \frac{61}{490} \approx 0.1245$$

because $\theta < 1$ (Exercise 6.12). If $r < \frac{61}{490}$, on the other hand, then v^* is still an ESS if

$$(6.38) \qquad\qquad \theta \leq \frac{108+35r}{13(13-35r)}.$$

The critical value of θ increases with r from $\frac{108}{169}$ at $r = 0$ toward 1 as $r \to \frac{61}{490}$, in perfect agreement with (6.20) and (6.37). Note that $\phi(v^*, v^*)$ again increases with r.

Games among kin, which are games in which a mutant strategy has a greater—by r—than infinitesimal probability of interacting with itself, must be carefully distinguished from games *between* kin, which are games between specific individuals whose proportion of (variable) genes shared by descent is precisely r. For example, parents and their offspring, or two brothers, share $\frac{1}{2}$; an uncle and his nephew share $\frac{1}{4}$; and first cousins share $\frac{1}{8}$.[13] Although it is traditional in evolutionary game theory to use the same symbol r for both relatedness between kin and relatedness among kin, possible implications of the different interpretations should always be kept in mind. In particular, because a game between kin is not a population game, the concept of ESS is—strictly speaking—inappropriate; but we can still apply the concept of Nash equilibrium.[14] For $k = 1, 2$, let $f_k(u, v)$ denote Player k's reward in a game between non-relatives, and let $\phi_k(u, v)$ denote Player k's reward in the corresponding game between individuals whose relatedness is r. Then, because fraction r of either's reproductive success counts as reproductive success for the other, we have

$$(6.39) \qquad \begin{aligned} \phi_1(u, v) &= f_1(u, v) + r f_2(u, v) \\ \phi_2(u, v) &= f_2(u, v) + r f_1(u, v). \end{aligned}$$

These rewards are usually known as the players' *inclusive fitnesses*.

Although in general we must allow for asymmetry (Exercise 6.16), it is sometimes reasonable to assume that the reward structure is symmetric, i.e., $f_2(u, v) = f_1(v, u)$. Then the game can be analyzed in terms of a single reward or inclusive-fitness function ϕ defined by

$$(6.40) \qquad \phi(u, v) = f(u, v) + r f(v, u).$$

Comparing (6.40) with (6.27), we see that a symmetric game between kin differs from the corresponding population game among kin not only in terms of interpretation, but also in terms of mathematical

---

[13] See, e.g., Chapter 9 of [**32**].

[14] This is not to say that a Nash equilibrium between kin could not correspond to an ESS of a larger population game in which it specifies a conditional strategy; then Player 1 and Player 2 in the game between kin would correspond to roles that any individual could have in the population game. This point is illustrated by Exercises 1.32 and 2.25 (with regard to ordinary ESSes).

structure. Nevertheless, their properties often overlap. For example, if f is bilinear, then v^* is an ESS among kin only if (v^*, v^*) is a Nash equilibrium of the corresponding symmetric game between kin (although the first may be strong but the second weak, as illustrated by Exercise 6.15); and sometimes an ESS v^* among kin corresponds to a symmetric Nash equilibrium between kin, even if f is not bilinear (as illustrated by Exercise 6.13). In general, however (and as Exercise 6.14 illustrates), there is no such correspondence—which is unsurprising, because the games are different![15]

6.4. Information and strategy: a mating game

The games we have studied so far in this chapter are separable in a degenerate sense because $p = 1$ in (6.1). Moreover, although we have already seen a population game that is separable with $p = 2$—namely, Four Ways (§1.4)—its reward function is bilinear, and so its ESS is weak. Here is a game with $p = 2$ whose ESS is strong.

Sperm competition is competition between the ejaculates of different males for fertilization of a given set of eggs [175]. For example, male 13-lined ground squirrels routinely "queue" for mating with oestrous females, typically in pairs, with the first male having an advantage [202]. If ejaculates are costly, in the sense that increased expenditure of reproductive effort on a given ejaculate reduces the number of matings that can be achieved, should a male expend more on a mating in the disfavored role to compensate for his disadvantage—or should he expend more on a mating in the favored role to capitalize on his advantage? We explore this question in terms of game theory.

Let α denote the *fairness* of the competition between males, i.e., the effectiveness ratio of a unit of disfavored sperm compared to a unit of favored sperm: as α increases from 0 to 1, the unfairness of the favored male's advantage in the "raffle" for fertilization decreases, until the raffle becomes fair when $\alpha = 1$. Consider a game among a population of males who mate randomly either with unmated or with once mated females from a set of females who mate at most twice. The proportion of unmated females may be any positive number less than 1. Thus each male may occupy either of two roles, and a strategy

---

consists of an amount of sperm to ejaculate in each. Unless the raffle is fair, one of these roles is favored; let us denote it by A, and the disfavored role by B. Let the amounts ejaculated in roles A and B be v_1 and v_2, respectively, for the population strategy but u_1, u_2 for a potential mutant. That is, a mutant's strategy is a two-dimensional vector $u = (u_1, u_2)$ in which u_1 is the amount of sperm ejaculated when that male is in role A, and u_2 is the amount of sperm ejaculated when he is in role B. Similarly, the population strategy is $v = (v_1, v_2)$, where v_1 and v_2 are the amounts of sperm ejaculated in roles A and B, respectively.

For any given female, let K denote her maximum potential future reproductive success from a given set of eggs, let X denote the favored male's sperm expenditure, let Y denote the disfavored male's expenditure and let T denote the female's effective total number of sperm. Then the proportion of her reproductive success that accrues to the favored male is X/T, and the proportion that accrues to the disfavored male is $1 - X/T$. To obtain T, we devalue the disfavored male's expenditure by the fairness α. Thus

$$(6.41) \qquad\qquad T = X + \alpha Y.$$

Let $Kg(T)$ denote the female's EFRS as a function of effective total sperm number. Then it is reasonable to assume that the proportion g is a concave increasing function of T, with $g(0) = 0$ and $g(\infty) = 1$.[16] For the sake of simplicity, we satisfy these conditions by assuming

$$(6.42) \qquad\qquad g(T) = \frac{T}{\epsilon + T}$$

throughout, where ϵ—the number of sperm that would fertilize half of a female's eggs—is relatively small, in a sense to be made precise.

Let $W_A(X, Y)$ denote the favored male's expected reproductive gain from the female, and let $W_B(X, Y)$ denote that of the disfavored male. Then, from (6.41)-(6.42),

$$(6.43a) \qquad W_A(X, Y) = \frac{X}{T} \cdot Kg(T) = \frac{KX}{\epsilon + X + \alpha Y}$$

$$(6.43b) \qquad W_B(X, Y) = \left(1 - \frac{X}{T}\right) Kg(T) = \frac{K\alpha Y}{\epsilon + X + \alpha Y}$$

---

[16]For a discussion of this point see [**147**], on which §6.4 is based.

Note that $W_B(X, X)/W_A(X, X) = \alpha$: whenever sperm expenditures are equal, the disfavored male's gain is lower than that of the favored male by factor α, the fairness. Note also that $W_A(X, Y) + W_B(X, Y) = Kg(T)$: because a female mates at most twice, the sum of reproductive gains for two mates must equal her EFRS.

The cost of a mating in terms of EFRS increases with sperm expenditure, ultimately at a prohibitive rate; i.e., denoting sperm expenditure by s (which may be either X or Y) and mating cost by $Kc(s)$, so that $c(s)$ is a dimensionless quantity, $c(0) = 0$, $c'(s) > 0$ and $c''(s) \geq 0$. For simplicity's sake, we satisfy these conditions with

$$(6.44) \qquad c(s) = \gamma s,$$

where γ has dimensions SPERM^{-1}, so that $\delta = \sqrt{\epsilon\gamma}$ is a dimensionless measure of uncertainty of fertilization in the absence of competition: the lower the value of δ, the lower the amount of sperm with which an uncontested male can expect to fertilize a given egg set. We now quantify our assumption that the egg set can be fertilized by a relatively small amount of sperm by requiring $\delta < 1$.

Let p_A denote the probability that a mutant focal male is in role A, allocating u_1 against a male who allocates v_2 in role B, so that the focal male's expected reproductive gain from the female is $W_A(u_1, v_2)$. The cost of this mating is $Kc(u_1)$, by assumption, so that the mutant's net increase of EFRS is $W_A(u_1, v_2) - Kc(u_1)$. Similarly, let $p_B = 1 - p_A$ denote the probability that the mutant is in role B, allocating u_2 against a male who allocates v_1 in role A, so that his net reward is $W_B(v_1, u_2) - Kc(u_2)$. Note that the proportion of unmated females is p_A when mating first is favored but p_B when mating second is favored (because the favored role is always role A). On multiplying the reward from each role by the probability of occupying that role and adding, the reward to a u-strategist in a population of v-strategists— the expected payoff from the next female, who is analogous to the next customer in Store Wars (§1.5)—becomes

$$(6.45) \quad f(u, v) = p_A\{W_A(u_1, v_2) - Kc(u_1)\}$$
$$+ \ p_B\{W_B(v_1, u_2) - Kc(u_2)\}$$
$$= Kp_A u_1 \left(\tfrac{1}{\epsilon + u_1 + \alpha v_2} - \gamma\right) + Kp_B u_2 \left(\tfrac{\alpha}{\epsilon + v_1 + \alpha u_2} - \gamma\right)$$

Note that this expression is a special case of (6.1) with $p = 2$.

Straightforward partial differentiation reveals that

$$(6.46) \qquad \frac{\partial f}{\partial u_1} = \frac{\epsilon + \alpha v_2}{(\epsilon + u_1 + \alpha v_2)^2} - \gamma, \qquad \frac{\partial f}{\partial u_2} = \frac{\alpha(\epsilon + v_1)}{(\epsilon + r u_2 + v_1)^2} - \gamma$$

Hence the unique best reply to $v = (v_1, v_2)$ is $u = (\hat{u}_1, \hat{u}_2)$, where

$$(6.47a) \qquad \hat{u}_1 = \begin{cases} \sqrt{\frac{\epsilon + \alpha v_2}{\gamma}} - \epsilon - \alpha v_2 & \text{if } \alpha \gamma v_2 < 1 - \delta^2 \\ 0 & \text{if } \alpha \gamma v_2 \geq 1 - \delta^2 \end{cases}$$

$$(6.47b) \qquad \hat{u}_2 = \begin{cases} \sqrt{\frac{\epsilon + v_1}{\alpha \gamma}} - \frac{\epsilon + v_1}{\alpha} & \text{if } \gamma v_1 < \alpha - \delta^2 \\ 0 & \text{if } \gamma v_1 \geq \alpha - \delta^2 \end{cases}$$

(Exercise 6.17). If v is the best reply to itself, i.e., if $\hat{u}_1 = v_1$ and $\hat{u}_2 = v_2$, then v is the unique strong ESS. We cannot satisfy $\hat{u}_1 = v_1, \hat{u}_2 = v_2$ with $v_1 = 0$, because (6.47) would then require $\alpha \gamma v_2 \geq 1 - \delta^2$ with $v_2 = \sqrt{\epsilon / \alpha \gamma} - \epsilon / \alpha$ and $\alpha > \delta^2$ or $\delta \sqrt{\alpha} \geq 1$ and $\sqrt{\alpha} > \delta$, the first of which contradicts $\delta < 1$ even if the second is satisfied (because $0 < \alpha \leq 1$). So we must have $v_1 > 0$. But it is possible to have either $v_2 = 0$ or $v_2 > 0$ (Exercise 6.17). The upshot is that the unique ESS for $\epsilon > 0$ is $v = (v_1^*, v_2^*)$, where

$$(6.48a) \qquad \gamma v_1^* = \begin{cases} \delta(1 - \delta) & \text{if } 0 \leq \alpha \leq \delta \\ \dfrac{\sqrt{\alpha^2 + 4\delta^2 \alpha(1 + \alpha)} - 2\delta^2 \alpha(1 + \alpha) + \alpha}{2(1 + \alpha)^2} & \text{if } \delta < \alpha \leq 1 \end{cases}$$

$$(6.48b) \qquad \gamma v_2^* = \begin{cases} 0 & \text{if } 0 \leq \alpha \leq \delta \\ \dfrac{2\delta^2(\alpha^2 - \delta^2)}{\alpha\left\{2\delta^2(1 + \alpha) - \alpha^2 + \alpha\sqrt{\alpha^2 + 4\delta^2 \alpha(1 + \alpha)}\right\}} & \text{if } \delta < \alpha \leq 1 \end{cases}$$

(and both are dimensionless, because γ has dimensions SPERM^{-1}). For $\epsilon = 0$, however, we obtain the somewhat surprising result that

$$(6.48c) \qquad v_1^* = v_2^* = \frac{\alpha}{\gamma(1 + \alpha)^2}$$

(Exercise 6.17). Because (6.48) implies $\alpha(v_1^* - v_2^*) = \epsilon(1 - \alpha)$ whenever v_1^* and v_2^* are both positive, we also have $v_1^* = v_2^*$ when the raffle is fair, i.e., when $\alpha = 1$; but otherwise $v_1^* > v_2^*$.[17]

Sperm expenditures at the ESS are plotted against α in Figure 6.4 in units of γ^{-1}, i.e., the right-hand sides of (6.48) are plotted. In

---

[17]Note that this result requires the assumption that costs are independent of role: if γ is replaced by γ_A and γ_B in roles A and B, respectively, then $\alpha(\gamma_A v_1^* - \gamma_B v_2^*) = \epsilon(\gamma_B - \alpha \gamma_A)$ when $v_1^* > 0, v_2^* > 0$ at the ESS, implying $v_1^* < v_2^*$ if $\gamma_B < \alpha \gamma_A$.

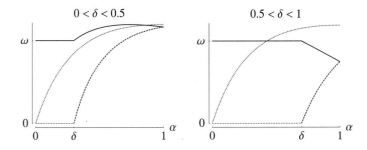

Figure 6.4. Sperm expenditure versus fairness α at the ESS when roles are certain for fixed $\delta = \sqrt{\epsilon\gamma}$, where ϵ is the number of sperm that would fertilize half of the egg set, γ is the marginal-cost parameter and $\omega = \delta(1 - \delta)$. The solid curve shows expenditure in favored role A; the dashed curve shows expenditure in disfavored role B; and the dotted curve shows expenditure in either role when $\delta = 0$. Sperm expenditure is measured in units of γ^{-1} along the vertical axis. (Units of ϵ would preclude a comparison with the result for $\epsilon = 0$.)

each diagram, the solid curve corresponds to favored role A and the dashed curve to disfavored role B. The dotted curve corresponds to $\delta = 0$, when sperm expenditure is the same in either role, by (6.48c).

The form of the ESS depends on whether δ exceeds $\frac{1}{2}$. If not, i.e., if $0 < \delta \leq \frac{1}{2}$, then the maximum possible expenditure at the ESS is always $\frac{1}{4\gamma}$, precisely the same value as when fertilization is assured $(\delta = 0)$; but the maximum occurs where $\alpha = 1/(1 + 4\delta^2)$ instead of at $\alpha = 1$ (as for $\delta = 0$). If $\frac{1}{2} < \delta < 1$, on the other hand, then expenditure is always lower than when fertilization is assured, the maximum occurring where $\alpha \leq \delta$, i.e., in the absence of competition. In both cases, $v_1 - v_2$ is a strictly decreasing function of α when $\delta < \alpha \leq 1$. That is, the divergence between expenditures in roles A and B increases with unfairness in the de facto competitive domain.

All of the above assumes, however, that animals know which role they are in. A different result emerges if they know the proportion of unmated females but do not know (with certainty) whether they are first or second to mate. Then sperm expenditure is of necessity role-independent, say $\overline{v}$ for the population, $\overline{u}$ for a mutant; the reward

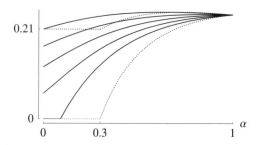

Figure 6.5. Sperm expenditure versus fairness at the ESS when roles are uncertain (solid curves) for $p \to 0$ (lowest), $p = 0.25$, $p = 0.5$, $p = 0.75$, $p \to 1$ (highest) and $\delta = 0.3$. The separate expenditures when roles are certain are also shown for comparison (dotted curves). As in Figure 6.4, sperm expenditure is measured in units of γ^{-1} along the vertical axis.

to a $\overline{u}$-strategist in a population of $\overline{v}$-strategists becomes

$$(6.49) \qquad f(\overline{u}, \overline{v}) = p_A \{ W_A(\overline{u}, \overline{v}) - Kc(\overline{u}) \}$$
$$+ \; p_B \{ W_B(\overline{v}, \overline{u}) - Kc(\overline{u}) \}$$
$$= K\overline{u} \left(\frac{p}{\epsilon + \overline{u} + \alpha \overline{v}} + \frac{(1-p)\alpha}{\epsilon + \overline{v} + \alpha \overline{u}} - \gamma \right),$$

on using p in place of p_A to indicate that roles are now uncertain (thus, for consistency, $0 < p < 1$); and the unique ESS $\overline{v}^*$ is given by

$$(6.50) \quad \gamma \overline{v}^* = \begin{cases} 0 & \text{if } \alpha \leq \alpha_c \\ \dfrac{2\delta^2 \{ p + \alpha(1-p) - \delta^2 \}}{2\delta^2 (1+\alpha) - \alpha + \sqrt{\alpha^2 + 4\delta^2 (1+\alpha) \{ p + \alpha^2 (1-p) \}}} & \text{if } \alpha > \alpha_c \end{cases}$$

where $\alpha_c = \frac{\delta^2 - p}{1 - p}$ (Exercise 6.18). This ESS is plotted in Figure 6.5 for $\delta = 0.3$—i.e., for the same value of δ as in the left-hand panel of Figure 6.4. Note that the ESS for uncertain roles agrees with the ESS for certain roles if the raffle is fair ($\alpha = 1$): if expenditures are equal anyway, then it makes no difference whether animals are aware of their roles. For the most part, however, the ESSes differ. Typically, expenditure when the role is uncertain is higher than when the mating is known to be disfavored but lower than when it is known to be favored; in particular, the threshold α_c, below which there is zero expenditure, is lower than the corresponding threshold δ for a mating that is known to to be disfavored. Increasing p increases the sperm expenditure. Note, however, that v_1^* is not the limit of $\overline{v}^*$ as $p \to 1$, and v_2^* is not the limit of $\overline{v}^*$ as $p \to 0$—unsurprisingly, because each allocation is a response to itself when roles are uncertain, but to a separate allocation when roles are known.

The result that $\overline{v}^* = 0$ if $\alpha \leq \alpha_c$ is at first paradoxical, because no eggs are fertilized if $\overline{v}^* = 0$. But we have to remember that our game is a game among males, and that γK is the (marginal) *opportunity* cost of a unit of sperm expenditure—i.e., it is the cost of the unit in terms of future mating opportunities that the male thereby forgoes. These forgone opportunities need not exist among the present set of females: because the males are out for themselves, they should have no qualms about jeopardizing the reproductive success of the present females if their own mating prospects elsewhere are brighter thereby. Moreover, if the current set of females were the only set available, then the cost of forgoing future matings would have to be close to zero; and $\alpha \leq \alpha_c$ cannot be satisified if $\delta \approx 0$.

All the same, it is perhaps unsatisfactory to resolve a paradox in terms of that which is not explicitly modelled. Sperm competition is a complex affair, requiring more sophisticated game-theoretic models than we have developed here. For pointers to the literature, see §6.9.

6.5. Roving ravens: a recruitment game

It is not unusual for an animal who finds a major new source of food to recruit others to the site. But it is comparatively rare for an animal to delay recruitment until, e.g., the following day. The best known example concerns juvenile ravens, who, for the sake of simplicity, are assumed to be male. These animals will often return to their communal roost after spotting a large carcass and wait until the following morning before leading others to it. Why? Bernd Heinrich [95], who has conducted long-term studies of ravens in the forests of Western Maine, suggests two possible answers. The first, the status-enhancement hypothesis, is that delayed recruitment is favored because a recruiter's social status (and hence attractiveness to females) increases with the number of followers he leads to a food source. The second possible answer, the posse hypothesis, is that aggregation is favored because larger groups are more likely to usurp a

defended carcass. Here we use a game-theoretic model to explore the
logic of delayed recruitment in the light of these hypotheses.

Consider a population of unrelated overwintering juveniles who
forage independently by day, but roost together at night in groups of
size $N + 1$. They search during a period of L daylight hours, from
early morning until dusk, for a "bonanza"—i.e., an opened carcass.
Bonanzas are rare and ephemeral, but bountiful. For the sake of
simplicity, we assume that a bonanza lasts only a day (before, e.g.,
it is irretrievably buried by a heavy snowfall), but is ample enough
to satiate any juvenile who exploits it. Moreover, we assume that
bonanzas reappear at a rate of precisely one per day. Thus the object
of the search is to find the lone bonanza, wherever in a huge area it
may have randomly appeared.

We scale EFRS (defined on p. 7) so that feeding at a bonanza
increases it by 1 unit of fitness. Also, we assume that juveniles feed
solely at bonanzas, and we ignore the effect of predation. Thus any
additional fitness increment is due to a rise in social status. We
assume that the fitness of a discoverer increases by α units for every
animal he recruits, and we will refer to this assumption as the *status-
enhancement* effect.

Each day, an isolated individual locates the bonanza with probability b, and hence with probability $1-b$ finds no bonanza. We assume
that $0 < b < 1$. The time until a focal individual locates the bonanza
by himself is a random variable Z, continuously distributed between
0 and ∞ with probability density function (p.d.f) g and cumulative
distribution function (c.d.f) G defined by $G(z) = \mathrm{Prob}(Z \leq z) =
\int_0^z g(\xi)\,d\xi$, so that $G(0) = 0$, $G(L) = b$ and $g(z) = G'(z)$. Because
Z has a continuous distribution, two individuals cannot discover the
bonanza simultaneously (although any number of individuals may
discover the bonanza at some time during the day).

Individuals cease to search either at the end of the day or when
they become privy to the bonanza. But the focal individual need not
discover the bonanza for himself. Let W denote the time until one
of the N non-focal individuals locates the bonanza. Then W is also
continuously distributed between 0 and ∞, but with a different c.d.f.
and p.d.f., say H and h, respectively. Because $1-\mathrm{Prob}(W \leq w)$ is the
probability that none of N independent foragers locates the bonanza

by time w, we have

(6.51a) $\qquad H(w) \;=\; \mathrm{Prob}(W \le w) \;=\; 1 - (1 - G(w))^N$

(6.51b) $\qquad h(w) \;=\; H'(w) \;=\; N(1 - G(w))^{N-1}G'(w).$

If $Z < W < L$, then the focal individual ceases to search at time Z. If $Z > W$, however, then his stopping time depends on the strategy of the other individuals.

For the sake of tractability, we assume each individual in the population to be either an immediate recruiter (strategy 1, or IR) or a delayed recruiter (strategy 2, or DR). Both types respond to recruitment by others: this aspect of behavior is assumed to be fixed. If an IR-strategist discovers the bonanza, then he immediately recruits all other individuals within range. He then remains at the carcass with his gang of recruits until all return to the roost at dusk, i.e., at time L. If another individual subsequently discovers the carcass himself, independently of recruitment, then he neither rises in status nor joins the gang; but he is able to satiate himself if the gang acquires access to the food. (If necessary, we can think of such a late discoverer as lurking in the vicinity of the carcass until the gang has left and gorging himself rapidly before likewise returning to the roost.)

Each juvenile will have access to the food if it is not defended, or if it is defended but the gang contains enough recruits to repulse the resident adults. Let δ denote the probability that the bonanza is defended by a resident pair, and let $\rho(I)$ denote the probability that a gang of I juveniles (including its leader) will repulse the residents. For the sake of simplicity, we assume that a lone animal has no chance of repulsing them, but that the probability of a gang's success thereafter increases in direct proportion to the number of recruits, with maximum probability σ (when the entire roost is at the bonanza):

(6.52) $\qquad\qquad \rho(I) \;=\; (I - 1)\sigma/N.$

With this assumption, the probability of access for a gang of size I is $a(I) = (1 - \delta)\cdot 1 + \delta\rho(I)$, or

(6.53) $\qquad\qquad a(I) \;=\; 1 - \delta\{1 - (I - 1)\sigma/N\}.$

We will refer to the assumption embodied in (6.52) as the *posse* effect; it holds whenever bonanzas are defended with positive probability, but is absent when bonanzas are undefended. In other words, there is a posse effect if and only if $\delta > 0$.

We are now ready to begin our calculation of the reward matrix

$$(6.54) \qquad A = \begin{bmatrix} R & S \\ T & P \end{bmatrix}$$

in which a_{ij} is the increase in EFRS to an individual playing strategy i against N individuals playing strategy j; for example, T is the reward to an IR-strategist among N DR-strategists. For this calculation we need some combinatorial results associated with the binomial distribution, namely,

$$(6.55a) \qquad \sum_{k=0}^{N} \binom{N}{k} p^k (1-p)^{N-k} = 1$$

$$(6.55b) \qquad \sum_{k=0}^{N} k \binom{N}{k} p^k (1-p)^{N-k} = Np$$

$$(6.55c) \qquad \sum_{k=0}^{N} k^2 \binom{N}{k} p^k (1-p)^{N-k} = Np + N(N-1)p^2$$

$$(6.55d) \qquad \sum_{k=0}^{N} \frac{1}{k+1} \binom{N}{k} p^k (1-p)^{N-k} = \frac{1 - (1-p)^{N+1}}{(N+1)p}$$

for $0 < p < 1$, where $\binom{N}{k}$ denotes the number of ways of choosing k objects from N (Exercise 6.19).

First we calculate R. Although, in an IR population, at most one animal per day—the first discoverer—can rise in status, in a DR population several animals may rise in status by locating the carcass independently; in which case, the corresponding increase of fitness is shared equally among them. A DR-strategist who discovers the bonanza delays recruitment to the following day at dawn. If he is the sole discoverer, then the other N juveniles follow him at dawn from the roost to the site of the carcass; but if several animals find the bonanza, then each is equally likely to lead. A juvenile who fails to discover the bonanza will fail to rise in status, but his fitness will still increase by 1 unit if some DR-strategist found the bonanza and the flock gains access to the food. So, conditional upon access, the increase of EFRS to a DR-strategist in a DR population is

$$(6.56) \qquad b\,\mathrm{E}\left[1 + \frac{\alpha(N-D)}{1+D}\right] + (1-b)\,\mathrm{Prob}(D \geq 1),$$

where E denotes expected value and D is the number of other discoverers. Thus, on using (6.55) and multiplying by the probability of access (Exercise 6.20), we find that the increase of EFRS to a DR-strategist in a population of DR-strategists is

$$(6.57) \qquad R = (1 - \delta\{1 - \sigma\})\left\{(1 + \alpha)\{1 - (1 - b)^{N+1}\} - \alpha b\right\}.$$

Next we calculate T. Suppose that the population contains a mutant IR-strategist (in addition to N DR-strategists). If this focal individual finds a bonanza at time Z, then he will immediately recruit the other individuals—all of them DR-strategists—within his range of attraction. Let $F(Z)$ denote the number of other individuals still foraging at time Z, and let r denote the recruitment probability for each. In ravens, immediate recruitment appears to be predominantly vocal, although visual cues may play a lesser role. Thus, because individuals are assumed to search at random over a large area, we can interpret r as the ratio of the area of the animal's call range to the total search area for the roost. Then the expected number of recruits is $rF(Z)$. The focal individual's fitness will increase by 1 unit for access to the food plus $\alpha rF(Z)$ units for the status of a discoverer; and the size of the gang will be $rF(Z) + 1$, so that the probability of access will be $a(rF(Z) + 1)$, where a is defined by (6.53). Even if the IR-strategist fails to discover the bonanza, however, i.e., if $Z > L$, his fitness will increase by 1 if one of the DR-strategists discovers it (i.e., if $W < L$) and the group as a whole gains access the following dawn. But a payoff of 1 with conditional probability $1 - \delta(1 - \sigma)$ is equivalent to a payoff of $1 - \delta(1 - \sigma)$ with conditional probability 1. So the IR-strategist's payoff against the N DR-strategists is $\psi_T(Z, W) =$

$$(6.58) \qquad \begin{cases} \left(1 - \delta\left\{1 - \frac{\sigma rF(Z)}{N}\right\}\right)(1 + \alpha rF(Z)) & \text{if } Z < L \\ 1 - \delta(1 - \sigma) & \text{if } Z > L, W < L \\ 0 & \text{if } \min(Z, W) > L \end{cases}$$

and the increase of EFRS to an IR-strategist in a population of DR-strategists is its expected value,

$$(6.59) \qquad T = \mathrm{E}\left[\psi_T(Z, W)\right].$$

To calculate this expression, we must average over the distributions of both $F(Z)$ and Z. Because $F(Z) = k$ when $N - k$ others have

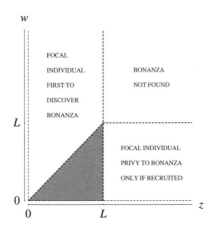

Figure 6.6. Joint sample space of Z and W.

ceased to search, and because—in a population of DR-strategists—
an animal has ceased to search at time Z with probability $G(Z)$,
$\mathrm{Prob}(F(Z) = k) = \binom{N}{k}\{1 - G(Z)\}^k\{G(Z)\}^{N-k}$. Thus, if we temporarily fix the value of Z and denote the average of (6.58) over the
distribution of $F(Z)$ by $\phi_T(Z)$, for $Z < L$ we obtain

$$(6.60) \quad \phi_T(Z) = 1 - \delta + r\{N\alpha(1-\delta) + (1+\alpha r)\delta\sigma\}\{1 - G(Z)\}$$
$$+ (N-1)\alpha\delta\sigma r^2\{1 - G(Z)\}^2$$

(Exercise 6.20). Now, on using the result that

$$(6.61) \quad \int_0^L \{1 - G(z)\}^i g(z)\, dz = \int_{1-b}^1 x^i\, dx = \frac{1-(1-b)^{i+1}}{(i+1)}$$

with $i = 0, 1, 2$ and noting that Z and W are independent random
variables, and hence are distributed over the infinite quadrant in Figure 6.6 with joint probability density $g(z)h(w) = G'(z)H'(w)$ per unit
area, we deduce from (6.58)-(6.59) that

$$(6.62) \quad T = \int_0^\infty h(w) \int_0^L \phi_T(z)g(z)\, dz\, dw$$
$$+ (1 - \delta(1 - \sigma)) \int_0^\infty h(w) \int_L^\infty g(z)\, dz\, dw$$

$$
\begin{aligned}
= \ & b(1-\delta) + r\left\{N\alpha(1-\delta) + (1+\alpha r)\delta\sigma\right\} b \left(1 - \tfrac{1}{2}b\right) \\
& + (N-1)\alpha\delta\sigma r^2 b \left(1 - b + \tfrac{1}{3}b^2\right) \\
& + (1-b)\{1 - (1-b)^N\}\left(1 - \delta(1-\sigma)\right).
\end{aligned}
$$

Next we calculate P. As remarked above, if the focal IR-strategist is not a mutant but belongs instead to a population of IR-strategists, then we must allow for the possibility that he ceases to search before the end of the day, not because he has discovered the bonanza himself, but because another IR-strategist has discovered the bonanza and the focal individual is near enough to be recruited. In that case, the focal individual's status will not rise, but his fitness will still increase by 1 unit if (immediate) recruitment yields a large enough gang for access to the carcass. If the focal individual is first to discover the carcass himself, then his fitness increases by 1 plus α times the expected number of recruits, or $1 + \alpha r N$. If he is not the first discoverer of the carcass (implying $Z > W$), then his fitness will increase by 1 if either he is recruited, or subsequently he discovers the carcass for himself. In the triangle shaded in Figure 6.6, the focal individual is recruited with conditional probability r; and with conditional probability $1 - r$ he is not recruited, but subsequently finds the bonanza himself. In either case, his reward is 1. In the open rectangle to the right of this triangle, however, the focal individual enjoys a reward of 1 only if he is recruited. But a payoff of 1 with conditional probability r is equivalent to an expected payoff of r with conditional probability 1. Thus the focal individual's payoff, conditional upon access to the carcass, is

$$
(6.63) \qquad \psi_P(Z,W) \ = \ \begin{cases}
1 + \alpha r N & \text{if } Z < \min(L, W) \\
1 & \text{if } W < Z < L \\
r & \text{if } W < L < Z \\
0 & \text{if } \min(Z, W) > L,
\end{cases}
$$

and so his increase in EFRS, conditional on access to the carcass, is

$$
\begin{aligned}
(6.64) \quad \mathrm{E}\left[\psi_P(Z,W)\right] \ = \ & (1 + \alpha r N)\mathrm{Prob}(Z < \min(L, W)) \\
& + 1 \cdot \mathrm{Prob}(W < Z < L) + r \cdot \mathrm{Prob}(W < L < Z).
\end{aligned}
$$

Whenever the reward is positive—regardless of whether the focal individual is the discoverer, a recruit, or neither—the expected size of the gang of juveniles is $rN + 1$. So the probability of access is $a(rN + 1) = 1 - \delta(1 - \sigma r)$, and the (unconditional) increase of EFRS to an IR-strategist in an R population is

$$(6.65) \qquad P = \{1 - \delta(1 - \sigma r)\} \, \mathrm{E}\left[\psi_P(Z, W)\right],$$

which (Exercise 6.20) readily reduces to $P =$

$$(6.66) \quad (1 - \delta + \delta\sigma r)\left\{\left(1 + \tfrac{\alpha N}{N+1}\right)r\{1 - (1 - b)^{N+1}\} + (1 - r)b\right\}.$$

Finally, to complete the reward matrix, we calculate S. Let the population consists of N IR-strategists and a mutant DR-strategist. This focal individual obtains an immediate reward only if he responds to an IR-strategist's call; if he finds the bonanza himself, then his reward is delayed to the following dawn. Thus we must allow for the fact that if the focal DR-strategist is first discoverer, then his expected number of recruits is no longer rN, but rather the number of IR-strategists that were not made privy to the bonanza the previous day (all others being satiated, by assumption). An IR-strategist is not made privy to the bonanza if he fails to find it, and it is not the case that both another IR-strategist finds it and the first animal is recruited. The probability of this event is

$$(6.67) \qquad \begin{aligned} q &= (1 - G(L)) \left(1 - r\{1 - (1 - G(L))^{N-1}\}\right) \\ &= (1 - b)(1 - r) + r(1 - b)^N. \end{aligned}$$

The corresponding expected number of recruits is qN. Thus although the probability of access remains $1 - \delta(1 - \sigma r)$ when $Z > W$, for $Z < W$ it changes to $a(Nq + 1) = 1 - \delta(1 - \sigma q)$. So, by analogy with (6.63)-(6.66), the increase of EFRS to a DR-strategist in an IR population is

$$(6.68) \quad \begin{aligned} S = \;&\{1 - \delta(1 - \sigma q)\}(1 + \alpha qN)\mathrm{Prob}(Z < \min(L, W)) \\ &+ \{1 - \delta(1 - \sigma r)\}\,\mathrm{Prob}(W < Z < L) \\ &\qquad + r\{1 - \delta(1 - \sigma r)\}\,\mathrm{Prob}(W < L < Z), \end{aligned}$$

from which (Exercise 6.20)

$$(6.69) \qquad S = P + \frac{\left(1 - \{1 - b\}^{N+1}\right)(q - r)\{\delta\sigma + \alpha N(1 - \delta + \delta\sigma\{q + r\})\}}{N + 1}.$$

Depending on the signs of $R - T$ and $S - P$, either IR or DR is potentially an ESS. If $R - T$ and $S - P$ are both positive, then DR is the only ESS; if both are negative, then IR is the only ESS; if $R - T$ is positive but $S - P$ is negative, then both strategies are evolutionarily stable; and if $R - T$ is negative but $S - P$ is positive, then neither strategy is a monomorphic ESS but each can infiltrate the other, so that a mixture of both will yield a polymorphic ESS. As in Chapter 5, we will refer to the strategy that yields the higher reward to the population as the cooperative strategy. Thus IR is the cooperative strategy if $P > R$, and DR is the cooperative strategy if $R > P$. In practice, however, it seems extremely unlikely that b and r are large enough to make $P > R$ for ravens; furthermore, mutualistic food sharing via immediate recruitment has been adequately explained elsewhere, and is not the subject of our investigation.[18] We therefore assume that $R > P$. So our game is an $(N + 1)$-player analogue of the cooperator's dilemma, the games being equivalent for $N = 1$. But because we are concerned with larger group sizes, $2R > S + T$ in (5.2) no longer applies.

First we ask whether a posse effect alone suffices for the evolution of delayed recruitment.[19] Setting $\alpha = 0$ in (6.62) and (6.69) yields

$$(6.70) \qquad R - T \;=\; b\delta\sigma\big\{1 - r\big(1 - \tfrac{1}{2}b\big)\big\}$$

$$(6.71) \qquad S - P \;=\; \frac{\delta\sigma(1-b)\{1-(1-b)^{N+1}\}}{N+1}\left(1 - \frac{r}{r_1}\right)$$

where

$$(6.72) \qquad r_1 \;=\; \frac{(1-b)}{2-b-(1-b)^N}.$$

By inspection, $R - T$ is invariably positive, but the sign of $S - P$ depends on r. Thus, in the absence of a status-enhancement effect, DR is always an evolutionarily stable strategy. Moreover, if $r < r_1$, then DR is the only evolutionarily stable strategy. If, on the other hand, $r > r_1$, then both IR and DR are evolutionarily stable; but if the population consists of immediate recruiters initially, then the noncooperative strategy will prevail. We can now answer the question of interest. If the immediate recruitment probability r exceeds

---

[18]For details, see [**155**], on which §6.5 is based.

[19]Note that in this case $R > P$ holds regardless: $\alpha = 0$ in (6.57) and (6.66) yields
$R - P = (1 - r)\{1 - \delta(1 - \sigma r)\}(1 - b)\{1 - \{1 - b\}^N\} + \delta\sigma(1 - r)\{1 - (1 - b)^{N+1}\}$,
which is invariably positive because b, δ, r and σ are all probabilities between 0 and 1.

Table 6.1. Evolutionarily stable strategies in the absence of
a posse effect. The critical values of r, namely, r_1 and r_2
are defined by (6.72) and (6.76), respectively, and invariably
satisfy $\max(r_1, r_2) < 1 - b < r_0$. The critical value of b,
namely, b_c, is the value of b for which $r_1 = r_2$.

$0 < r < \min(r_1, r_2)$	DR unique ESS
$0 < b < b_c,\ r_1 < r < r_2$	Two ESSes, DR and IR
$b_c < b < 1,\ r_2 < r < r_1$	Polymorphic ESS
$\max(r_1, r_2) < r < r_0$	IR unique ESS

r_1, then the posse effect alone does not suffice for the evolution of
cooperation. If, however, r lies below this critical value, then the cooperative strategy DR is a dominant strategy and will invade an IR
population. In sum, the posse effect alone suffices for the evolution of
delayed from immediate recruitment if r is sufficiently small. In fact,
such data as are available do tentatively suggest that $r < r_1$ [**155**].

Next we ask whether the status-enhancement effect alone suffices
for the evolution of delayed recruitment. Setting $\delta = 0$ in (6.57),
(6.62), (6.66) and (6.69) we find that

$$(6.73) \qquad R - T = Nb\alpha \left\{ 1 - \tfrac{1}{2}b \right\} (r_2 - r)$$

$$(6.74) \qquad S - P = \frac{\alpha N(1-b)\{1-(1-b)^{N+1}\}}{N+1} \left(1 - \frac{r}{r_1} \right)$$

and $R - P = (1 + \alpha)Nbr_2(1 - b/2)(1 - r/r_0)$, where

$$(6.75) \qquad r_0 = (1 + \alpha) \left\{ 1 + \frac{\alpha N}{N+1} \left(1 + \frac{b}{(1-b)(1-\{1-b\}^N)} \right) \right\}^{-1}$$

$$(6.76) \qquad r_2 = \frac{2(1-b)\{1-(1-b)^N\}}{Nb(2-b)}$$

and r_1 is defined by (6.72). Thus for $N \geq 3$, which always holds in
practice [**155**, p. 382], whether IR or DR is an ESS depends on r
according to Table 6.1, and as illustrated by Figure 6.7 for $\alpha = 0.5$
and various values of N. Note that $R > P$ excludes the shaded region,
whose curvilinear boundary is $r = r_0$. Note also that two ESSes seem
quite unlikely to arise in practice.

We can now answer the question of whether a status-enhancement
effect by itself suffices for the evolution of delayed recruitment. The
answer is yes, and for at least moderately large roosts the conditions
for DR to infiltrate an IR population are essentially the same; but

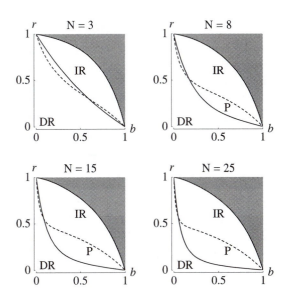

Figure 6.7. ESS regimes for $\delta = 0$ and $\alpha = 0.5$. The shaded region has boundary $r = r_0$; the other solid curve is $r = r_2$; the dashed curve is $r = r_1$; and P denotes a polymorphism. In general, r_1 and r_2 are independent of α but decrease with respect to N and b, whereas r_0 decreases with respect to all three parameters (and approaches $1 - b$ in the limit as $\alpha \to \infty$, $N \to \infty$); however, the dependence of r_0 on N is weak.

the conditions for DR to invade and become a monomorphic ESS are potentially much more stringent. Nevertheless, they must always be satisfied for sufficiently small b and r.

What happens if there is both a status-enhancement and a posse effect? See Exercise 6.21.

6.6. Cooperative wildlife management

What can an African government do to deter illegal hunting of wildlife, and hence promote its conservation? In strictly protected areas, such as large national parks, it can employ professional scouts to discourage and intercept poachers. But much of the Africa's most valuable wildlife is found elsewhere in communities where people are living and hunting, such as Zambia's Luangwa Valley, and there aren't nearly enough professional scouts to go round. So, as a practical matter, governments must find ways to create an incentive for communities to police themselves. Several such schemes, through which governments promise financial rewards to communities for refraining from illegal hunting and to individuals for reporting poachers, have been implemented over the years; yet they have met with only limited success. For example, improved law enforcement in Zambia curbed the decimation of large mammals that was rampant in the 1970s and 1980s, but mainly because poachers switched to targeting smaller animals to reduce the risk of being detected [**69**]. Could these schemes have been better designed? Here we attempt to answer this question by constructing a discrete population game in which each player chooses whether or not to poach, and whether or not to monitor the resource.

Let a community consist of $n + 1$ decision-making families or individuals, to whom a government is offering incentives to conserve a wildlife resource.[20] For the sake of simplicity, we assume that incentives are offered in cash (and thus avoid the issue of how non-cash benefits could be equitably distributed among deserving individuals). The maximum potential community benefit per period for conserving the resource is B. It is paid in full if no individual is hunting illegally, but reduced proportionately to zero as the number of poachers rises from zero to $n + 1$;[21] in any event, it is distributed equally among all individuals. Thus the community "wage" per period to each individual for conserving wildlife when m individuals are poaching is

$$(6.77) \qquad W_m = \frac{B}{n+1}\left(1 - \frac{m}{n+1}\right).$$

A choice of hunting technologies is available to residents. For example, in the Luangwa Valley of Zambia, they may either snare surreptitiously or hunt with guns. Before recent increases in law enforcement levels, hunting with guns was preferred. But snaring has now become more prevalent, due to its lower probability of detection. These two options exemplify the tradeoffs a poacher may face. We

---

[20]We use n in place of N to be consistent with [**147**], on which §6.6 is based.

[21]We implicitly assume that the government has a means of estimating the total number of poachers, even when infractions are not reported by the community.

use L to denote hunting with guns, a long-range technology, and S to denote hunting with snares, a short-range technology.

Let V_Z be the expected value of returns per period from using type-Z technology, let D_{iZ} denote the associated probability of being detected if i individuals are monitoring; and let C_Z denote the corresponding expected cost per period of a conviction for poaching. Then the expected benefit per period to each individual who poaches with type-Z technology, if i individuals are monitoring, is

$$(6.78) \qquad Z_i = V_Z \cdot (1 - D_{iZ}) + (V_Z - C_Z) \cdot D_{iZ} = V_Z - C_Z D_{iZ}$$

where Z is either L or S.[22] For the sake of simplicity, we assume that each monitor has a constant probability q_Z per period of observing while type-Z technology is used. So the probability that some monitor is observing—i.e., that not all monitors are not observing—is

$$(6.79) \qquad D_{iZ} = 1 - (1 - q_Z)^i,$$

where Z again is either L or S. Long-range technologies typically have a higher probability of detection, and so we assume that $q_L > q_S$.

The existence of a community wage ensures that the net value of poaching with type-Z technology is less than V_Z. When m individuals are poaching, an additional poacher reduces the community wage by $W_m - W_{m+1} = \frac{B}{(n+1)^2}$. Thus the net expected value per period of poaching with type-Z technology is

$$(6.80) \qquad v_Z = V_Z - \frac{B}{(n+1)^2}.$$

If v_L and v_S were both negative, then there would exist no incentive to poach. The persistence of poaching suggests, however, that one of them is positive, and so we assume that

$$(6.81) \qquad \max(v_L, v_S) > 0.$$

Expected returns V_Z depend not only on the value of captured animals, but also on the numbers taken per period with type-Z technology. Accordingly, it will be convenient to denote the higher-value technology by H, regardless of whether $H = L$ or $H = S$, and the other technology by O (so that $H = L$ implies $O = S$, and vice versa). Thus (6.81) is equivalent to $v_H > 0$.

---

[22]Note that (6.78) does not imply that convicted poachers keep their ill-gotten gains, because C_Z may greatly exceed V_Z.

We assume that each individual who monitors is paid a scouting fee σ by the government. We envisage that this payment would be remuneration for a variety of monitoring duties (e.g., carrying out a wildlife census), of which game law enforcement is only one; thus monitors would have useful work to do, even if no one were breaking the law. We also assume that monitors would always report observed violations to the government, for which there would be an additional reward R_Z to the first informant of a type-Z infraction. Monitoring is not without its costs, however; even if no one is poaching, there is an opportunity cost c_0 associated with the remaining duties. Furthermore, reporting poachers to the government incurs opprobrium, whose intensity increases with the number of poachers. For the sake of simplicity, we assume that the total expected cost of monitoring is $c_0 + c_L j + c_S k$ per period, where j and k are the numbers of type-L and type-S poachers, respectively. Given that hunters gain social status from providing their lineage dependents with meat and goods exchanged for meat [**69**, p. 943], c_L and c_S are probably much larger than c_0, which is probably very small. Accordingly, we assume that

$$(6.82) \qquad c_L \gg c_0, \quad c_S \gg c_0, \quad c_0 \approx 0.$$

Observing a violation is not the same thing as being rewarded for it, however, because there may be more than one observer, and only the first to report an offence is assumed to be remunerated. Thus, for a monitoring individual, the probability of obtaining a reward from a specific violation is the probability that the violation is detected times the conditional probability that the monitor is first to report it. For the sake of simplicity, we assume that all monitors are equally likely to obtain a reward from each violation. Then the reward probability for type-Z technology is D_{iZ}/i, where i is the number of monitors. Note that this quantity is approximately equal to q_Z when q_Z is small (that is, when there is effectively no competition in the race to report). Combining our results, we find that the expected benefit per period to each monitoring individual, when j other individuals are engaged in type-L poaching, k other individuals are engaged in type-S poaching and i individuals are monitoring, is

$$(6.83) \qquad M_i^{jk} \;=\; \frac{jR_L D_{iL} + kR_S D_{iS}}{i} + \sigma - c_0 - c_L j - c_S k.$$

Table 6.2. The strategy set

	Strategy	Definition
1	LM	Poach with type-L technology, monitor and report (other) lawbreakers
2	LX	Poach with type-L technology, do not monitor or report lawbreakers
3	SM	Poach with type-S technology, monitor and report
4	SX	Poach with type-S technology, do not monitor or report
5	NM	Do not poach, monitor and report
6	NX	Do not poach, do not monitor or report

Table 6.3. The reward matrix

	LM	LX	SM	SX	NM	NX
LM	$W_{n+1}+L_n$ $+M_n^{n0}$	$W_{n+1}+L_0$ $+M_1^{n0}$	$W_{n+1}+L_n$ $+M_n^{0n}$	$W_{n+1}+L_0$ $+M_1^{0n}$	W_1+L_n $+M_{\cdots}^{00}$	W_1+L_0 $+M_{\cdots}^{00}$
LX	$W_{n+1}+L_n$	$W_{n+1}+L_0$	$W_{n+1}+L_n$	$W_{n+1}+L_0$	W_1+L_n	W_1+L_0
SM	$W_{n+1}+S_n$ $+M_n^{n0}$	$W_{n+1}+S_0$ $+M_1^{n0}$	$W_{n+1}+S_n$ $+M_n^{0n}$	$W_{n+1}+S_0$ $+M_1^{0n}$	W_1+S_n $+M_{\cdots}^{00}$	W_1+S_0 $+M_{\cdots}^{00}$
SX	$W_{n+1}+S_n$	$W_{n+1}+S_0$	$W_{n+1}+S_n$	$W_{n+1}+S_n$	W_1+S_n	W_1+S_0
NM	$W_n+M_n^{n0}$	$W_n+M_1^{n0}$	$W_n+M_n^{0n}$	$W_n+M_1^{0n}$	$W_0+M_{\cdots}^{00}$	$W_0+M_{\cdots}^{00}$
NX	W_n	W_n	W_n	W_n	W_0	W_0

We assume that poachers may use either type-L or type-S technology, but not both. Thus each individual has six possible strategies, defined in Table 6.2. Decimation of large mammals without law enforcement broadly corresponds to a population at LX; reduced exploitation, in which improved law enforcement induces residents to take small mammals only, broadly corresponds to a population at SM; and conservation with law enforcement broadly corresponds to a population at NM. In essence, therefore, the wildlife management scheme in Zambia's Luangwa Valley failed to achieve a desideratum of NM, but instead evolved from LX in the 1970s and 1980s to SM in more recent times. The goal of our analysis is to indicate how the desired outcome could yet be attained. Note that there is no direct feedback between individuals: any cooperation is mutualistic (§5.8).

The matrix of rewards per period to a focal individual, using a given row strategy against a population using a given column strategy, can now be determined from (6.77)-(6.80) and (6.83); see Table 6.3. We denote this matrix by A, so that a_{IJ} is the reward to an individual using strategy I against n individuals using strategy J. For example, the reward to strategy NX against population strategy LM is a_{61} = W_n because strategy NX neither poaches nor monitors; the only benefit is the community wage, which is W_n, because there are n +

$0 = n$ type-L poachers and no type-S poachers among the $n + 1$ individuals in the entire population. Similarly, the reward to strategy SM against population strategy SX is $a_{34} = W_{n+1} + S_0 + M_1^{0n}$, which we obtain as follows. First, because there are $n + 1$ type-S poachers when strategy SM confronts n individuals using strategy SX, the community wage is W_{n+1} $(= 0)$. Second, because SX does not monitor and SM does not monitor itself, no one monitors the SM-strategist, so that the expected benefit of poaching is S_0. Finally, the benefit from monitoring is M_1^{0n}, because the focal individual is the only monitor; and although all $n + 1$ individuals are poaching, the focal individual does not monitor itself. Note that the benefits of monitoring to the focal individual are independent of the number of monitors if no one else is poaching, as indicated by the subscripted dots in columns 5 and 6 of Table 6.3.

From §2.6, population strategy J is (strongly) stable if a_{JJ} exceeds a_{IJ} for all $I \neq J$; or equivalently, if $p_{JJ} = 0$ is the only non-positive term in column J of the stability matrix P defined by

$$(6.84) \qquad p_{IJ} = a_{JJ} - a_{IJ}, \quad 1 \leq I, J \leq 6.$$

The first two columns of P are shown in Table 6.4(a), the next two columns in Table 6.4(b) and the last two columns in Table 6.4(c).

Some general results emerge at once. First, because $p_{21} + p_{12} = R_L\big(1 - (1 - q_L)^n - nq_L\big) \leq 0$ for $n \geq 1$, if p_{12} is positive then p_{21} is negative, and vice versa. Thus if either LM or LX is a stable strategy, then the other is not: if conditions do not favor a switch to monitoring in a no-monitoring population, then they also must favor a switch to no monitoring in a monitoring population, and vice versa. Similarly, SM and SX or NM and NX or LX and SX cannot both be stable strategies because $p_{43} + p_{34} \leq 0$ or $p_{65} + p_{56} = 0$ or $p_{42} + p_{24} = 0$, respectively; and only one of LM, SM and NM can be stable because $p_{51} + p_{15} = 0$ while both $p_{31} + p_{13} = 0$ and $p_{53} + p_{35} = 0$. Furthermore, and unsurprisingly, NX cannot be a stable strategy, because (6.81) implies that either p_{26} or p_{46} is negative: lack of vigilance favors poaching. The upshot is that at most two strategies can be stable strategies; and if there are two stable strategies, then one must be a monitoring strategy $(LM, SM$ or $NM)$ while the other must be a no-monitoring strategy $(LX$ or $SX)$. Indeed if one of these two

Table 6.4. The population stability matrix

(a) Type-L technology columns

I	p_{I1}	p_{I2}
1	0	$-\sigma+c_0+n(c_L-q_LR_L)$
2	$\sigma-c_0-nc_L+D_{nL}R_L$	0
3	$V_L-D_{nL}C_L-V_S+D_{nS}C_S$	$V_L-V_S-\sigma+c_0+n(c_L-q_LR_L)$
4	$V_L-D_{nL}(C_L-R_L)-V_S+D_{nS}C_S+\sigma-c_0-nc_L$	V_L-V_S
5	$v_L-(1-(1-q_L)^n)C_L$	$v_L-\sigma+c_0+n(c_L-q_LR_L)$
6	$v_L-D_{nL}(C_L-R_L)+\sigma-c_0-nc_L$	v_L

(b) Type-S technology columns

I	p_{I3}	p_{I4}
1	$V_S-D_{nS}C_S-V_L+D_{nL}C_L$	$V_S-V_L-\sigma+c_0+n(c_S-q_SR_S)$
2	$\sigma-c_0-nc_L+D_{nS}(R_S-C_S)+V_S-V_L+D_{nL}C_L$	V_S-V_L
3	0	$-\sigma+c_0+n(c_S-q_SR_S)$
4	$\sigma-c_0-nc_S+D_{nS}R_S$	0
5	$v_S-(1-(1-q_S)^n)C_S$	$v_S-\sigma+c_0+n(c_S-q_SR_S)$
6	$\sigma-c_0-nc_S+D_{nS}(R_S-C_S)+v_S$	v_S

(c) No-poaching columns

I	p_{I5}	p_{I6}
1	$(1-(1-q_L)^n)C_L-v_L$	$-\sigma+c_0-v_L$
2	$\sigma-c_0+(1-(1-q_L)^n)C_L-v_L$	$-v_L$
3	$(1-(1-q_S)^n)C_S-v_S$	$-\sigma+c_0-v_S$
4	$\sigma-c_0+(1-(1-q_S)^n)C_S-v_S$	$-v_S$
5	0	$-\sigma+c_0$
6	$\sigma-c_0$	0

no-monitoring strategies is stable, then it must be to use the higher-value technology, i.e., strategy HX; for if everyone else were using the lower-value technology (strategy OX), then it would pay to switch.

Nevertheless, our principal purpose is to find conditions for NM to be stable, which can happen only if $p_{65} > 0$ or

$$(6.85) \qquad \sigma > c_0.$$

In particular, σ must be positive: each individual must be paid to monitor even if no one poaches, or else the agreement is unsustainable. On the other hand, (6.82) implies that σ need not be large.

Let us assume that (6.85) holds. Then $p_{45} > p_{35}$, and $p_{25} > p_{15}$. So NM is a stable strategy if, and only if, p_{15} and p_{35} are both positive or, on using (6.80) in Table 6.4(c),

$$(6.86) \qquad (1-(1-q_H)^n)C_H + B/(n+1)^2 > V_H$$

and

$$(6.87) \qquad \left(1 - (1 - q_O)^n\right) C_O \; + \; B/(n+1)^2 \; > \; V_O$$

(where H denotes the higher-value technology and O the other type); by (6.80), the second inequality is automatically satisfied if v_O is negative. As remarked earlier, the only other potentially stable strategy is HX. From Table 6.4(a), if $H = L$, i.e., $V_L > V_S$, then to exclude the possibility that LX is stable we require $p_{12} < 0$ or $n\left(c_L - q_L R_L\right) < \sigma - c_0$; whereas, if $H = S$, then to exclude the possibility that SX is stable we require $p_{34} < 0$ or $n\left(c_S - q_S R_S\right) < \sigma - c_0$. So NM is the only stable strategy if, in addition to (6.85) and (6.86),

$$(6.88) \qquad n\left(c_H - q_H R_H\right) \; < \; \sigma - c_0.$$

Ignoring the fanciful circumstance that $q_H R_H$ almost exactly balances c_H, (6.88) cannot be satisfied for positive $c_H - q_H R_H$ unless either $\sigma - c_0$ is infeasibly large or n is unreasonably small. In practice, therefore, NM is the only stable strategy if (6.85), (6.86) and $q_H > c_H/R_H$ all hold. Because $1 - (1 - q_H)^n$ cannot be less than q_H for $n \geq 1$, (6.86) must hold if $q_H C_H/V_H > 1$. Similarly, (6.87) must hold if $q_O C_O/V_O > 1$. But it also holds if $v_O < 0$. Thus sufficient conditions for NM to be the only stable strategy are $\sigma > c_0$,

$$(6.89) \qquad q_H \; > \; \max\left(V_H/C_H, c_H/R_H\right)$$

(which requires $V_H < C_H$ and $c_H < R_H$) and

$$(6.90) \qquad \text{EITHER} \quad q_O \; > \; V_O/C_O \quad \text{OR} \quad v_O \; < \; 0.$$

Otherwise, the stability of NM depends on group size: for NM to be stable, there is a critical value, say $n_{\text{crit}} + 1$, that $n+1$ must exceed. Suppose, e.g., that (6.89) holds and $V_O/C_O < 1$, but (6.90) is false. Then the effect of B/V_O on (6.87) is negligible, except to the extent of precluding (6.90), and so critical group size is well approximated by ensuring that $\left(1 - (1 - q_O)^{n_{crit}}\right) C_O/V_O$ exceeds 1:

$$(6.91) \qquad n_{crit} + 1 \; \approx \; \left[\frac{\ln(1 - V_O/C_O)}{\ln(1 - q_O)} \right] + 2,$$

where $[z]$ denotes the integer part of z. The effect of group size is illustrated by Table 6.5.[23]

---

[23] And is potentially significant; e.g., Child [**37**] has argued that, for a community-based scheme to work, the community should be small enough to meet under a tree, which he interprets as having no more than 200 households.

Table 6.5. Some critical group sizes when $q_O < V_O/C_O = 0.5$

q_O	$n_{crit} + 1$	q_O	$n_{crit} + 1$
0.005	140	0.025	29
0.01	70	0.03	24
0.015	47	0.05	15
0.02	36	0.1	8

In general, if a strategy is the only stable one, then we can expect it to emerge as the community norm; whereas if a second strategy is also stable, then we can expect the first to emerge only if it yields a higher community reward. Thus, for a conservation agreement to hold (when $v_H > 0$, as we have assumed), either the government must ensure that NM is the only stable strategy; or, if HX is also stable, then the government must ensure both that NM is stable and that it yields a higher community reward than HX. Now, in terms of the reward matrix A defined by Table 6.3, strategy J yields a higher community reward when it yields a higher value of a_{JJ} (i.e., when it yields a higher reward to each individual if the whole population adopts it). So, because NM is the fifth strategy in Table 6.2 whereas HX is either the second or fourth, if HX is stable then the government must ensure that $a_{55} > \max(a_{22}, a_{44})$ or, on using Table 6.4,

$$(6.92) \qquad \sigma - c_0 + \frac{B}{n+1} > V_H.$$

The higher the value of c_H, or the lower the value of R_H, the greater the significance of the above inequality. If the conservation strategy NM is the only stable one, then the relatively high value of R_H/c_H that guarantees (6.88) will coerce the community into conserving the resource, and (6.92) need not hold. If NM is stable but R_H/c_H too small to destabilize the anti-conservation strategy HX, however, then the emergence of NM will require the community's voluntary cooperation, which can be induced by the government only if it pays a community benefit high enough to make NM yield a higher reward than HX; or, on rearranging (6.92), if $B > (n+1)\{V_H - \sigma + c_0\}$. Then $(n+1)(V_H - \sigma + c_0)$ is the minimum cost to the government of ensuring that NM has the higher reward. But it also costs the government σ per individual, or $(n+1)\sigma$ in all, to render NM stable.

Including this cost makes $(n + 1)(V_H + c_0)$ the total minimum price tag for inducing conservation through community self-monitoring.

Suppose, however, that the government neglects to pay the community for monitoring per se. In other words, suppose that $\sigma = 0$. Then although, by (6.92), NM still yields a higher reward than HX, it is no longer a stable strategy, because p_{65} is negative: if no one is poaching and no one is being paid to monitor, then it pays to switch to not monitoring, thus avoiding the opportunity cost c_0. But although NX would conserve the resource, it is not a stable strategy. From Table 6.4(c), to render it stable requires $v_H < 0$, or $B > (n + 1)^2 V_H$: when $\sigma = 0$, the minimum price tag for inducing conservation is raised to $(n + 1)^2 V_H$. In other words, the minimum cost to the government of an effective community agreement to conserve the resource when $\sigma = 0$ is greater than when $\sigma > c_0$ by a factor $(n + 1)V_H / (V_H + c_0)$. If $c_0 \ll V_H$, as $c_0 \approx 0$ would imply, then the relevant factor is approximately $n + 1$, which could never be less than 10 and is typically far greater [**157**]. So an agreement among residents to conserve a wildlife resource through community self-monitoring may be cheaper by at least an order of magnitude for a government to sustain if its community incentive structure separates benefits for not poaching and bonuses for arrests made from payments for monitoring per se.

Our analysis clarifies the important distinction between the value of an agreement and its strategic stability; community benefits may strongly influence the former, yet scarcely influence the latter. For example, if $B/(n+1)$ is very much greater than $\sigma - c_0$, then the (individual) value of an agreement to adhere to NM, $a_{55} = \sigma - c_0 + B/(n+1)$, is dominated by the magnitude of B; the influence of $\sigma - c_0$ is negligible. But the magnitude of B need have no effect on the stability of the agreement; whereas the effect of $\sigma - c_0$ on the agreement's stability can never be ignored because no self-monitoring agreement is sustainable without a payment to each individual that exceeds the opportunity cost of monitoring—even if no one is poaching. Otherwise, it pays to switch to neither poaching nor monitoring, which is not a stable strategy. Thus, to answer the question that motivated our analysis: it appears that community-based wildlife management schemes could, in fact, be better designed—residents could always be paid to monitor, even if no one is breaking the law.

6.7. Winner and loser effects

We often speak of someone having a winning streak or being "on a roll" after a string of consecutive successes. It seems that if you win today then you have a higher probability of winning tomorrow (though you could still lose), and similarly for a losing streak; if you lose today then you have a higher probability of losing tomorrow (though you could still win). We will refer to these effects as winner or loser effects, or collectively as prior-experience effects.

Winner and loser effects are not restricted to humans: they have also been observed in laboratory experiments with beetles, fishes, snakes and spiders.[24] Some of these experiments have produced only a loser effect, and some have produced both a loser and a winner effect; but so far as we know, no experiment has ever produced a winner effect without a loser effect. Why? Here we attempt to make sense of these observations with the help of a game-theoretic model.

Consider a game between unrelated animals who interact in triads chosen randomly from a large population. An interaction consists of three pairwise contests. The outcome of each contest is determined by the difference between the contestants' perceptions of their strengths, which are known only to themselves, and may be revised in the light of experience. The cost of a contest is determined by the extent to which an animal overestimates his strength, that is, by the difference between his actual strength—which we assume he does not know—and his perception of that strength. Both perceived and actual strengths are assumed to be numbers between 0 (weakest) and 1 (strongest); actual strengths are drawn at random from a continuous distribution on $[0, 1]$ with probability density function g and cumulative distribution function G (defined by $G(s) = \int_0^s g(\xi)\,d\xi$). Although there is no direct assessment of strength, animals can respond to the distribution of strength among the population at large.

In general, the greater an animal perceives his strength, the harder he fights, and hence the greater his probability of winning; however, as remarked above, it is costly for an animal to overestimate his strength, and the cost increases with the magnitude of the overestimate. Thus the fundamental tradeoff to be captured by our model is between the

---

[24] See references cited in [149], on which §6.7 is based. For further discussion and references to other examples of winner and loser effects, see [184, pp. 255-258].

benefits to an animal of raising his strength perception above that
of his opponent and the costs of overestimating his actual strength.
These costs are assumed to arise principally from excessive depletion of energy reserves when strength is overestimated. They could
also, in principle, arise partially from increased risk of injury with
long-term consequences; however, we assume that an animal's actual
strength does not change between contests, so that the short-term
consequences of any such injury would have to be negligible.

Specifically, let an animal whose strength perception is S_1 defeat
an opponent whose strength perception is S_2 with probability

$$(6.93) \qquad W(S_1, S_2) = \tfrac{1}{2} + \tfrac{1}{2}(S_1 - S_2).$$

Note that W increases with $S_1 - S_2$, $W(1, 0) = 1$, $W(S, S) = \tfrac{1}{2}$ and
$W(0, 1) = 0$: animals with equal strength perceptions are equally
likely to win a fight, and a perception of maximum strength is guaranteed to defeat one of minimum strength. Moreover, if $L(S_1, S_2)$ denotes the probability that an animal with strength perception S_1 loses
against an opponent with strength perception S_2, then $L(S_1, S_2) =
W(S_2, S_1) = \tfrac{1}{2} + \tfrac{1}{2}(S_2 - S_1)$. For an individual whose strength is X
but whose perception of it is S, let k_0 denote the fixed cost of fighting
per contest and r the maximum variable cost (which would be paid
by the worst possible fighter if he perceived himself as best). Then
the animal's contest cost, denoted by $c(S, X)$, is k_0 when $S \leq X$ but
increases with respect to $S - X$ when $S > X$; moreover, $c(S, S) = k_0$
and $c(1, 0) = k_0 + r$. For the sake of simplicity, we satisfy these conditions by taking $c(S, X) = k_0 + r(S - X)$ if $S > X$, so that r becomes
the marginal cost of overestimating strength perception (with respect
to the magnitude of that overestimate). Every contestant pays the
fixed cost k_0: this parameter reduces each animal's excess over that
of the gamma individual by the same amount $2k_0$, and hence has no
strategic effect. So, without loss of generality, we set $k_0 = 0$:

$$(6.94) \qquad c(S, X) = \begin{cases} 0 & \text{if } S \leq X \\ r(S - X) & \text{if } S > X. \end{cases}$$

A strategy in this game consists of a triple of numbers, an initial
strength perception and a pair of revised perceptions, one to adopt in
the event of a win and another to adopt in the event of a loss. All are

numbers between 0 and 1. For the population, let v_0 denote initial strength perception, v_1 the level to which strength perception rises after a win, and v_2 the level to which strength perception falls after a loss. Thus the population's strategy is a three-dimensional vector $v = (v_0, v_1, v_2)$ satisfying $0 \leq v_2 \leq v_0 \leq v_1 \leq 1$. Correspondingly, let $u = (u_0, u_1, u_2)$ denote the strategy of a focal individual who is a potential mutant and changes his initial strength perception of u_0 to u_1 after a win but to u_2 after a loss, where $0 \leq u_2 \leq u_0 \leq u_1 \leq 1$. Both of the focal individual's opponents play the population strategy v. Thus a triad consists of a u-strategist and a pair of v-strategists. All orders of interaction within a triad are assumed to be equally likely, and no animal observes the outcome of the contest between the other two individuals.

After a win, a v-strategist's probability of a win against an opponent with the same initial perception increases from $W(v_0, v_0) = \frac{1}{2}$ to $W(v_1, v_0) = \frac{1}{2} + \frac{1}{2}(v_1 - v_0)$, i.e., by $\frac{1}{2}(v_1 - v_0)$. When $v_1 > v_0$, we will say that there is a winner effect of magnitude $v_1 - v_0$ (omitting the factor $\frac{1}{2}$ for the sake of simplicity). Correspondingly, when $v_0 > v_2$ there is a loser effect of magnitude $v_0 - v_2$, because the v-strategist's probability of a loss against an opponent with the same initial perception increases by $L(v_2, v_0) - L(v_0, v_0) = \frac{1}{2}(v_0 - v_2)$.

Because there are eight possible ways in which any set of three pairwise contests can be won or lost and three possible orders of interaction between a u-strategist and two v-strategists, there are 24 possible outcomes in all; see Table 6.6, where four of the outcomes have been combined in pairs to yield a modified total of 22. An animal is called *naive* if he has yet to engage in a contest, and otherwise *experienced*. If the focal individual faces two naive individuals, then his reward is denoted by $f_{nn}(u, v)$; if he faces a naive individual followed by an experienced individual, then his reward is $f_{ne}(u, v)$; and if he faces two experienced individuals, then his reward is $f_{ee}(u, v)$. These three possibilities are equally likely, by assumption. Thus

$$(6.95) \qquad f(u, v) = \tfrac{1}{3} f_{nn}(u, v) + \tfrac{1}{3} f_{ne}(u, v) + \tfrac{1}{3} f_{ee}(u, v)$$

is the unconditional reward to a u-strategist against v-strategists.

There are two possible outcomes for the triad overall as a result of this interaction. The first is that one animal wins twice, implying

Table 6.6. Possible outcomes for a focal individual, F; $O1$ and $O2$ are his first and second opponents, respectively, and parentheses indicate a contest in which he is not involved.

CASE k	1st	WINNERS 2nd	3rd	CASE k	1st	WINNERS 2nd	3rd
1	F	F	(O1)	11	F	(O1)	O2
	F	F	(O2)	12	O1	(O2)	F
2	F	O2	(O2)	13	O1	(O1)	O2
3	O1	F	(O1)	14	O1	(O2)	O2
4	F	O2	(O1)	15	(O1)	F	F
5	O1	F	(O2)	16	(O2)	F	F
6	O1	O2	(O1)	17	(O1)	O1	F
	O1	O2	(O2)	18	(O2)	F	O2
7	F	(O1)	F	19	(O1)	F	O2
8	F	(O2)	F	20	(O2)	O1	F
9	F	(O2)	O2	21	(O1)	O1	O2
10	O1	(O1)	F	22	(O2)	O1	O2

that another loses twice and the third both wins and loses. In this case, we shall say that there is a linear dominance hierarchy with the double winner or "alpha" male on top, the double loser or "gamma" male at the base and the remaining contestant or "beta" male at an intermediate rank. The second possibility is a circular arrangement, which arises if each animal both wins and loses. In the first case, we assume that the alpha male's reproductive benefits from its position exceed those of the gamma male by 1 unit, and that the beta male's benefits exceed those of the gamma male by b units, where $0 \leq b < 1$. In the second case, we assume that the total excess of $1 + b$ is divided equally among the triad. Thus b is an inverse measure of reproductive inequity. The middle rank of a linear hierarchy yields a higher benefit than an equal share from a circular triad if $\frac{1}{2} < b < 1$, but a lower benefit if $0 \leq b < \frac{1}{2}$; and if $b = \frac{1}{2}$, then additional benefits are directly proportional to number of contests won.

We now discuss each interaction order in turn. Let $\phi_{nn}(u, v, X)$ denote the payoff to a u-strategist with strength X when both opponent v-strategists are naive, so that each has strength perception v_0. The u-strategist's strength perception for his opening contest is

u_0, and so he incurs a fighting cost $c(u_0, X)$. He wins with probability $W(u_0, v_0)$, and he loses with probability $L(u_0, v_0) = W(v_0, u_0)$. For the subsequent contest, the u-strategist's perception will increase from u_0 to u_1 following a win, but decrease from u_0 to u_2 following a loss. After a win, he wins again with probability $W(u_1, v_0)$, but he loses with probability $L(u_1, v_0)$; in either case, the cost is $c(u_1, X)$. After a loss, he loses again with probability $L(u_2, v_0)$, but he wins with probability $W(u_2, v_0)$; in either case, the cost is $c(u_2, X)$. The u-strategist becomes the alpha individual, with payoff $1 - c(u_0, X) - c(u_1, X)$, if he wins both contests, i.e., with probability $W(u_0, v_0)W(u_1, v_0)$; see Table 6.7 (Case 1). Similarly, the u-strategist becomes the gamma individual, with payoff $0 - c(u_0, X) - c(u_2, X)$, if he loses both contests, i.e., with probability $L(u_0, v_0)L(u_2, v_0)$; again see Table 6.7 (Case 6).

In either of these cases, i.e., if the focal individual either wins or loses twice, then the outcome of the third contest does not affect his payoff. If the focal individual is both a winner and a loser, however, then the remaining contest determines whether he occupies the middle rank in a linear hierarchy or belongs to an egalitarian triad. Each of these outcomes can arise in two ways. The u-strategist will occupy the middle rank of a linear hierarchy if he wins before he loses and his second opponent prevails over his first (Table 6.7, Case 2), or if he loses before he wins and his first opponent prevails over his second (Case 3). The benefit is b in either case but the costs differ, as shown in Table 6.7. Similarly, the u-strategist will belong to an egalitarian triad he wins before he loses and his first opponent prevails over his second (Case 4), or if he loses before he wins and his second opponent prevails over his first (Case 5). The benefit is now $\frac{1}{3}(1 + b)$ in either case and the costs again differ, as shown in Table 6.7.

Let $\omega_k(u, v)$ denote the probability of Event k in Table 6.7 and let $P_k(u, X)$ denote the associated payoff to the focal individual. Then

$$(6.96) \qquad \phi_{nn}(u, v, X) = \sum_{k=1}^{6} \omega_k(u, v) P_k(u, X),$$

and integration over the sample space of X yields

$$(6.97) \qquad f_{nn}(u, v) = \mathrm{E}\left[\phi_{nn}(u, v, X)\right] = \int_0^1 \phi_{nn}(u, v, x)g(x)\, dx$$

Table 6.7. Payoff to focal individual, conditional on

Cases 1-6: Two naive opponents

Cases 7-14: A naive and an experienced opponent

Cases 15-22: Two experienced opponents.

CASE k	PROBABILITY $\omega_k(u,v)$	PAYOFF $P_k(u,X)$
1	$W(u_0,v_0)\,W(u_1,v_0)$	$1 - c(u_0,X) - c(u_1,X)$
2	$W(u_0,v_0)\,L(u_1,v_0)\,W(v_1,v_2)$	$b - c(u_0,X) - c(u_1,X)$
3	$L(u_0,v_0)\,W(u_2,v_0)\,W(v_1,v_2)$	$b - c(u_0,X) - c(u_2,X)$
4	$W(u_0,v_0)\,L(u_1,v_0)\,W(v_2,v_1)$	$\frac{1}{3}(1+b) - c(u_0,X) - c(u_1,X)$
5	$L(u_0,v_0)\,W(u_2,v_0)\,W(v_2,v_1)$	$\frac{1}{3}(1+b) - c(u_0,X) - c(u_2,X)$
6	$L(u_0,v_0)\,L(u_2,v_0)$	$-c(u_0,X) - c(u_2,X)$
7	$W(u_0,v_0)\,W(v_2,v_0)\,W(u_1,v_2)$	$1 - c(u_0,X) - c(u_1,X)$
8	$W(u_0,v_0)\,W(v_0,v_2)\,W(u_1,v_1)$	$1 - c(u_0,X) - c(u_1,X)$
9	$W(u_0,v_0)\,W(v_0,v_2)\,L(u_1,v_1)$	$b - c(u_0,X) - c(u_1,X)$
10	$L(u_0,v_0)\,W(v_1,v_0)\,W(u_2,v_2)$	$b - c(u_0,X) - c(u_2,X)$
11	$W(u_0,v_0)\,W(v_2,v_0)\,L(u_1,v_2)$	$\frac{1}{3}(1+b) - c(u_0,X) - c(u_1,X)$
12	$L(u_0,v_0)\,W(v_0,v_1)\,W(u_2,v_1)$	$\frac{1}{3}(1+b) - c(u_0,X) - c(u_2,X)$
13	$L(u_0,v_0)\,W(v_1,v_0)\,L(u_2,v_2)$	$-c(u_0,X) - c(u_2,X)$
14	$L(u_0,v_0)\,W(v_0,v_1)\,L(u_2,v_1)$	$-c(u_0,X) - c(u_2,X)$
15	$W(v_0,v_0)\,W(u_0,v_1)\,W(u_1,v_2)$	$1 - c(u_0,X) - c(u_1,X)$
16	$W(v_0,v_0)\,W(u_0,v_2)\,W(u_1,v_1)$	$1 - c(u_0,X) - c(u_1,X)$
17	$W(v_0,v_0)\,L(u_0,v_1)\,W(u_2,v_2)$	$b - c(u_0,X) - c(u_2,X)$
18	$W(v_0,v_0)\,W(u_0,v_2)\,L(u_1,v_1)$	$b - c(u_0,X) - c(u_1,X)$
19	$W(v_0,v_0)\,W(u_0,v_1)\,L(u_1,v_2)$	$\frac{1}{3}(1+b) - c(u_0,X) - c(u_1,X)$
20	$W(v_0,v_0)\,L(u_0,v_2)\,W(u_2,v_1)$	$\frac{1}{3}(1+b) - c(u_0,X) - c(u_2,X)$
21	$W(v_0,v_0)\,L(u_0,v_1)\,L(u_2,v_2)$	$-c(u_0,X) - c(u_2,X)$
22	$W(v_0,v_0)\,L(u_0,v_2)\,L(u_2,v_1)$	$-c(u_0,X) - c(u_2,X)$

in (6.95), where E denotes expected value.

A similar analysis yields the payoff $\phi_{ne}(u,v,X)$ or $\phi_{ee}(u,v,X)$ to a u-strategist against a naive opponent and an experienced one, or against two experienced opponents. The relevant expressions are obtained from Table 6.7 and analogues of (6.96) in which k is summed between 7 and 14 for a naive and an experienced opponent and between 15 and 22 for two experienced opponents. Then $f_{ne}(u,v)$ and $f_{ee}(u,v)$ are calculated by analogy with (6.97); and, on using $W(v_0,v_0) = \frac{1}{2}$, it follows from (6.95) that the reward $f(u,v)$ to a

Table 6.8. Reward function coefficients.

$$
\begin{aligned}
72J_0(v) &= (1-2b)\left(2\{2v_1 - 2v_2 + 3\}v_0^2 + 7v_1v_2\right) - (2-b)(1-v_0)v_2^2 \\
&\quad + \{24 + (1+v_0)v_1^2\}(1+b) - 12\{v_0 + (1-b)v_1 + bv_2\} \\
&\quad - 3\{v_1 + v_2 + (v_1 - v_2 + v_1v_2 + 2)v_0\} \\
72J_1(v) &= 3(1-2v_0) - (1-2b)v_0(v_0 + v_1 - 2v_2 + 3) \\
&\quad - (v_1 - v_2)(1+b) + 6(1-b)(2-v_2) \\
72J_2(v) &= (1-2b)v_0(v_0 - 2v_1 + v_2 + 3) + (v_2 - v_1)(2-b) \\
&\quad + 6b(2v_0 + v_1 + 2) + 3 \\
72K_0(v) &= 18 - (1+b)v_1^2 + 3v_1v_2 - (2-b)v_2^2 \\
&\quad - (1-2b)(2v_0\{2v_1 - 2v_2 + 3\} + 3\{v_1 + v_2\}) \\
72K_1(v) &= (1-2b)(6 + v_0 + v_1 - 2v_2) + 9 \\
72K_2(v) &= (1-2b)(6 - v_0 + 2v_1 - v_2) - 9
\end{aligned}
$$

u-strategist against v-strategists is

$$
(6.98) \quad f(u,v) = \frac{1}{3}\sum_{k \in N_1} \omega_k(u,v) + \frac{1}{3}b \sum_{k \in N_2} \omega_k(u,v)
$$

$$
+ \frac{1}{9}(1+b) \sum_{k \in N_3} \omega_k(u,v) - \int_0^1 c(u_0,x)\,g(x)\,dx
$$

$$
- \frac{1}{6}\left\{4W(u_0,v_0) + W(u_0,v_1) + W(u_0,v_2)\right\} \int_0^1 c(u_1,x)\,g(x)\,dx
$$

$$
- \frac{1}{6}\left\{4L(u_0,v_0) + L(u_0,v_1) + L(u_0,v_2)\right\} \int_0^1 c(u_2,x)\,g(x)\,dx
$$

where the index sets N_1, N_2 and N_3 are defined by

$$
(6.99) \quad \begin{aligned} N_1 &= \{1,7,8,15,16\}, \ N_2 = \{2,3,9,10,17,18\}, \\ N_3 &= \{4,5,11,12,19,20\}. \end{aligned}
$$

It now follows from (6.94) that

$$
(6.100) \quad f(u,v) = J_0(v) + J_1(v)u_1 + J_2(v)u_2
$$

$$
+ \{K_0(v) + K_1(v)u_1 + K_2(v)u_2\}\,u_0 - rH(u_0)
$$

$$
- \frac{1}{12}r\left\{6 + 6u_0 - 4v_0 - v_1 - v_2\right\}H(u_1)
$$

$$
- \frac{1}{12}r\left\{6 - 6u_0 + 4v_0 + v_1 + v_2\right\}H(u_2)
$$

with J_0, J_1, J_2, K_0, K_1 and K_2 defined by Table 6.8 and H by

$$
(6.101) \quad H(s) = \int_0^s (s - \xi)\,g(\xi)\,d\xi = \int_0^s G(\xi)\,d\xi.
$$

Table 6.9. The gradient of f.

$$\frac{\partial f}{\partial u_0} = K_0(v) + K_1(v)u_1 + K_2(v)u_2 - rG(u_0) - \tfrac{1}{2}r\{H(u_1) - H(u_2)\}$$

$$\frac{\partial f}{\partial u_1} = J_1(v) + K_1(v)u_0 - \tfrac{1}{12}r\{6 + 6u_0 - 4v_0 - v_1 - v_2\}G(u_1)$$

$$\frac{\partial f}{\partial u_2} = J_2(v) + K_2(v)u_0 - \tfrac{1}{12}r\{6 - 6u_0 + 4v_0 + v_1 + v_2\}G(u_2)$$

Table 6.10. The gradient of f, evaluated for $u = v$.

$$\frac{\partial f}{\partial u_0}(v_0, v_1, v_2) = \tfrac{1}{24}\{6(1 + v_1) - (1 - 2b)(v_1 - v_2 + 2)v_0 + (v_1 - v_2)v_2\}$$
$$- \tfrac{1}{24}\{2(1 + b) + b(v_1 - v_2)\}(v_1 + v_2)$$
$$- \tfrac{1}{2}r\{H(v_1) - H(v_2) + 2G(v_0)\}$$

$$\frac{\partial f}{\partial u_1}(v_0, v_1, v_2) = \tfrac{1}{72}\{6(1 - b)(v_0 - v_2 + 2) - (1 + b)(v_1 - v_2) + 3\}$$
$$- \tfrac{1}{12}r(6 - v_1 - v_2 + 2v_0)G(v_1)$$

$$\frac{\partial f}{\partial u_2}(v_0, v_1, v_2) = \tfrac{1}{72}\{6b(v_1 - v_0 + 2) - (2 - b)(v_1 - v_2) + 3\}$$
$$- \tfrac{1}{12}r(6 + v_1 + v_2 - 2v_0)G(v_2)$$

The gradient of f is also readily calculated, and its components are recorded in Table 6.9. For $u = v$ the expressions reduce to those in Table 6.10, where $\frac{\partial f}{\partial u_i}(v_0, v_1, v_2)$ denotes $\frac{\partial f}{\partial u_i}\big|_{u=v}$ for $i = 0, 1, 2$.

All of these results hold for an arbitrary distribution of actual strength, but such generality allows only limited further progress. Accordingly, we assume henceforward that strength has a uniform distribution—i.e., $g(\xi) = 1$, as in (2.23)—and hence that

$$(6.102) \qquad G(s) = s, \quad H(s) = \tfrac{1}{2}s^2, \qquad 0 \le s \le 1.$$

More realistic, nonuniform distributions are discussed in [**149**].

To any population strategy v, the best reply—u^*, say— is u that maximizes $f(u, v)$ subject to $0 \le u_2 \le u_0 \le u_1 \le 1$. This best reply is unique, because it follows from Table 6.9 that $\partial^2 f/\partial u_0^2 = -rg(u_0)$, $\partial^2 f/\partial u_1^2 = -r(6 + 6u_0 - 4v_0 - v_1 - v_2)g(u_1)/12$ and $\partial^2 f\partial u_2^2 = -r(6 - 6u_0 + 4v_0 + v_1 + v_2)g(u_2)/12$, all of which are negative in the interior of the constraint set. If $u^* = v$, then v is a (strong) ESS. Note, however, that the uniqueness of u^* does not by itself imply a unique ESS (Exercise 6.28), although for a uniform distribution there is indeed a unique ESS. We now proceed to discover how it depends on b and r.

The ESS is most readily found by applying the Kuhn-Tucker conditions of constrained optimization theory; see, e.g., Chapter 10 of [**123**]. Adapted to the purpose at hand, these conditions state that if u^* maximizes f subject to constraints of the form $h_i(u) \geq 0, i = 0, \dots, 3$ then there must exist nonnegative Lagrange multipliers $\lambda_i, i = 0, \dots, 3$ such that the Lagrangean $L = f + \sum_{i=0}^{3} \lambda_i h_i(u)$ satisfies $\nabla L = (0,0,0)$ and $\lambda_i h_i(u^*) = 0, i = 0, \dots, 3$, with the gradient evaluated at $u = u^*$. If these conditions are satisfied with $h_i(u^*) = 0$, then we will say that Constraint i is *active*. Writing $h_0(u) = 1 - u_1$, $h_1(u) = u_1 - u_0$, $h_2(u) = u_0 - u_2$ and $h_3(u) = u_2$, it follows at once that u^* maximizes f subject to $0 \leq u_2 \leq u_0 \leq u_1 \leq 1$ if nonnegative $\lambda_0, \lambda_1, \lambda_2$ and λ_3 exist such that

$$(6.103a) \qquad \frac{\partial f}{\partial u_0}\Big|_{u=u^*} - \lambda_1 + \lambda_2 = 0$$

$$(6.103b) \qquad \frac{\partial f}{\partial u_1}\Big|_{u=u^*} - \lambda_0 + \lambda_1 = 0$$

$$(6.103c) \qquad \frac{\partial f}{\partial u_2}\Big|_{u=u^*} - \lambda_2 + \lambda_3 = 0$$

and

$$(6.104a) \qquad \lambda_0(1 - u_1^*) = 0 \qquad \text{(Constraint 0)}$$

$$(6.104b) \qquad \lambda_1(u_1^* - u_0^*) = 0 \qquad \text{(Constraint 1)}$$

$$(6.104c) \qquad \lambda_2(u_0^* - u_2^*) = 0 \qquad \text{(Constraint 2)}$$

$$(6.104d) \qquad \lambda_3 u_2^* = 0.$$

To determine the ESS, we must solve these equations for $u^* = v$.

We first establish that there is no ESS with $v_2 = 0$. For suppose that $v_2 = 0$. Then $\lambda_2 - \lambda_3 = \frac{1}{72}\{3 - 2v_1 + (12 - 6v_0 + 7v_1)b\}$ is positive, from (6.103c) and Table 6.10; hence λ_2 is positive, because $\lambda_3 \geq 0$. So Constraint 2 can be satisfied with $u^* = v$ only if also $v_0 = 0$, in which case it follows from Table 6.10 and (6.103a)-(6.103b) that $\lambda_1 - \lambda_2 = \frac{1}{24}\{2(2v_1 + 3) - (2 + v_1)bv_1\} - \frac{1}{4}rv_1^2$ and $\lambda_0 - \lambda_1 = \frac{1}{72}\{12(1-b)-(1+b)v_1+3\} - \frac{1}{12}rv_1(6-v_1)$, the second of which implies $\lambda_0 > 0$ and contradicts (6.104a) if $v_1 = 0$; hence $v_1 > 0$, requiring $\lambda_1 = 0$, by (6.104b). Now, from $\lambda_2 > 0, \lambda_1 = 0$ and $\lambda_0 \geq 0$, we require $2(2v_1+3)-(2+v_1)bv_1 < 6rv_1^2$ and $12(1-b)-(1+b)v_1+3 \geq 6rv_1(6-v_1)$, implying $3(1 + b)v_1^2 > bv_1^3 + 3v_1 + 36$, which is clearly false. So we must have $v_2 > 0$ at the ESS, i.e., the last constraint is never active.

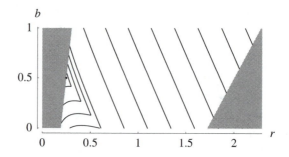

Figure 6.8. Variation of $\zeta = \min(\lambda_0, \lambda_2)$ with r and b for $\frac{1}{36}(4b+7) \leq r \leq \frac{1}{18}(10b+31)$ when $v = (v_0, 1, v_0)$. Contours are drawn for (right to left) $\zeta = -0.87, \ldots, -0.07, \ldots -0.01$, with the difference in heights between contours decreasing from 0.1 to 0.02 at the ninth. The dot represents the global maximum at $(r, b) = (\frac{1}{4}, \frac{1}{2})$, where $\zeta = 0$.

It follows from (6.104d) that $\lambda_3 = 0$, and hence from (6.103) that

(6.105a) $\lambda_0 = \frac{\partial f}{\partial u_0}(v_0, v_1, v_2) + \frac{\partial f}{\partial u_1}(v_0, v_1, v_2) + \frac{\partial f}{\partial u_2}(v_0, v_1, v_2)$

(6.105b) $\lambda_1 = \frac{\partial f}{\partial u_0}(v_0, v_1, v_2) + \frac{\partial f}{\partial u_2}(v_0, v_1, v_2)$

(6.105c) $\lambda_2 = \frac{\partial f}{\partial u_2}(v_0, v_1, v_2)$

for any ESS v.

Next we establish that there is no ESS of the form $v = (v_0, v_1, v_0)$ with $v_1 > v_0$. For suppose that such an ESS exists. Then (6.104b) implies $\lambda_1 = 0$, so that (6.102),(6.105) and Table 6.10 imply

(6.106a) $\{10 - 6r(3v_1 + 4v_0) + b\}(v_1 - v_0)$

$$+ \; 3(7 - 36rv_0 + 4b) = 3b(v_1 - v_0)^2$$

(6.106b) $\lambda_0 = \frac{1}{24}(5 - 12rv_1 - 4b) + \frac{1}{72}(6rv_1 - 1 - b)(v_1 - v_0)$

(6.106c) $\lambda_2 = \frac{1}{24}(1 - 12rv_0 + 4b) + \frac{1}{72}(7b - 6rv_0 - 2)(v_1 - v_0)$.

If $v_1 < 1$, then (6.104a) implies that $\lambda_0 = 0$. Thus v_0 is determined by (6.106a) if $v_1 = 1$, whereas v_0 and v_1 are jointly determined by (6.106a) and (6.106b) if $v_1 < 1$.

First suppose that $v_1 = 1$. Then, from Exercise 6.22, (6.106a) has a solution satisfying $0 < v_0 < 1$ only if

(6.107) $\frac{4b+7}{36} < r < \frac{10b+31}{18}$, i.e., if (r, b) lies in the unshaded region of Figure 6.8, and the only such solution is

$$(6.108) \quad v_0 = \frac{2(10b-18r+31)}{114r-5b+10+\sqrt{(114r-5b+10)^2+12(b-8r)(10b-18r+31)}}.$$

Substitution of v_0 into (6.106b)-(6.106c) yields

$$(6.109) \quad \begin{aligned} \lambda_0 &= \frac{14-30r-13b-(6r-b-1)v_0}{72} \\ \lambda_2 &= \frac{36r^2+132br-19b^2-54r-b-(108r^2+24br-7b^2-36r+2b)v_0}{72(8r-b)} \end{aligned}$$

with v_0 given by (6.108). The dependence of $\zeta = \min(\lambda_0, \lambda_2)$ on r and b for $\frac{4b+7}{36} \leq r \leq \frac{10b+31}{18}$ can now be determined from a contour map of ζ, and the result is shown in Figure 6.8. The unique global maximizer of ζ is $(r, b) = \left(\frac{1}{4}, \frac{1}{2}\right)$, at which $\zeta = 0$. So it is impossible to have $\lambda_0, \lambda_2 \geq 0$ with $v = (v_0, 1, v_0)$ and $v_0 < 1$.

Next suppose that $v_1 < 1$ and hence $\lambda_0 = 0$, or Constraint 0 fails to be satisfied. Then (6.106b) implies

$$(6.110) \quad (1 - 6rv_1 + b)(v_1 - v_0) + 12(3rv_1 + b) = 15.$$

Together, 6.106a) and (6.110) yield a pair of simultaneous quadratic equations for v_0 and v_1, which in principle can be solved analytically. Unless $b = \frac{1}{2}$, yielding $v_0 = v_1 = \frac{1}{4r}$, the result is far too cumbersome to be of practical use. For $b \neq \frac{1}{2}$, however, (6.106a) and (6.110) are easy to solve numerically with standard desktop mathematical software (e.g., [**239**]). In some regions of the r-b plane, no solution satisfies $0 < v_0 < v_1 < 1$; for example, inspection shows that (6.110) is inconsistent with $0 < v_1 - v_0 < 1$ if $36r + 13b < 14$. Whenever $0 < v_0 < v_1 < 1$ is satisfied, however, the corresponding value of λ_2 is readily obtained from (6.106c), and so the dependence of v_0, v_1 and λ_2 on b and r can be determined. It is found that $\lambda_2 \geq 0, v_1 > v_0$ is never satisfied: as illustrated by Figure 9, λ_2 is always negative if $b < \frac{1}{2}$, and $v_1 - v_0$ is always negative if $b > \frac{1}{2}$. It follows that no ESS of the form $v = (v_0, v_1, v_0)$ with $v_1 > v_0$ can exist.

A similar analysis demonstrates that an ESS of the form $v = (v_0, v_0, v_2)$ with $v_0 > v_2$ is possible only if $v_0 = 1$. For if $v_0 < 1$ then neither Constraint 0 nor Constraint 2 is active, requiring $\lambda_0 = 0 = \lambda_2$. These equations determine v_0 and v_2; the corresponding value of $\lambda_1 = \frac{\partial f}{\partial u_0}(v_0, v_0, v_2)$ is determined by setting $v_1 = v_0$ in Table 6.10; and it is found that $\lambda_1 < 0$ if $b < \frac{1}{2}$ and $v_0 - v_2 < 0$ if $b > \frac{1}{2}$ whenever

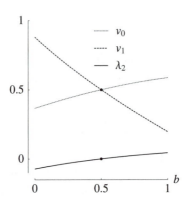

Figure 6.9. Variation of λ_2 (solid curve) with b for $r = \frac{1}{2}$ when $v = (v_0, v_1, v_0)$; v_0 is shown dotted, v_1 dashed. The dots indicate that $v_0 = v_1$ and $\lambda_2 = 0$ when $b = \frac{1}{2}$.

Table 6.11. Types of evolutionarily stable strategy.

TYPE	CONSTRAINTS	LOSER EFFECT?	WINNER EFFECT?
Ia	$1 = v_1 = v_0 = v_2$	No	No
Ib	$1 > v_1 = v_0 = v_2$	No	No
II	$1 = v_1 = v_0 > v_2$	Yes	No
IIIa	$1 = v_1 > v_0 > v_2$	Yes	Yes
IIIb	$1 > v_1 > v_0 > v_2$	Yes	Yes

$v_2 < v_0 < 1$. So $\lambda_1 \geq 0, v_2 < v_0 < 1$ can never be satisfied. The upshot is that the only possible ESSes are of the types defined in Table 6.11. We discuss each type in turn.

At a Type-I ESS there is neither a winner nor a loser effect, i.e., $v = (v_0, v_0, v_0)$, and so Constraints 1 and 2 are both active, with $\lambda_1, \lambda_2 \geq 0$. From Table 6.10) with $v_1 = v_0 = v_2$, (6.105) reduces to

$$(6.111) \qquad (\lambda_0, \lambda_1, \lambda_2) = \left(\tfrac{1-4rv_0}{2}, \tfrac{7-36rv_0+4b}{24}, \tfrac{1-12rv_0+4b}{24} \right).$$

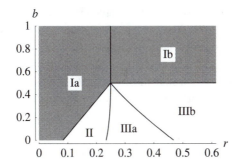

Figure 6.10. Type of evolutionarily stable strategy as a function of r and b. See Table 6.11 for definitions of types.

At an ESS of Type Ia, where $v_0 = 1$, Constraint 0 is also active, with $\lambda_0 \geq 0$. Thus, from (6.111) with $v_0 = 1$,

$$(6.112) \qquad r \leq \tfrac{1}{12}\min(1 + 4b, 3).$$

At an ESS of Type Ib, $v_0 < 1$, so that (6.104a) implies $\lambda_0 = 0$. Then it follows from (6.111) that $r \geq \tfrac{1}{4}$, $b \geq \tfrac{1}{2}$ and

$$(6.113) \qquad v_0 = \tfrac{1}{4r}.$$

The Type-I region of the r-b plane is shaded in Figure 6.10.

At a Type-II ESS, a loser effect exists without a winner effect; Constraints 0 and 1 are active, but (6.104c) is satisfied with $\lambda_2 = 0$, so that (6.102), (6.105) and Table 6.10 with $v_1 = 1 = v_0$ imply

$$(6.114a) \qquad 6rv_2(5 + v_2) = (2 - b)v_2 + 13b + 1$$

$$(6.114b) \qquad \lambda_0 = \tfrac{41-10b-5(1+b)v_2-3(1-b)v_2^2}{72} - \tfrac{(7-v_2)r}{12} - \tfrac{(5-v_2^2)r}{4}$$

$$(6.114c) \qquad \lambda_1 = \tfrac{7-v_2^2+(3-v_2)(1-v_2)b}{24} - \tfrac{(5-v_2^2)r}{4}.$$

From Exercise 6.22, (6.114a) has a solution satisfying $0 < v_2 < 1$ only if $12r > 4b + 1$, and the only such solution is

$$(6.115) \qquad v_2 = \tfrac{1}{12r}\left\{\sqrt{(30r+b-2)^2+24(1+13b)r}-30r-b+2\right\}.$$

Because (6.114b) implies that $\lambda_0 < \tfrac{41-10b}{72} - \tfrac{3r}{2}$ for $0 < v_2 < 1$, (6.115) can satisfy $\lambda_0 \geq 0$ only if $r < \tfrac{41-10b}{108}$. Thus, if $b \geq \tfrac{1}{2}$, the only possibility for $\lambda_0 \geq 0$ is at a point in the triangle in the r-b plane with vertices at $\left(\tfrac{1}{4}, \tfrac{1}{2}\right)$, $\left(\tfrac{1}{3}, \tfrac{1}{2}\right)$ and $\left(\tfrac{29}{92}, \tfrac{16}{23}\right)$. But substitution of (6.115) into (6.114b)-(6.114c) yields expressions for λ_0 and λ_1 in terms of b and r alone, from which it is readily established that $\lambda_0 < 0$ throughout the triangle (Exercise 6.22). It follows that a Type-II ESS exists only if $b < \frac{1}{2}$. Although the equations of the curves on which $\lambda_0 = 0$ and $\lambda_1 = 0$—which intersect at $\left(\frac{1}{4}, \frac{1}{2}\right)$—can in principle be found analytically, the resultant expressions are too unwieldy to be of practical use. Instead, therefore, we proceed numerically. We find that $\lambda_0 > \lambda_1$ for $b < \frac{1}{2}$, and so the right-hand boundary of the region in which $(1, 1, v_2)$ is an ESS for $v_2 < 1$ is determined by $\lambda_1 = 0$ (Exercise 6.22). In Figure 6.10 it is the solid curve joining $\left(\frac{1}{4}, \frac{1}{2}\right)$ to $(r_2, 0)$, where $r_2 \approx 0.233$ is the larger positive root of the quartic equation $43200r^4 - 23040r^3 + 4332r^2 - 344r + 9 = 0$.

To the right of the Type-II region there exists both a winner and a loser effect, i.e., $v_1 > v_0 > v_2$, so that (6.104b) and (6.104c) imply $\lambda_1 = 0 = \lambda_2$. The ESS is of Type IIIa if $v_1 = 1$. Then the equations

$$(6.116) \qquad 6r\{4v_0 - v_2^2 + 1\} + \{2(1+b) + b(1-v_2)\}(1+v_2)$$
$$+ (1 - 2b)(3 - v_2)v_0 = (1 - v_2)v_2 + 12$$

$$(6.117) \quad 6r(7 - 2v_0 + v_2)v_2 + (2 - b)(1 - v_2) = 6b(3 - v_0) + 3$$

jointly determine v_0 and v_2, and $(v_0, 1, v_2)$ is an ESS as long as

$$(6.118) \qquad \lambda_0 = \frac{(v_0 - v_2 + 2)(1-b) - (2v_0 - v_2 + 5)r}{12} - \frac{(1-v_2)(1+b)}{72} + \frac{1}{24}$$

is nonnegative. As in the case of a Type-II ESS, for any b there is an upper bound on r, which is found by solving (6.116)-(6.117) simultaneously with $\lambda_0 = 0$ for v_0, v_2 and r. The resultant boundary is sketched in Figure 6.10.

Finally, to the right of this boundary, there is always an ESS of Type IIIb with $1 > v_1 > v_0 > v_2 > 0$, so that (6.104a)-(6.104c) imply $\lambda_i = 0$ for all i. That is, v_0, v_1 and v_2 are jointly determined by

$$(6.119) \quad 6r\{v_1^2 - v_2^2 + 4v_0\} + \{2(1+b) + b(v_1 - v_2)\}(v_1 + v_2)$$
$$= 6(1 + v_1) - (1 - 2b)(v_1 - v_2 + 2)v_0 + (v_1 - v_2)v_2$$

$$(6.120) \quad 6r(6 - v_1 - v_2 + 2v_0)v_1 + (1 + b)(v_1 - v_2)$$
$$= 6(1 - b)(v_0 - v_2 + 2) + 3$$

$$(6.121) \quad 6r(6 - 2v_0 + v_1 + v_2)v_2 + (2 - b)(v_1 - v_2)$$
$$= 6b(v_1 - v_0 + 2) + 3.$$

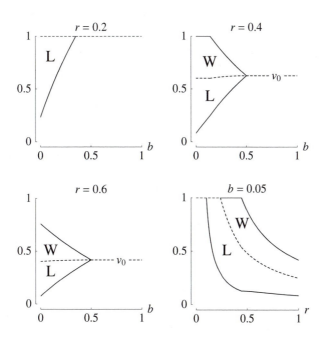

Figure 6.11. Evolutionarily stable strategy $v = (v_0, v_1, v_2)$ as a function of b for **(a)** $r = 0.2$, **(b)** $r = 0.4$, **(c)** $r = 0.6$ and **(d)** as a function of r for $b = 0.05$; v_0 is shown dashed, v_1 and v_2 (where different from v_0) are shown solid. A loser effect (of magnitude $v_0 - v_2$) is indicated by L; a winner effect (of magnitude $v_1 - v_0$) is indicated by W.

These equations agree with (6.113) along the common boundary $b = \frac{1}{2}$ between regions Ib and IIIb.

Note that there is a unique ESS, as determined by Figure 6.10. The magnitude of any winner or loser effect decreases with b and is greatest at intermediate values of r, as illustrated by Figure 6.11. The uniqueness of the ESS turns out to be a special property of the uniform distribution; for nonuniform distributions, there may be small transition regions where different types of ESS overlap (Exercise 6.28). But the main prediction, namely, that winner effects cannot evolve without loser effects, has been shown to hold for a very general class of distributions. As remarked at the beginning of the section, this prediction corroborates experimental results. Intuitively, if costs are too low to support a loser effect, then they are also low enough to support such a high initial strength perception that there is no advantage to raising it after a win. For further details, see [149].

6.8. Stomatopod strife: a threat game

Several years ago, Eldridge Adams and Roy Caldwell [1] observed a series of contests between stomatopods, or mantis shrimps. These crustaceans occupy cavities in coral rubble. If one intrudes upon another, the resident often defends his cavity by threatening with a pair of claw-like appendages. These threat displays often deter intruders, so that contests are settled without any physical contact. A surprising observation is that when stomatopods are weakened by molting, so that they are completely unable to fight, they threaten more frequently than animals that are between molts. Moreover, threats by weaklings often deter much stronger intruders, who would easily win a fight if there were one; that is, weaklings often bluff. But if the very weakest members of the population can threaten profitably, why don't all animals threaten? If the display can be given by animals that cannot back it up, why do their opponents respect it?

To explore this paradox, we develop a game in which a resident possessing a resource of value V can either threaten or not threaten in defense of it, and an intruder responds by either attacking or fleeing.[25] We assume that, if there's a fight, then the stronger animal wins. Furthermore, both contestants pay a combat cost, which is higher for the weaker animal; specifically, an animal of strength s pays $C(s)$, where $C'(s) < 0$. Threats increase the vulnerability of a resident to injury inflicted by an intruder. Thus a threat is a display of bravado that bears no cost if the resident is not attacked, but which carries a cost T (for threat) in addition to the combat cost if the resident is attacked by a stronger opponent. Because the molt condition of stomatopods is not externally visible, we assume (as in the war-of-attrition game) that each contestant is unaware of his opponent's strength. So his own strength must determine his behavior.

---

[25]§6.8 is based on [2, 150].

Let fighting strengths be continuously distributed between 0 and 1 with probability density function g (as in §6.7); and consider a focal individual or protagonist with fighting strength X, called Player 1 for convenience. In the role of resident, Player 1 threatens if either $X < u_1$ or $X > u_2$, but does not threaten if $u_1 < X < u_2$.[26] In the role of intruder, on the other hand, Player 1 attacks when $X > u_4$ if his opponent threatens but when $X > u_3$ if his opponent does not threaten; correspondingly, Player 1 flees when $X < u_4$ or when $X < u_3$ according to whether his opponent threatens or not. Thus Player 1's strategy is a four-dimensional vector $u = (u_1, u_2, u_3, u_4)$, whose first two components govern his behavior as resident, while its last two components govern his behavior as intruder. The corresponding fighting strength and strategy of Player 1's opponent, called Player 2 for convenience, are denoted by Y and $v = (v_1, v_2, v_3, v_4)$, respectively; e.g., Player 2 threatens as resident if either $Y < v_1$ or $Y > v_2$. Thus threats occur only when a resident's strength strength is either above or below a certain threshold. Nevertheless, potential ESSes include ones such that only the strongest residents threaten ($v_2 \gg v_1 = 0$), such that the weakest residents also threaten ($v_2 \gg v_1 > 0$) or such that residents always threaten ($v_1 = v_2$). In the first case, threats would carry reliable information; in the second case, threats could be either honest or deceitful; and in the third case, threats would carry no information at all.

Using notation that temporarily suppresses dependence on u and v, let $F(X, Y)$ denote the payoff to a u-strategist (Player 1) against a v-strategist (Player 2); let $F_k(X, Y)$ denote the payoff to u against v in role k; and let p_k be the probability of occupying role k. Then, if r stands for resident and i for intruder, we have $p_r + p_i = 1$ and

$$(6.122) \qquad F(X, Y) = p_r F_r(X, Y) + p_i F_i(X, Y).$$

Note that F, F_r and F_i are random variables, because X and Y are random variables; and it follows from (6.122) that the reward to a u-strategist in a population of v-strategists is

$$(6.123) \qquad f(u, v) = \mathrm{E}[F(X, Y)] = p_r f_r(u, v) + p_i f_i(u, v)$$

---

[26]Because X is continuously distributed, the event that $X = u_1$ or $X = u_2$ occurs with zero probability, and so we ignore it. Similarly for $X = u_3$ or $X = u_4$.

where E denotes expected value and, for $k = r$ or $k = i$,

$$(6.124) \qquad f_k(u,v) = \int_0^1 \int_0^1 F_k(x,y)\, dA,$$

with dA denoting $g(x)\, g(y)\, dx\, dy$ as in (2.22).

It is convenient at this juncture to define ρ, σ and τ as follows:

$$(6.125a) \qquad \sigma(X,Y) = \begin{cases} V - C(X) & \text{if } X > Y \\ -C(X) & \text{if } X < Y \end{cases}$$

$$(6.125b) \qquad \tau(X,Y) = \begin{cases} 0 & \text{if } X > Y \\ -T & \text{if } X < Y \end{cases}$$

$$(6.125c) \qquad \rho(X,Y) = \sigma(X,Y) + \tau(X,Y).$$

Thus ρ or σ, respectively, is the payoff to a threatening or non-threating resident protagonist against an attacking opponent. The protagonist's payoff from any contest is now defined by Table 6.12, in which the first, second, fifth and sixth rows correspond to threatening behavior by the resident, and the remaining four rows correspond to non-threatening behavior.

For $u_1 \leq u_2$, substitution from (6.125) into (6.124) now yields

$$(6.126) \quad f_r(u,v) = \int_0^{u_1} \int_{v_4}^1 \rho(x,y)\, dA + \int_{u_2}^1 \int_{v_4}^1 \rho(x,y)\, dA + \int_0^{u_1} \int_0^{v_4} V\, dA$$

$$+ \int_{u_2}^1 \int_0^{v_4} V\, dA + \int_{u_1}^{u_2} \int_{v_3}^1 \sigma(x,y)\, dA + \int_{u_1}^{u_2} \int_0^{v_3} V\, dA$$

and[27]

$$(6.127) \quad f_i(u,v) = \left\{ \int_{u_4}^1 \int_0^{v_1} + \int_{u_4}^1 \int_{v_2}^1 + \int_{u_3}^1 \int_{v_1}^{v_2} \right\} \sigma(x,y)\, dA.$$

In each case, the first integral sign corresponds to integration variable x and the second to variable y. If $u_1 > u_2$, however, then the u-strategist always threatens, and in place of (6.126) we have

$$(6.128) \qquad f_r(u,v) = \int_0^1 \int_{v_4}^1 \rho(x,y)\, dA + \int_0^1 \int_0^{v_4} V\, dA = Q,$$

---

[27] $\left\{ \int_{\xi_1}^{\xi_2} \int_{\eta_1}^{\eta_2} + \int_{\xi_3}^{\xi_4} \int_{\eta_3}^{\eta_4} \right\} \sigma(x,y)\, dA \equiv \int_{\xi_1}^{\xi_2} \int_{\eta_1}^{\eta_2} \sigma(x,y)\, dA + \int_{\xi_3}^{\xi_4} \int_{\eta_3}^{\eta_4} \sigma(x,y)\, dA.$

Table 6.12. Payoff to a protagonist of strength X with strategy $u = (u_1, u_2, u_3, u_4)$ against an opponent of strength Y with strategy $v = (v_1, v_2, v_3, v_4)$. The role of resident or intruder is denoted by r or i; ρ and σ are defined by (6.125).

k	Relative magnitudes of X and Y	$F_k(X,Y)$
r	$X < u_1$ or $X > u_2$ and $Y > v_4$	$\rho(X,Y)$
r	$X < u_1$ or $X > u_2$ and $Y < v_4$	V
r	$u_1 < X < u_2$ and $Y > v_3$	$\sigma(X,Y)$
r	$u_1 < X < u_2$ and $Y < v_3$	V
i	$Y < v_1$ or $Y > v_2$ and $X > u_4$	$\sigma(X,Y)$
i	$Y < v_1$ or $Y > v_2$ and $X < u_4$	0
i	$v_1 < Y < v_2$ and $X > u_3$	$\sigma(X,Y)$
i	$v_1 < Y < v_2$ and $X < u_3$	0

say, where Q is independent of u. In other words, any strategy satisfying $u_1 > u_2$ is equivalent mathematically to any strategy satisfying $u_1 = u_2$. Therefore, from now on we constrain u to satisfy

$$(6.129) \qquad 0 \le u_1 \le u_2 \le 1, \ 0 \le u_3 \le 1, \ 0 \le u_4 \le 1.$$

It is a moot point whether the strategies thus excluded are all equivalent biologically: An animal who always threatens because his "high" threshold (for reliable communication) is normal but his "low" threshold (for deceitful communication) is abnormally high may be said to behave very differently from an animal who always threatens because his low threshold is normal but his high threshold is abnormally low, whereas our game does not distinguish between them. Nevertheless, the question becomes irrelevant, because $u_1 < u_2$ at the only ESS.

To calculate the rational reaction set R we must maximize f defined by (6.123) with respect to u. We first observe that the game is separable, because (6.123) may be written as (6.1) with $p = 4$ and

$$(6.130a) \qquad f_1(u_1, v) \;=\; p_r \int_0^{u_1} \int_{v_3}^{v_4} \{V - \sigma(x, y)\} \, dA$$

$$+ \; p_r \int_0^{u_1} \int_{v_4}^{1} \tau(x, y) \, dA$$

$$(6.130b) \qquad f_2(u_2, v) \;=\; p_r Q - f_1(u_2, v)$$

Table 6.13.   Quantities that appear in Tables 6.14-6.15.
Note that all but the last two depend on $v = (v_1, v_2, v_3, v_4)$.

$$\delta = \frac{(a+b)(v_1-v_2+1)-v_1}{v_1-v_2+1} \qquad \omega_1 = \frac{(1+a+b)v_4-(a+b)v_3-t(1-v_4)}{1+b(v_4-v_3)}$$

$$\omega_4 = \frac{(a+b)(v_1-v_2+1)}{1+b(v_1-v_2+1)} \qquad \gamma_4 = \frac{(a+b)(v_1-v_2+1)-v_1+v_2}{1+b(v_1-v_2+1)}$$

$$\theta_1 = \frac{(1+a+b)(v_4-v_3)-t(1-v_4)}{b(v_4-v_3)} \qquad \theta_2 = 1 - \frac{a(v_4-v_3)}{t-b(v_4-v_3)}$$

$$\theta_3 = \frac{v_1+(a+b)(v_2-v_1)}{1+b(v_2-v_1)} \qquad \theta_4 = \frac{(a+b)(v_1-v_2+1)-v_1}{b(v_1-v_2+1)} = \frac{\delta}{b}$$

$$\Delta = \frac{t(1-v_4)}{v_4-v_3} \qquad \omega_3 = 1 - \frac{1-a}{b} \qquad \gamma_3 = \frac{a+b}{1+b}$$

$$(6.130c) \quad f_3(u_3, v) = p_i \int_{u_3}^{1} \int_{v_1}^{v_2} \sigma(x, y)\, dA$$

$$(6.130d) \quad f_4(u_4, v) = p_i \int_{u_4}^{1} \int_{0}^{v_1} \sigma(x, y)\, dA + p_i \int_{u_4}^{1} \int_{v_2}^{1} \sigma(x, y)\, dA$$

where Q is defined by (6.128). Thus maximization with respect to u_3 may be performed separately from that with respect to u_4, and both independently of that with respect to u_1 or u_2. This separability of the reward function makes the game tractable analytically.

Some general features of R require only that V, C, T and g are all positive, which we assume. From (6.125a) and (6.130c), $\frac{\partial f_3}{\partial u_3} = -p_i \int_{v_1}^{v_2} \sigma(u_3, y)g(u_3)g(y)\, dy = p_i C(u_3)g(u_3) \int_{v_1}^{v_2} g(y)\, dy0$ is positive for $u_3 < v_1 < v_2$; so the maximum of f_3 for $0 \le u_3 \le 1$ must occur where $v_1 \le u_3 \le 1$. Again, (6.125a)-(6.125b) and (6.130a) imply that if $v_4 \le v_3$ then $\frac{\partial f_1}{\partial u_1} < 0$ for all $0 < u_1 < 1$ unless $v_4 = v_3 = 1$. Thus if $v_4 \le v_3$ then the maximum of f_1 must occur at $u_1 = 0$; unless $v_4 = v_3 = 1$, in which case f_1 is independent of u_1. Correspondingly, (6.130b) yields $\frac{\partial f_2}{\partial u_2} > 0$ for all $0 < u_2 < 1$ unless $v_4 = v_3 = 1$, and so the maximum of f_2 must occur at $u_2 = 1$; unless $v_4 = v_3 = 1$, in which case f_2 is independent of u_2.

Nevertheless, we cannot fully calculate R until we specify C and g in (6.125)-(6.128). In this regard we make two assumptions. First, combat cost decreases linearly with fighting strength according to

$$(6.131) \qquad C(s) = V\{a + b(1-s)\},$$

with $0 < a < 1$ and $b > 0$; thus $V > C$ for the strongest animal, and $V > C$ for every animal in the limit as $B \to 0$, but in general there may be (weaker) animals for which $C > V$. Second, as in the previous section, fighting strength is uniformly distributed between 0 and 1, i.e., $g(x) = g(y) = 1$ or $dA = dx\,dy$ in (6.130). Furthermore, it is convenient to introduce a dimensionless threat-cost parameter

$$(6.132) \qquad\qquad t = T/V.$$

We can now proceed to calculate R.[28] We find that $f_3(u_3, v) = \frac{1}{2}V(v_2-v_1)(1-u_3)\{2(1-a)-b(1-u_3)\} - \frac{1}{2}V(v_2-u_3)^2$ if $v_1 \le u_3 \le v_2$, whereas the last (squared) term must be omitted to obtain the correct expression for f_3 if $v_2 \le u_3 \le 1$. It follows from Exercise 6.23 that the maximum of f_3 for $v_1 \le u_3 \le 1$ (and hence also for $0 \le u_3 \le 1$) occurs at $u_3 = \theta_3$ if $b(1-v_2) \le 1-a$ but at $u_3 = \omega_3$ if $b(1-v_2) > 1-a$, where θ_3 and ω_3 are defined in Table 6.13.

Also, $f_4(u_4, v) = \frac{1}{2}V(1-u_4)\{2v_1 - (2a+b\{1-u_4\})(v_1-v_2+1)\} + \frac{1}{2}V(1-v_2)^2 - \frac{1}{2}V(v_1-u_4)^2$ when $0 \le u_4 \le v_1$, but the last (negative squared) term must be omitted to obtain the correct expression for f_4 when $v_1 \le u_4 \le v_2$; and when $v_2 \le u_4 \le 1$ we obtain $f_4(u_4, v) = \frac{1}{2}V(1-u_4)\{2v_1 - 2v_2 + u_4 + 1 - (2a+b\{1-u_4\})(v_1-v_2+1)\}$. Provided $v_1 - v_2 + 1 \ne 0$, the maximum of f_4 for $0 \le u_4 \le 1$ can now be shown to occur at $u_4 = \omega_4$ if $\delta < bv_1$, at $u_4 = \theta_4$ if $bv_1 \le \delta \le bv_2$ and at $u_4 = \gamma_4$ if $\delta > bv_2$, where δ, γ_4, θ_4 and ω_4 are defined in Table 6.13. If $v_1 - v_2 + 1 = 0$, which can happen only if $v_1 = 0$ and $v_2 = 1$, then f_3 is maximized at $u_3 = \gamma_3$ (defined in Table 6.13) and any u_4 maximizes f_4. These results imply that the maximum of f_i—defined by (6.127)—is given by Table 6.14.

We have already seen that when $v_3 \ge v_4$, f_r is maximized for $0 \le u_1 \le u_2 \le 1$ where $u_1 = 0$, $u_2 = 1$ (unless $v_3 = v_4 = 1$, in which case both u_1 and u_2 are arbitrary). Moreover, it is clear from (6.125a) and (6.130a)-(6.130b) that when $v_3 < v_4 = 1$, f_r is maximized where $u_1 = u_2$. Let us therefore assume that $v_3 < v_4 < 1$, and hence that

$$(6.133) \qquad b(1-v_4) < a+b(1-v_4) < a+b(1-v_3).$$

Then $f_1(u_1, v) = \frac{1}{2}Vu_1\{2(1+a+b)(v_4-v_3) - 2t(1-v_4) - b(v_4-v_3)u_1\}$ for $0 \le u_1 \le v_3$; $\frac{1}{2}V(u_1 - v_3)^2$ must be subtracted to obtain the

---

[28]This calculation is rather complicated; readers who would prefer to take its outcome on trust are advised to skip ahead to p. 296.

Table 6.14. Maximizers u_3 and u_4 of f_i for $0 \leq u_3, u_4 \leq 1$.

Constraints on v_1, v_2 $(\geq v_1)$	u_3	u_4	Constraints on u_3, u_4
$\delta < bv_1, b(1 - v_2) \leq 1 - a$	θ_3	ω_4	$v_1 \leq u_3 \leq v_2, u_4 < v_1$
$bv_1 \leq \delta \leq bv_2, b(1 - v_2) \leq 1 - a$	θ_3	θ_4	$v_1 \leq u_3, u_4 \leq v_2$
$\delta > bv_2, b(1 - v_2) \leq 1 - a$	θ_3	γ_4	$v_1 \leq u_3 \leq v_2, u_4 > v_2$
$\delta < bv_1, b(1 - v_2) > 1 - a$	ω_3	ω_4	$u_3 > v_2, u_4 < v_1$
$bv_1 \leq \delta \leq bv_2, b(1 - v_2) > 1 - a$	ω_3	θ_4	$u_3 > v_2, v_1 \leq u_4 \leq v_2$
$\delta > bv_2, b(1 - v_2) > 1 - a$	ω_3	γ_4	$u_3 > v_2, u_4 > v_2$
$v_1 = 0, v_2 = 1$	γ_3	u_4	u_4 arbitrary

Table 6.15. Maximizers u_1 and u_2 of f_r for $0 \leq u_1 \leq u_2 \leq 1$.

Constraints on v_3, v_4 $(> v_3)$	u_1	u_2	Constraints on u_1, u_2
$\Delta \geq 1 + a + b$	0	θ_2	$u_1 \leq v_3, u_2 > v_4$
$1 + a + b > \Delta \geq 1 + a + b(1 - v_3)$	θ_1	θ_2	$u_1 \leq v_3, u_2 > v_4$
$1 + a + b(1 - v_3) > \Delta > a + b(1 - v_4)$	ω_1	θ_2	$v_3 < u_1 \leq v_4, u_2 > v_4$
$\Delta \leq a + b(1 - v_4), v_4 > v_3$	u_1	u_2	$u_1 = u_2, u_2$ arbitrary
$v_3 \geq v_4, v_4 \neq 1$	0	0	
$v_3 = v_4 = 1$	u_1	u_2	u_1, u_2 both arbitrary

correct expression for $v_3 \leq u_1 \leq v_4$; and, for $v_4 \leq u_1 \leq 1$, $f_1(u_1, v) = \frac{1}{2}V\{(v_4 - v_3)(v_4 + v_3 + u_1\{2(a + b) - bu_1\}) - 2tu_1(1 - v_4) + t(u_1 - v_4)^2\}$. From Exercise 6.23, f_1 varies between $u_1 = 0$ and $u_1 = 1$ as follows. If $\Delta \geq 1 + a + b$, then f_1 decreases between $u_1 = 0$ and $u_1 = \theta_2$ $(> v_4)$ but increases again between $u_1 = \theta_2$ and $u_1 = 1$. If $1 + a + b > \Delta \geq 1 + a + b(1 - v_3)$, then f_1 increases between $u_1 = 0$ and $u_1 = \theta_1$ $(\leq v_3)$, decreases between $u_1 = \theta_1$ and $u_1 = \theta_2$, and increases again between $u_1 = \theta_2$ and $u_1 = 1$. If $1 + a + b(1 - v_3) > \Delta > a + b(1 - v_4)$, then f_1 increases between $u_1 = 0$ and $u_1 = \omega_1$ (which satisfies $v_3 < \omega_1 < v_4$), decreases between $u_1 = \omega_1$ and $u_1 = \theta_2$, and increases again between $u_1 = \theta_2$ and $u_1 = 1$.[29] Finally, if $\Delta \leq a + b(1 - v_4)$ then f_1 increases monotonically between $u_1 = 0$ and $u_1 = 1$; its concavity is always downward for $0 \leq u_1 \leq v_4$, but it is upward or downward for $v_4 \leq u_1 \leq 1$ according to whether $\Delta > b(1 - v_4)$ or $\Delta < b(1 - v_4)$.

---

[29] Note that $\Delta > a + b(1 - v_4)$ and (6.133) imply $t > b(v_4 - v_3)$ in θ_2 (and hence, eventually, that $\eta'(s) > 0$ for $L < s < 1$ in (6.137)).

Correspondingly, from (6.130b), f_2 varies between $u_2 = 0$ and $u_2 = 1$ as follows. If $\Delta \geq 1 + a + b$, then f_2 increases between $u_2 = 0$ and $u_2 = \theta_2$, and decreases again between $u_2 = \theta_2$ and $u_2 = 1$. If $1+a+b > \Delta \geq 1+a+b(1-v_3)$, then f_2 decreases between $u_2 = 0$ and $u_2 = \theta_1$, increases between $u_2 = \theta_1$ and $u_2 = \theta_2$, and decreases again between $u_2 = \theta_2$ and $u_2 = 1$. If $1+a+b(1-v_3) > \Delta > a+b(1-v_4)$, then f_2 decreases between $u_2 = 0$ and $u_2 = \omega_1$, increases between $u_2 = \omega_1$ and $u_2 = \theta_2$, and decreases again between $u_2 = \theta_2$ and $u_2 = 1$. Finally, if $\Delta \leq a + b(1 - v_4)$ then f_2 decreases monotonically between $u_2 = 0$ and $u_2 = 1$. Thus the maximum of f_r—defined by (6.126)—is given by Table 6.15. Note that the maximum corresponds to unconditional threatening if $\Delta \leq a + b(1 - v_4)$.

Now, if v is an ESS, then the maximum in Table 6.14 must occur where $u_3 = v_3$ and $u_4 = v_4$; the maximum in Table 6.15 must occur where $u_1 = v_1$ and $u_2 = v_2$; and all conditions on u must be satisfied. Let us first of all look for a strong ESS. Then v must be the only best reply to itself. This immediately rules out the fourth and sixth rows of Table 6.15, where u_1 and u_2 do not yield a unique best reply to v_3 and v_4; and although the fifth row of Table 6.15 does yield a unique best reply, it corresponds to the bottom row of Table 6.14, where u_4 is not unique. Accordingly, we restrict our attention to the first three rows of Table 6.15. Then, for the maximum to occur at $u_2 = v_2$, each possibility requires $v_2 > v_4$. Thus the maximum at $u_4 = v_4$ in Table 6.14 must satisfy $v_4 < v_2$, excluding the third and sixth row of that table. Again, the relative magnitudes of v_3 and v_4 in the first three rows of Table 6.15 all imply $v_3 < v_4 < 1$, so that the maximum at $u_3 = v_3$ in Table 6.14 cannot satisfy $v_3 \geq v_4$, and hence (because $v_4 < v_2$) cannot satisfy $v_3 \geq v_2$; thus the fourth and fifth rows of Table 6.14 are excluded. The first row of the table is likewise excluded, because the maximum where $u_3 = v_3$ and $u_4 = v_4$ would have to satisfy $v_4 < v_1 \leq v_3$, which is impossible because $v_4 > v_3$. Only the second row of Table 6.14 now remains. Because the maximum at $u_3 = v_3$ must therefore satisfy $v_1 \leq v_3$, we have to exclude the third row of Table 6.15. But the maximum where $u_1 = v_1$ and $u_2 = v_2$ in Table 6.15 cannot now occur where $u_1 = 0$ and $u_2 = \theta_2$ because the second row of Table 6.14 would then imply $0 \leq a+b \leq b\theta_2$, which is impossible for $a > 0$. We have thus excluded the top row of Table 6.15, and only the second remains. We conclude that a strong ESS must correspond to the second row in each table.

Let us now set $v = (I, J, K, L)$ in Table 6.13, so that θ_3, θ_4, δ, ω_4 and γ_4 depend on I and J, whereas θ_1, θ_2, Δ and ω_1 depend on K and L. Then what we have shown is that (I, J, K, L) is a strong ESS if it satisfies the equations $I = \theta_1$, $J = \theta_2$, $K = \theta_3$ and $L = \theta_4$. The last two equations yield

$$(6.134) \qquad K = \frac{I + (a+b)(J-I)}{1 + b(J-I)}, \qquad L = \frac{(a+b)(I-J+1)-I}{b(I-J+1)}.$$

Substituting into $I = \theta_1$ and $J = \theta_2$, we obtain a pair of equations for I and J. The first has the form

$$(6.135) \qquad tab(1-J)^2 + d_1(1-J) + d_0 = 0,$$

where $d_0 = (1-a)\{(1+t)(1-bI+b) + a\}I$ is quadratic in I and $d_1 = -(a+b+at)(1-bI+b) - bt(1-a)I - a(a+b)$ is linear in I. The second equation is cubic in J, and can be used in conjunction with the first to express J as a quotient of cubic and quadratic polynomials in I. Substitution back into the first equation yields a sextic equation for I, of which three solutions—namely, $I = 0$, $I = 1 + a/b$ and $I = 1 + (1+a)/b$—can be found by inspection. None of these solutions satisfies $0 < I < 1$. Thus, removing the appropriate linear factors, we find that I must satisfy the cubic equation

$$(6.136) \qquad c_3 I^3 + c_2 I^2 + c_1 I + c_0 = 0,$$

whose coefficients are defined by

$$
\begin{aligned}
c_0 &= -a\{(a+b)(1+2a+b) + at(1+a+b)\} \\
c_1 &= (1+a+b)\{(1+t)\{a + (1+b)(1+t) + b^2\} + 2ab\} \\
&\qquad + a(1+t)\{1+b+b(3a+2b)\} + a^2 \\
c_2 &= -b\{(1+t)\{(2+b)t + 2b^2 + (3b+2)(1+a)\} + a(1+b)\} \\
c_3 &= b^2(1+b)(1+t).
\end{aligned}
$$

Because $c_0 < 0$ and $c_0 + c_1 + c_2 + c_3 > 0$, it is clear at once that there is always a real solution satisfying $0 < I < 1$. It is not difficult (but a bit tedious) to show that this solution is the only solution satisfying $0 < I < 1$; the other two solutions are either complex conjugates or, if they are real, satisfy $I > 1$. Moreover, only one solution of quadratic equation (6.135) for J satisfies $J > I$. Thus the strategy $v = (I, J, K, L)$ defined by (6.134)-(6.136) is the only strong ESS.

Nevertheless, there are several candidates for a weak ESS. First, from the last row of Table 6.14 and the fifth row of Table 6.15, we find that $v^* = (0, 1, \gamma_3, \lambda)$ satisfies (2.12a) for any $\lambda \leq \gamma_3$ (where γ_3 is defined in Table 6.13); however, v^* fails to satisfy (2.12b), because $f(u, v^*) = f(v^*, v^*)$ for $u = (0, 1, \gamma_3, u_4)$, $0 \leq u_4 \leq 1$. From (6.126)-(6.127), we then find that $f(v^*, u) = f(u, u) = 0$, so that (2.12c) fails to hold. Thus v^* is not a weak ESS. Intuitively, never threatening cannot be an evolutionarily stable behavior because in equilibrium the threshold λ is irrelevant; even if the population strategy v satisfies $v_4 \leq \gamma_3$ to begin with, there is nothing to prevent v_4 from drifting to $v_4 > \lambda$, in which case never threatening is no longer a best reply. (In particular, there is nothing to prevent v_4 from drifting to 1, and never threatening cannot be a best reply to an opponent who never attacks when threatened.)

Second, from the last row of Table 6.15, we must investigate the possibility that there is a weak ESS of the form $v = (v_1, v_2, 1, 1)$. Because $a < 1$, however, we see from Table 6.14 that this possibility requires $\theta_3 = 1$ or $(1 - a)v_1 + av_2 = 1$, which implies $v_1 = v_2 = 1$. But then, from the first three rows of Table 6.14, either $v_4 = \gamma_3$ or $v_4 = \omega_3$, contradicting $v_4 = 1$. Hence there is no such ESS.

The remaining possibility for a weak ESS is an always-threatening equilibrium with $v_1 = v_2 = \lambda$, say, which corresponds to the fourth row of Table 6.15, and therefore satisfies $v_4 > v_3$. This equilibrium cannot correspond to the first row of Table 6.14, because $v_1 \leq v_3 \leq v_2$ and $v_4 < v_1$ then imply $v_4 < \lambda \leq v_3$, contradicting $v_4 > v_3$. For similar reasons, the equilibrium cannot correspond to either the second row of Table 6.14 (which would require $v_4 = \lambda = v_3$) or the fourth or fifth row (each of which would require $v_3 > v_4$). Thus the equilibrium must correspond to either the third or sixth row of Table 6.14, and hence have the form $v^* = (\lambda, \lambda, \zeta, \gamma_3)$, where $\zeta = \max(\lambda, \omega_3)$ satisfies $\zeta < \gamma_3$. Then $f(u, v^*) = f(v^*, v^*)$ and $f(v^*, u) = f(u, u)$ for any u such that $u_1 = u_2$ and $u_4 = \gamma_3$: although v^* satisfies (2.12a), it fails to satisfy (2.12b)-(2.12c), and so is not a weak ESS. Intuitively, always threatening cannot be an evolutionarily stable behavior because in equilibrium the threshold ζ is irrelevant; even if the population

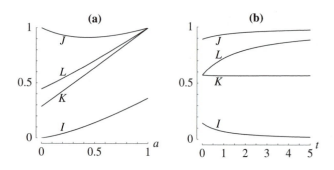

Figure 6.12. The effect of varying **(a)** the fixed cost of fighting or **(b)** the threat cost on the ESS thresholds. In these examples $b = 0.4$ and **(a)** $t = 0.4$ or **(b)** $a = 0.4$.

strategy v satisfies $v_3 < \gamma_3$ to begin with, there is nothing to prevent v_3 from drifting to $v_3 \geq \gamma_3$, in which case always threatening is no longer a best reply. (In particular, there is nothing to prevent v_3 from drifting to 1, and always threatening cannot be a best reply to an opponent who never attacks when not threatened.)

The upshot is that the sole ESS is the strong ESS, for which $J > L > K > I$ (Figure 6.12). At this ESS, the weakest and strongest animals both threaten when resident, whereas those of intermediate strength do not. The proportion of intruders deterred by threats is $L - K$ (because an intruder attacks if his strength exceeds K when not threatened, but only if his strength exceeds L when threatened). All such intruders would lose against a resident whose strength exceeds J (because $J > L$); the threats of the strongest residents are therefore honest. On the other hand, all deterred intruders would win against a resident whose strength does not exceed I (because $I < K$); the threats of the weakest residents are therefore deceptive—but they cannot be distinguished from honest threats without escalation. The proportions of residents who threaten deceptively, who do not threaten and who threaten honestly are I, $J - I$ and $1 - J$, respectively; and it is readily shown that $J - I > \frac{1}{2}$, so that fewer than half of the residents threaten (Exercise 6.24). Nevertheless, the proportion of threats that are deceptive can be considerable (Exercise

6.24). In other words, not only is bluffing a part of the ESS, but also it can persist at high frequency.

To see why the weakest and strongest animals both threaten when resident while those of intermediate strength do not, it is instructive to compute the expected difference in net gain between threatening and not threatening to a resident of known strength s against an intruder whose unknown strength S is drawn at random from the uniform distribution (so that $\text{Prob}(S \leq z) = z$). We compute this quantity by subtracting the expected difference in total cost—i.e., combat cost plus threat cost— from the expected difference in benefit.

On the one hand, the resident's combat cost—if paid, i.e., if $S > L$ or if $S > K$, according to whether the resident threatens or not— is $C(s)$. Thus the expected difference in combat cost between threatening and not threatening is $C(s)\{\text{Prob}(S > L) - \text{Prob}(S > K)\} = \{K - L\}C(s)$. Also, the resident's threat cost is T if $S > \max(s, L)$ but zero otherwise, with expected value $T\text{Prob}(S > \max(s, L)) = T\{1 - \max(s, L)\}$—which is also the expected *difference* in threat cost, because the cost is avoided by not threatening. Adding, we find that the expected difference in total cost is $\{K - L\}C(s) + T\{1 - \max(s, L)\}$.

On the other hand, the benefit to a threatening resident, who wins if the intruder is weaker or does not attack, is V if $S < \max(s, L)$ but 0 if $S > \max(s, L)$, with expected value $V\text{Prob}(S < \max(s, L))$. For a non-threatening resident, the corresponding expected value is $V\text{Prob}(S < \max(s, K))$; and their difference is the expected difference in benefit. So $V\{\text{Prob}(S < \max(s, L)) - \text{Prob}(S < \max(s, K))\} - \{K - L\}C(s) - T\{1 - \max(s, L)\}$ is the expected difference in net gain. Let us denote this quantity by $V\eta(s)$, so that η is dimensionless. Then, from (6.131)-(6.132) and Exercise 6.25, we have $\eta(s) =$

$$
(6.137) \quad
\begin{cases}
\{1 + a + b(1 - s)\}(L - K) - t(1 - L) & \text{if } 0 \leq s \leq K \\
L - s + \{a + b(1 - s)\}(L - K) - t(1 - L) & \text{if } K \leq s \leq L \\
\{a + b(1 - s)\}(L - K) - t(1 - s) & \text{if } L \leq s \leq 1.
\end{cases}
$$

Furthermore, $\eta(I) = 0 = \eta(J)$; $\eta(s) < 0$ if $I < s < J$; and $\eta(s) > 0$ if $s < I$ or $s > J$. So the strongest and the weakest residents both threaten because the expected net gain from doing so exceeds that from not threatening; however, their threats are profitable for different reasons. The strongest residents threaten because, although their expected benefit (of avoiding combat costs) is low, their expected cost is even lower—they are very unlikely to meet an opponent strong enough to inflict the threat cost. At the other extreme, the weakest residents threaten because, although their expected cost is high—an intruder who attacks invariably inflicts the threat cost—their expected benefit from threatening is even higher: they are able thereby to deter some considerably stronger intruders (who would win a fight if there were one), and to do so without the cost of combat (which is highest for the weakest animals).

Note, finally, that our partial-bluffing ESS arises only in special circumstances. The ESS does not persist in the limit as $b \to 0$, or if we change the reward structure so that a threatening resident pays the threat cost either regardless of whether he is attacked; or only if he is attacked, but regardless of whether he wins or loses.[30] From (6.131), we have $Vb = -C'(s)$ where s is strength and C is cost of combat. Thus our model predicts a partial-bluffing ESS only if the combat cost is higher for weaker animals *and* a threatening resident pays an additional cost only when he is attacked and loses. And this is a strength of the model: it helps to identify the particular circumstances in which we might expect to observe high-frequency bluffing in nature.

6.9. Commentary

In Chapter 6, we have used continuous or discrete population games to study topics in behavioral ecology and resource management, namely, sex allocation (§6.1), contest behavior (§6.2 and §§6.7-6.8), kinship (§6.3), sperm competition (§6.4) and mutualism in social foraging (§6.5) or wildlife conservation (§6.6). The list is by no means exhaustive but exemplifies the scope and variety of applications. Other topics include habitat selection [**31, 102**], seed dispersal [**119**] and parent-offspring conflict [**160**]. For overviews, see [**83, 177**].

Evolutionary game theory did not fully emerge as a field of study in its own right until 1982, when Maynard Smith's definitive monograph [**132**] consolidated the advances that he and others had made during the 1970s. But the application of game-theoretic reasoning to

---

[30] For details, see Exercise 6.26.

the study of sex ratios (§6.1) is significantly older, and can be traced through Hamilton [**82**] all the way back to Fisher [**63**] in 1930. For an introduction to this topic, see [**36**]. For more recent developments, see [**65**].

The application of game theory to contest behavior (§6.2, §6.7 and §6.8) began with two basic models introduced by Maynard Smith [**129, 134**]. The first of these models was the Hawk-Dove game (§1.2, §2.3 and §2.7), in which use of weaponry determines the winner of any actual fight; costs result from injury, and are discrete with respect to time. The Hawk-Dove game has since been developed in various ways by numerous authors; see, e.g., [**5, 44, 111**], Chapter 7 of [**100**], and references therein. The second of Maynard Smith's models was a war-of-attrition model, in which persistence determines who wins; costs result from depletion of energy reserves, and are continuous with respect to time. This model also has evolved considerably; see, e.g., [**17, 18, 85, 117, 211**], and references therein. (Although the extremes of costs that are discrete or continuous with respect to time are useful idealizations, many contests potentially involve both types of cost, as in §6.8's threat game.) Other models of contest behavior include Enquist and Leimar's sequential assessment game [**57**] and Payne's cumulative assessment [**179**] game, which consider how animals may assess opponents' fighting strengths (directly, as opposed to indirectly through their distribution). For earlier work on this issue, see [**178**]. For a recent review of the literature, see [**192**].

In matters of assessment, analysis of contest behavior dovetails with that of communication or signalling, where game theory has been central in recent years; see, e.g., [**26, 104, 103, 107, 108, 210**], and references therein. An early prediction was that honest threat displays are unlikely to be evolutionarily stable signals, because—unless it is physically impossible to be dishonest—they could be infiltrated by bluffs until it would no longer pay receivers (e.g., §6.8's intruders) to respect them [**131**]. A later view was that threat displays must be honest to be stable, with signal costs enforcing reliability [**75, 242**]; and similar logic applies to other signals, e.g., offspring's calls for food from their parents [**167**]. As we saw in §6.8, however, [**2**] has since shown that stable communication can involve both reliable and deceptive signals, and that the frequency of bluffs may be high.[31] Thus, in a sense, game theory has come full circle over this issue.

Winner and loser effects (§6.7) may help to explain the persistence of linear dominance hierarchies, in which no individuals are of equal or indeterminate rank. Such hierarchies can form in social groups through a round-robin tournament of modified Hawk-Dove games in which contestants are able to assess one another's fighting strength prior to any escalation (unlike in §6.7), with the winner of each pairwise contest dominating the loser [153]. Because relative strength is never a perfectly reliable predictor of the outcome of a contest, however, some games are won by the weaker contestant; thus, for a hierarchy to persist, stronger animals must sometimes consider themselves subordinate to weaker animals, instead of attempting to reverse the asymmetry at a subsequent encounter. Why? Enter winner and loser effects: one effect of victory by a weaker animal may be to raise his perception of his strength and/or to lower that of his opponent, so that after they fight the opponent perceives his relative strength to lie below some evolutionarily stable aggression threshold—whereas previously it lay above, provoking a fight. For a recent perspective on related issues, see [184].

Aspects of dominance are implicit in the recruitment game of §6.5 because a bonanza's discoverer may rise in status. This model focuses on the dichotomy of immediate versus delayed recruitment, but social foraging games more often focus on "producing" (i.e., finding food) versus "scrounging" (i.e., exploiting the discoveries of others). An aspect of dominance to consider here is that individuals of high rank may coerce those of low rank into producing [12]. For a recent synthesis of social foraging theory, see [70].

In terms of §5.8, the recruitment game showed that delayed recruitment to food bonanzas among juvenile ravens can arise mutualistically either as cooperation against the common enemy of adult ravens who may defend such bonanzas or as cooperation toward the

---

[31]See also [10, 220]; but beware of two false claims, namely, that [2] involves "the assumption that cheating must be rare in an evolutionarily stable signalling system" [220, p. 225] and that a bluffing frequency as high as 44% is "unanticipated by current signalling theory" [10, p. 719]. See Exercise 6.24.

common end of rising in social status (which is a property of the individual, albeit a relative one). Either way, cooperation is an incidental consequence of each individual's selfish goal. In the first case, that goal is access to the bonanza. In the second case, that goal is acquiring prestige; for discussion, see Chapter 12 of [**242**]. In neither case is there any direct feedback between specific individuals. In both cases, therefore, cooperation is mutualism [**152**, p. 270].

Despite that, cooperation via the status-enhancement effect in ravens is reminiscent of what has been called "indirect reciprocity" [**3**]. Here two remarks are in order. First, if one regards mutualism and reciprocity as the poles of a grand continuum, then the degree of feedback between cooperators should steadily increase between the first pole and the second; any behavior described as a form of reciprocity should be close to the second pole. Second, there is no central authority to act as supreme arbiter of names for sub-categories. As a result, "indirect reciprocity" means different things to different people: either that a well defined network of specific donors and recipients contains more than two individuals [**25**], or that individuals can enhance their status through generosity and be more generous to partners of higher status [**3, 172**]. The first interpretation is far closer to reciprocity, especially for a short network; and the second interpretation—which "does not require the same two individuals ever to meet again" [**172**, p. 573]—is so close to mutualism that, in one opinion, it "stretches the definition of reciprocity to the point of meaninglessness" [**242**, p. 149].

Sperm competition games began with Parker [**174**] in 1990. He and others have since developed a suite of models to deal with various strategic aspects of sperm expenditure, e.g., risk assessment [**11**]; for a review, see [**175**]. On the whole, this theory has developed in isolation from sex allocation theory (§6.1), although occasionally the two overlap; see, e.g., [**180**].

Games between and among kin derive, respectively, from Hamilton [**81**] and Grafen [**74**]. Grafen's conditions for an ESS were discovered independently by Fagen [**60**]. §6.3 is based on [**156**], which in turn is based on [**97**]. For a recent perspective, see [**189**]. For an application to sperm competition, see [**176**].

Although game theory has long been applied to the economics of fisheries management [**87, 142**], applications of game theory to wildlife conservation—where models can be used to predict the behavior either of humans who exploit a wildlife resource (as in §6.6) or of animals that are themselves the resource to be conserved—are among the least developed. Nevertheless, opportunity is often greatest where development is least, and recent literature suggests enormous potential for future applications of game theory in this area; see, e.g., [**72, 73, 159**].

Finally, as stated at the beginning of §6.2, we tacitly assume that behavior observed in a real population can be adequately approximated by the ESS of a game-theoretic model (at least for the purpose of resolving a paradox). Although this assumption is at least highly plausible—if the rate of deviation or mutation is significantly lower than the rate of selection, and regardless of whether the underlying dynamic is cultural or genetic (§2.3)—it has generated considerable skepticism,[32] especially among biologists who worry about its consistency with details of genetical inheritance in sexually reproducing populations. Game theorists have addressed this issue by showing that, under reasonable assumptions about the mapping between genotype and "behavioral phenotype" (= strategy)—which rarely is actually known [**76**, pp. 5-8]—ESSes correspond to stable, long-run equilibria of explicit genetic models; see [**58, 68, 84, 98, 127**], and references therein. For further discussion see [**190**], which systematically addresses the criticisms and argues powerfully in favor of evolutionary game theory.

Exercises 6

1. Sketch the rational reaction set R of the game whose reward is defined by (6.5), and verify that $v = \frac{1}{2}$ is the only ESS.
2. Verify (6.12)-(6.14). Where would it help to reverse the order of integration?

---

[32]See, e.g., [**158**, p. 70].

3. **(a)** Obtain the reward function for the war of attrition in which T_{max} is exponentially distributed with mean μ according to

$$g(t) = \tfrac{1}{\mu}e^{-t/\mu}, \quad 0 < t < \infty.$$

   (b) Calculate the rational reaction set R.

   (c) Find the ESS.

4. Verify (6.16)-(6.18). Why is it unnecessary to differentiate f for $u \le v$?

5. **(a)** Verify that (6.20) is a necessary and sufficient condition for (6.19) to be an ESS in the war of attrition defined by (6.11).

   (b) Verify that the ESS is weak if (6.20) is satisfied with equality, but otherwise strong.

   (c) Verify (6.21)-(6.22).

6. Verify Figure 6.2.

7. Show that if ω defined by

(i) $$\omega \int_0^\infty \{xg(x)\}^2 \, dx = \int_0^\infty g(y) \int_0^y xg(x) \, dx \, dy$$

satisfies

(ii) $$\mu \ge \tfrac{\omega(2+\{\omega+2\}\theta)}{1+\omega\theta} \int_0^\infty \{xg(x)\}^2 \, dx$$

then the unique ESS of the war of attrition is

(iii) $$v^* = \tfrac{1}{1+\omega\theta}.$$

8. **(a)** Verify that (iii) above yields the known ESS for the war of attrition defined by (6.11) with (ii) invariably satisfied.

   (b) Verify that (iii) above yields the known ESS for the war of attrition in Exercise 6.3 with (ii) invariably satisfied.

   (c) Verify that (iii) above yields the known ESS for the war of attrition defined by (6.15) with (ii) reducing to (6.20).

9. Suppose that initial reserves in a war of attrition have Weibull distribution with "scale" parameter s (> 0) and "shape" parameter c (≥ 1) defined by

$$g(t) = \tfrac{ct^{c-1}}{s^c}e^{-(t/s)^c}$$

What is the largest value of the cost/value ratio θ for which an ESS exists when **(a)** $c = 2$ and **(b)** $c = 3$? In each case, find the ESS (assuming that θ does not exceed its critical value).

10. **(a)** Verify (6.26).

    (b) Verify that (6.28) is merely (2.11) with ϕ in place of f.

11. **(a)** Verify (6.29) and (6.31), and hence that (6.30) is the unique ESS among kin of the war of attrition defined by (6.11).

    (b) Verify Figure 6.3. How does R change shape with r?

    (c) Verify (6.33).

12. **(a)** Verify (6.34).

    (b) Verify (6.36). Hence verify that (6.35) yields the ESS among kin for the war of attrition with parabolically distributed initial reserves if either (6.37) holds or $r < \frac{61}{490}$ and (6.38) holds.

13. **(a)** The war of attrition with exponentially distributed initial reserves (Exercise 6.3) has a unique ESS among kin, v^*. What is it?

    (b) Show that (v^*, v^*) is the unique symmetric Nash equilibrium of the corresponding game between kin.

14. Show that an ESS among kin of the war of attrition with uniformly distributed initial reserves need not correspond to a Nash equilibrium of the associated symmetric game between kin, in the sense described at the end of §6.3.

15. Assuming, for simplicity's sake, that (5.35) is satisfied with strict inequality, analyze the prisoner's dilemma as

    (a) a (continuous) game among kin and

    (b) a symmetric game between kin.

16. Suppose that two related females share a nest for breeding purposes. One of these animals dominates the other in the sense that she always enjoys a greater share of the pair's total reproductive success, even though the subordinate animal can increase her relative share by allocating more of the pair's resources to fighting over it (instead of to actually raising their young). Let the amounts of effort that the dominant and subordinate expend on this "tug-of-war" [191] have the effect of allocating fractions u and v, respectively, of the pair's resources to contesting relative shares, leaving fraction $1 - u - v$ for reproduction per se. Then it is reasonable to assume that total reproductive success is $K(1 - u - v)$ and that the relative shares are $u/(u + bv)$ for the dominant and $bv/(u + bv)$ for the subordinate, where $b < 1$; see

[**191**] (where it is assumed without loss of generality that $K = 1$). The interaction between these two animals can be analyzed as an asymmetric game between kin.

   (a) Obtain expressions for ϕ_1 and ϕ_2 in (6.39).

   (b) Sketch the rational reaction sets, and show that there is a unique strong Nash equilibrium (u^*, v^*).

   (c) Find (u^*, v^*), and discuss its dependence on b and r.

17. Verify (6.46)-(6.47), and hence that (6.48) yields the unique ESS for the sperm competition game. Does (6.48c) make sense?

18. Show that (6.50) yields the unique ESS of the game whose reward is defined by (6.49).

19. Verify (6.55).

20. (a) Verify (6.57).

   (b) Verify (6.60)-(6.61).

   (c) Verify (6.66) and (6.69).

21. How must the analysis of §6.5 be modified if there is both a status-enhancement and a posse effect?

22. (a) Verify that (6.106a) has a solution satisfying $0 \leq v_0 \leq 1$ for $v_1 = 1$ only if (6.107) is satisfied, and that (6.108) is the only such solution.

   (b) Verify (6.114a) has a solution satisfying $0 \leq v_2 \leq 1$ for $v_0 = v_1 = 1$ only if $12r > 4b + 1$, and that (6.115) is the only such solution.

   (c) Verify that λ_0 defined by (6.114b) and (6.115) is negative throughout the triangle in the r-b plane with vertices at $\left(\frac{1}{4}, \frac{1}{2}\right)$, $\left(\frac{1}{3}, \frac{1}{2}\right)$ and $\left(\frac{29}{92}, \frac{16}{23}\right)$.

   (d) Verify Figure 6.10.

23. Verify the calculations leading to Tables 6.14-6.15.

24. (a) Show that fewer than half of all residents threaten at the ESS of the game described in §6.8.

   (b) What is the proportion of threats that are deceptive? How large can it be?

25. (a) Verify (6.137).

   (b) Describe the graph of η, verifying that $\eta(I) = 0 = \eta(J)$.

26. (a) Show that strategy (I, J, K, L) defined by (6.134)-(6.136) does not remain evolutionarily stable in the limit as $b \to 0$.

    (b) Show that the ESS defined by (6.134)-(6.136) does not persist if a threat cost T is paid not only by animals that threaten and lose, but also by animals that threaten and win after an attack by the intruder.

    (c) Show that the ESS also does not persist if a threatening resident invariably pays a threat cost T, regardless of whether he is attacked.

27. In §6.1 we assumed that individuals mate at random across the entire population. Suppose instead that (small) proportions α of males and β of females mate within their brood.

    (a) Find the new evolutionarily stable sex ratio by suitably modifying §6.1.

    (b) Find the new ESS by some other method.

    (c) How does the sex ratio vary with α? With β? Interpret.

28. In §6.7 we assumed that (actual) fighting strength has a uniform distribution. Suppose instead that strength is distributed between 0 and 1 with probability density function g defined by $g(\xi) = 16224936\{\xi(1-\xi)\}^{11}$.

    (a) What equation must v_2 satisfy for $(1, 1, v_2)$ to be an ESS (with a loser but no winner effect) if $b = \frac{1}{20}$ and $r = \frac{199}{2000}$?

    (b) Does this equation have a unique solution?

Chapter 7

Appraisal

Our short introduction to game theory is almost over, but before concluding we pause to ask: How valuable are games? To answer this question we must first reflect on the purpose of mathematical models.

Crudely, we can classify models of natural or social phenomena as either *descriptive* or *prescriptive*, or as some combination of the two. A descriptive model is an attempt to say what things or animals *do* do, in the active sense of the verb. A celebrated example of a purely descriptive model is Newton's model of gravitation, which says that what planets do is to obey the inverse-square law $F = GMm/r^2$, where G is the gravitational constant, M the mass of the sun, m the mass of a planet, r the distance between their centers of mass and F their mutual force of attraction. By contrast, a prescriptive model is an attempt to say what decision makers *should* do, in the modal sense of the verb. An example of a purely prescriptive model is the simple, "lot-size" inventory model, which says that if dealers sell quantity Q of some product per year, and if they must pay a set-up cost c_1 to order a batch (of any size), and if they must pay storage cost c_2 per unit stock per year, and if demand for the product is uniform, and if they wish to minimize costs, then they should order the product in batches of size $u^* = \sqrt{2c_1 Q/c_1}$, because u^* is the value of u that minimizes the total annual cost, $c_1 Q/u + c_2 u/2$. Nevertheless, this model is purely prescriptive only because we have been careful to qualify its prediction by so many ifs. In particular, if we neglect the cost-minimization condition, then the model becomes at least partly descriptive—because if we still insist that dealers should order u^*, then we tacitly assume that what dealers do is to minimize costs of ordering. More generally, whenever we use an optimization model to say what decision makers should do, we tacitly assume at the very least that they are trying to maximize the thing we have called their reward. Often we assume much more. Thus every prescriptive model has at least some descriptive elements. The converse, however, is false—witness Newton's model of gravitation, which has no prescriptive elements.

Where do games lie on this spectrum of description versus prescription? Insofar as games are optimization models, it appears that their purpose is prescriptive. Four Ways tells us how drivers should behave at a 4-way junction, Store Wars tells us how managers should set their prices, and so on. If we were to observe decision makers following the recommendations of these or any other models, then our immediate reply to the question—why do these people behave in this way?—would be: Because this is the behavior that optimizes their rewards. Now, humans and other animals have interacted strategically for many thousands of years. Over the course of time, they have evolved behavior to deal with such interactions; and it is reasonable to suppose that, by a process of trial and error, the behavior they exhibit *in familiar situations* is already the behavior that optimizes their rewards. If so, then there is no more prescribing for a model to do. But what if the observed behavior is exceedingly curious? Shouldn't we then wonder: Why is this behavior optimal? In other words: What game are the players playing? In such circumstances, the purpose of games is purely descriptive.

To take a concrete example, we saw in §2.8 that the spider *Oecobius civitas* can behave most oddly. A disturbed spider may enter the lair of another spider. But the homeowner, far from shooing the intruder out, will scurry off to bump another spider; and that spider in turn will bump yet another spider; and so on, often until most of the spiders in a colony have been displaced from their homes. But the spider *O. civitas* has frequented rocks for countless generations, and must surely be familiar with getting disturbed; in which case, we would expect the spider's strategy to be evolutionarily stable. But what is the game for which the strategy is an ESS? Concerning this question, in §§2.7-2.8 we built a model of owner-intruder interactions. We postulated a reward and a strategy set, and we found a parameter regime with a unique ESS that resembles the spider's observed behavior. Thus the model suggests an answer to our question; and because we assumed that spider behavior is optimal (in the appropriate sense), our model is wholly descriptive.[1]

A further illustration of the use of games as descriptive models emerges from §4.7. At the end of that section we posed the question: Which is the fairer imputation in a characteristic function game— the nucleolus or the Shapley value? One approach to answering this question would be first to assume that decision makers who even bother to form the grand coalition must be fair-minded people; and second to collect some data on how, for example, various car pools— while blissfully ignorant of game theory—have in practice divided the benefits of their cooperations. Then we could regard as fairer the imputation that was closer on balance to the observed division of benefits. Our use of models would again be descriptive.

Now, in a given instance, the purpose of a game is always either descriptive or prescriptive. Nevertheless, in different instances, the same model can be used for different purposes. For example, Four Ways can also be used as a descriptive model. In Exercise 2.1's interpretation of this game, τ is the junction transit time, and δ and $\epsilon(< \delta)$ are the time penalties for selfishness and altruism, respectively. Different values of these parameters correspond to different traffic conditions: $\delta < \frac{1}{2}\tau$ to light traffic, in which the line of cars at a junction is short; and $\delta > \frac{1}{2}\tau$ to heavy traffic, in which the line of cars at a junction is long, and the time penalty for selfishness severe. The model's primary prediction is that G (selfishness) is optimal if $\delta < \frac{1}{2}\tau$; whereas a mixture of G, W (altruism) and C (impatient altruism) is optimal if $\delta > \frac{1}{2}\tau$. If we are prepared to assume that drivers who arrive simultaneously at a 4-way junction from opposite directions already behave so as to minimize delay, then the model predicts that they behave more selfishly when traffic is light than when

---

[1]For an extended discussion of the behavior of *O. civitas* in this context, see [**150**, pp. 339-341].

traffic is heavy. Do they? I suspect that, if anything, the opposite is true—but, either way, our use of Four Ways would be descriptive.

Although a model may have different purposes in different instances, it is only when its purpose has been declared that we can begin to assess its value. In this regard, it is widely accepted that the ultimate test of a descriptive model is its ability to predict observable data. As every student of calculus knows, Newton's model of gravitation predicts that planets have elliptical orbits; and because this prediction agrees so well with observations, we accept that $F = GMm/r^2$ is—for all practical purposes—correct. Nevertheless, the model would have been valuable even if it had predicted some other kind of orbit, because it would have told Newton that $F = GMm/r^2$ was wrong; and perhaps he would have found something else to replace it.

There is a world of difference, however, between the older science of astronomy and the newer natural and social sciences. One of those differences is that relevant data in the newer sciences are often inaccurately known. Therefore, if a model is going to make useful predictions, then it is at least desirable and often essential that the model should be robust, i.e., insensitive to small changes in parameter values. For example, estimates of the parameters δ and τ in Four Ways may be subject to considerable error. Nevertheless, the model's primary prediction depends—as we have just seen—only on the sign of $\delta - \frac{1}{2}\tau$, and that may be known with some degree of confidence. To further illustrate this property of robustness: We saw in §§2.7-2.8 that the ESS of Owners and Intruders is insensitive to the values of the parameters K, ϵ, λ and σ. If the anti-Bourgeois strategy DH were an ESS only for special values of these parameters, then our interpretation of $O.$ *civitas* behavior would carry no weight at all, because there would be no reason to suppose that K, ϵ, λ and σ actually take those special values. Because the model is robust, however, our interpretation requires only that ϵ and $1 - \lambda$ be small, and $1 - \sigma$ not too small; and that K be neither too small nor too large. Even if we cannot estimate these parameters accurately, we may have reason to believe that the constraints are satisfied. Likewise, the partial-bluffing ESS of §6.8 holds for arbitrary values of the threat-cost or combat-cost parameters; it can be shown that §6.7's prediction of no winner effect without a loser effect is robust to changes in the distribution of fighting strength or the cost or win-probability functions [**149**]; in order to conclude that community-based wildlife management schemes could be better designed in §6.6, we required only that duties associated with monitoring have a low opportunity cost; and so on.

Levins [**120**, p. 422] has characterized the models ecologists use as compromises between generality, realism and precision, reflecting compromises between the simultaneously unattainable goals of, respectively, understanding, predicting and modifying nature. From this perspective, our models are primarily general models. We hope that their assumptions are not unrealistic; but we acknowledge that some parameters (e.g., the marginal-cost parameter in §6.4's mating game) may be difficult or even impossible to measure, and so we cannot expect quantitative predictions. That does not mean that their qualitative predictions cannot be tested. For example, a test of the no-assessment ESS in §6.2 is whether loser reserves correlate positively with contest duration; and a test of the partial-bluffing ESS in §6.8 is whether animals that threaten and lose pay higher costs than those that lose without threatening. But even where tests are inconclusive, games are useful simply because they allow us to explore the logic of a verbal argument rigorously. They often demonstrate what is difficult to intuit; for example, that victory by stronger animals need not imply that strength is being assessed (§6.2), or that bluffing can persist at high frequency (§6.8).

We now consider prescriptive models—for which, however, there appears to be no widely accepted criterion of worth, although flexibility is at least an important factor [**144**, pp. 267-268]. Prescriptive models are flexible if they are easily altered to suit a specific instance. Thus, for example, the car pool games are flexible, because their cost functions are readily adapted to different locations; and Four Ways is flexible, because its solution is known for arbitrary values of δ, ϵ and τ. Flexibility makes models applicable to qualitatively similar but quantitatively different interactions—car pools are much the same wherever you go, but their cost functions differ numerically; various community-based wildlife management schemes have been tried in several African countries [**69, 159**], yet the insights from §6.6 could apply to any of them; and so on.

Indeed it is desirable to build prescriptive models with sufficient flexibility that two decision makers who agree over what to optimize, but disagree over parameter values, can both use the same model to derive an optimal decision. For example, two dealers who believe in minimizing ordering costs $c_1 Q/u + c_2 u/2$, but who disagree over values of the parameters Q, c_1 and c_2, can both use the lot-size inventory model to determine their optimal batch size, u^*, because subjective estimates of Q, c_1 and c_2 can be inserted into $\sqrt{2c_1 Q/c_1}$ to produce subjective values of u^*. There is absolutely no need for the first dealer to know anything about the second dealer's perceptions of Q, c_1 and c_2, or vice versa. Except in §1.7, however, we have had to assume throughout this book that players' rewards and decision sets are common knowledge—and relaxing this assumption makes game-theoretic analysis considerably more difficult.

Here two remarks are in order. First, the assumption of common knowledge may simply be quite reasonable. Then games are valuable as descriptive models if they make robust predictions, and games are valuable as prescriptive models if they are flexible enough to be widely applicable. Second, there is a large and growing literature on games of partial information, in which the assumption of common knowledge is relaxed. This literature distinguishes between incomplete information, i.e., partial knowledge of a conflict's structure, and imperfect information, i.e., partial knowledge of a conflict's history. To illustrate the first type of partial information, recall that in Store Wars Nan and San had a common perception of the maximum amount they could charge for their product. Thus Nan's price, p_1, and San's price, p_2, had a common upper bound, namely, $4\alpha c$. If Nan and San had different perceptions, however, then the constraints $p_1 \leq 4\alpha c$, $p_2 \leq 4\alpha c$ would have to be replaced by $p_1 \leq 4\alpha_1 c$, $p_2 \leq 4\alpha_2 c$, where $\alpha_1 \neq \alpha_2$; and there would be no reason to suppose that Nan knew α_2 and San knew α_1. Thus information would be incomplete. To illustrate the second type of partial information, consider once more the iterated prisoner's dilemma in a finite population. If the players interacted at random, and if a player met a particular opponent for the first time on the first move and for the second time on the third move, then she would have no idea what her opponent did on the second move—even though, by the third move, it would be part of the conflict's history (and very useful to know), and even though she had complete knowledge of the conflict's structure (random interaction, same payoff matrix for everyone, etc.). Thus information would be imperfect. Or to take another illustration, if animals did not know whether the sites they had just discovered in Owners and Intruders were occupied—in other words, if they were unaware of their roles—then information would again be imperfect.

Clearly, you cannot possibly know a conflict's history if you don't even know its structure, although you can know its structure without knowing its history. Thus incomplete information is always imperfect, although imperfect information can still be complete. The distinction between incomplete and imperfect information is less important in practice than it is in theory, however; as a practical matter, if there are things we don't know, then—regardless of whether our ignorance is an imperfection or an incompleteness—we either exclude them from our models, or else we call them random variables and assign them distributions. We did this several times in the preceding chapter, although we always assumed a common distribution. The characteristic feature of the literature on games of incomplete information—which generalizes Chapter 1's concept of Nash equilibrium to so-called Bayesian Nash equilibrium—is that distributions are allowed to be subjective: they are assigned, as it were, not by the modeller but by the players. Furthermore, they can be updated as the game progresses and new information is acquired (an inherently dynamical process). Thus games of incomplete information possess considerably more flexibility than games of complete information (at least in principle). But this is an introductory text, and we cannot cover everything. Instead we refer to the literature; see, e.g., Myerson [162], Phlips [181] or Rasmusen [187].

Von Neumann and Morgenstern's treatise on game theory is now more than half a century old. Yet despite much scholarly activity in recent years, difficulties surround almost every solution concept for any kind of game. How does one distinguish among Nash equilibria of noncooperative games? How does one distinguish among the nucleolus, the Shapley value and the egalitarian imputation as the solution of a characteristic function game? It is possible, of course, that different answers apply in different circumstances. Nevertheless, answers remain to be found, in large measure because game theory is still insufficiently dynamic—if we knew the algorithm by which players converged on a fair division of the benefits of cooperation, or if we knew the algorithm by which players converged on a Nash equilibrium, then we would also know which solution they arrived at. Perhaps in the future there will be a general theory of strategic behavior that fully and explicitly models the dynamics of interaction, including partnership formation and information transfer. In fact, some important steps in this direction have already been taken; see, e.g., Fudenberg and Levine [**66**], Greenberg [**77**], Kreps [**116**], Muthoo [**161**], Samuelson [**197**] and Young [**241**]. But animal behavior is extremely complicated, and valuable progress toward understanding it can still be made—especially in the short term—not by seeking a general dynamic, but rather by building a greater variety of explicit models of specific conflicts in which behavior has been observed. Levins [**120**, p. 423] has portrayed the rationale for this approach, in the context of population biology, as follows:

> ... we attempt to treat the same problem with several alternative models each with different simplifications but with a common biological assumption. Then, if these models, despite their different assumptions, lead to similar results we have what we can call a robust theorem which is relatively free of the details of the model. Hence our truth is the intersection of independent lies.

Or as Maynard Smith and Szathmáry have said, "complex systems can best be understood by making simple models" [**135**].

In conclusion: There is still trouble aplenty with games. But where there is trouble, there is also opportunity, and games are as susceptible to new ideas and applications as they were when I wrote the first edition of this book. In other words, games are still an attractive topic for research. Again I hope that this book will help to entice you toward them.[2]

---

[2] For economic perspectives on the problems of game theory and how to address them, see, e.g., [**16, 116, 196**]. For biological perspectives on the promise and pitfalls of game theory, see, e.g., [**71, 76, 177, 182, 190**].

Appendix A

The Tracing Procedure

The tracing procedure of Harsanyi [**92**] is a method for associating a unique Nash equilibrium with all tentative solutions of a noncooperative, n-person game. We will describe the method only as it applies to a 2-person game, and in particular to Crossroads. Further details are given by Harsanyi and Selten [**93**], who denote by priors what we have called tentative solutions because their method is closely related to Bayesian methods in decision theory. Thus (p, q) is the prior for Crossroads in §2.1. The Nash equilibrium selected by Harsanyi and Selten's theory always depends on the prior.

The tracing procedure obtains the solutions depicted in Figure A.1 by considering the infinite sequence of games with reward functions H_1, H_2 defined for $0 \leq t \leq 1$ by $H_1(u, v) = (1 - t)f_1(u, q) + tf_1(u, v)$ and $H_2(u, v) = (1 - t)f_2(p, v) + tf_2(u, v)$. Thus $H_k(u, v)$ is a convex linear combination of Player k's actual reward, $f_k(u, v)$, and the reward associated with naively assuming that the other player will select his prior strategy—$f_1(u, q)$ in the case of Player 1 (Ned), $f_2(p, v)$ in the case of Player 2 (Sed). As t increases from 0 to 1, the weight shifts continously from the reward associated with assuming that the other player selects his prior to the actual reward of the game. For Crossroads, it is readily shown that $H_1(u, v) = (\delta + \epsilon)(\theta_2 - (1 - t)q - tv)u + (\epsilon - \frac{1}{2}\tau_2)(tv + (1 - t)q) - \epsilon - \frac{1}{2}\tau_2$ and $H_2(u, v) = (\delta + \epsilon)(\theta_1 - (1 - t)p - tu)v + (\epsilon - \frac{1}{2}\tau_1)(tu + (1 - t)p) - \epsilon - \frac{1}{2}\tau_1$.

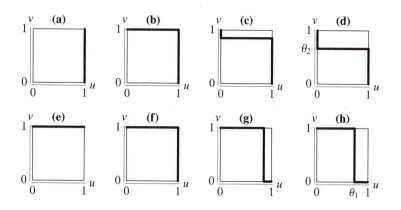

Figure A.1. $R_1(t)$ and $R_2(t)$ for $0 \leq t \leq 1, \theta_1 > p, \theta_2 > q$. $R_1(t)$ is shown for **(a)** $0 \leq t \leq (\theta_2 - q)/(1 - q)$, **(b)** $t = (\theta_2 - q)/(1 - q)$, **(c)** $(\theta_2 - q)/(1 - q) < t < 1$, **(d)** $t = 1$. $R_2(t)$ is shown for **(e)** $0 \leq t \leq (\theta_1 - p)/(1 - p)$, **(f)** $t = (\theta_1 - p)/(1 - p)$, **(g)** $(\theta_1 - p)/(1 - p) < t < 1$, **(h)** $t = 1$.

Because the rewards H_1 and H_2 depend upon t, the rational reaction sets R_1 and R_2 will also depend upon t; we therefore denote them by $R_1(t)$ and $R_2(t)$. Assuming, as in §2.1, that $\theta_2 > q$, $R_1(t)$ is depicted in the upper four diagrams of Figure A.1, for $0 \leq t \leq 1$. You can see that R_1 changes at $t = t_1$, where $t_1 = \frac{\theta_2 - q}{1 - q}$, from a vertical line to an inverted L; thereafter, there is a horizontal segment at $v = \{\theta_2 - (1 - t)q\}/t$, which moves downward from $v = 1$ to $v = \theta_2$ as t increases from t_1 to 1. Similarly, assuming $\theta_1 > p$, $R_2(t)$ is depicted for $0 \leq t \leq 1$ in the lower four diagrams of Figure A.1. It changes at $t = t_2$, where $t_2 = \frac{\theta_1 - p}{1 - p}$, from a horizontal line to an inverted L; thereafter, there is a vertical segment at $u = \{\theta_1 - (1 - t)p\}/t$, which moves leftward from $u = 1$ to $u = \theta_1$ as t increases from t_2 to 1.

Let us now assume, as in §2.1, that $p = \frac{1}{2} = q$. Then it is clear from Figure A.1 that the set of all Nash equilibria, $R_1(t) \cap R_2(t)$, evolves with t according to Figure A.2 if $\theta_1 > \theta_2$. For $t < t_1$ there is a unique equilibrium at $(1, 1)$; at $t = t_1$ there is a line segment of equilibria between $(1, 1)$ and $(0, 1)$; for $t_1 < t < t_2$ there is a unique equilibrium at $(0, 1)$; at $t = t_2$ there is an additional line segment of equilibria between $(1, \frac{\theta_1 + \theta_2 - 1}{2\theta_1 - 1})$ and $(1, 0)$; and thereafter

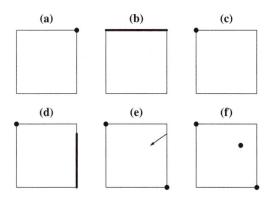

Figure A.2. Evolution of $R_1(t) \cap R_2(t)$ for $p = \frac{1}{2} = q, \theta_1 > \theta_2$ with **(a)** $0 \le t < 2\theta_2 - 1$, **(b)** $t = 2\theta_2 - 1$, **(c)** $2\theta_2 - 1 < t < 2\theta_1 - 1$, **(d)** $t = 2\theta_1 - 1$, **(e)** $2\theta_1 - 1 < t < 1$ and **(f)** $t = 1$.

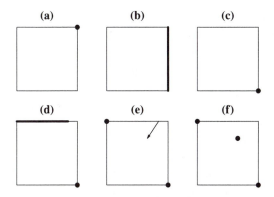

Figure A.3. Evolution of $R_1(t) \cap R_2(t)$ for $p = \frac{1}{2} = q, \theta_1 < \theta_2$ with **(a)** $0 \le t < 2\theta_1 - 1$, **(b)** $t = 2\theta_1 - 1$, **(c)** $2\theta_1 - 1 < t < 2\theta_2 - 1$, **(d)** $t = 2\theta_2 - 1$, **(e)** $2\theta_2 - 1 < t < 1$ and **(f)** $t = 1$.

there are three equilibria, one at $(1, 0)$, one at $(0, 1)$ and one at $\left(\frac{1}{2}, \frac{1}{2}\right) + t^{-1}\left(\theta_1 - \frac{1}{2}, \theta_2 - \frac{1}{2}\right)$, which migrates from $(1, \frac{\theta_1 + \theta_2 - 1}{2\theta_1 - 1})$ to (θ_1, θ_2) as t increases from t_2 to 1, as indicated by the arrow in Figure A.2. Likewise, when $\theta_2 > \theta_1, R_1(t) \cap R_2(t)$ evolves with t according to Figure A.3. For $t < t_2$ there is a unique equilibrium at $(1, 1)$; at $t = t_2$ there is a line segment of equilibria between $(1, 1)$ and $(1, 0)$;

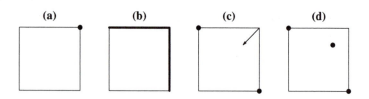

Figure A.4. Evolution of $R_1(t) \cap R_2(t)$ for $p = \frac{1}{2} = q, \theta_1 = \theta_2 = \theta$ with **(a)** $0 \le t \le 2\theta - 1$, **(b)** $t = 2\theta - 1$, **(c)** $2\theta - 1 < t < 1$ and **(d)** $t = 1$.

for $t_2 < t < t_1$ there is a unique equilibrium at $(1, 0)$; at $t = t_1$ there is an additional line segment of equilibria between $(\frac{\theta_1 + \theta_2 - 1}{2\theta_2 - 1}, 1)$ and $(0, 1)$; and thereafter there are three equilibria, one at $(1, 0)$, one at $(0, 1)$ and one at $\left(\frac{1}{2}, \frac{1}{2}\right) + t^{-1}\left(\theta_1 - \frac{1}{2}, \theta_2 - \frac{1}{2}\right)$, which migrates from $(\frac{\theta_1 + \theta_2 - 1}{2\theta_2 - 1}, 1)$ to (θ_1, θ_2) as t increases from t_1 to 1.

Now imagine that t is plotted on a third axis, perpendicular to the page. Then, as t increases from 0 to 1, $R_1(t) \cap R_2(t)$ will trace out a set of curves in three-dimensional space. Only one of the three Nash equilibria at $t = 1$ has the distinguishing property that it can be reached from the single Nash equilbrium $(1, 1)$ at $t = 0$ by moving continuously along one of the curves, as you can readily verify from Figures A.2-A.3. This distinguished equilibrium is Harsanyi's solution (which agrees with Figure 2.1).

If $\theta_1 = \theta_2$, however, then Figure A.4 shows that all three equilibria at $t = 1$ can be reached from $(1, 1)$ at $t = 0$ by moving continuously along the curves. For such contingencies, Harsanyi and Selten have proposed a modification of the tracing procedure, which they call the logarithmic tracing procedure. Broadly speaking, it resolves the indeterminacy by selecting the Nash equilibrium that is nearest to the center of the decision set; in particular, its prediction for the symmetric version of Crossroads agrees with Figure 2.2 [**93**, p. 165].

The general theory of Harsanyi and Selten does not consist solely of the tracing procedures, however; it also includes a rationale for selecting the prior. For details, see Chapter 5 of [**93**].

Appendix B

Solutions to Selected Exercises

Chapter 1

1. The game is equivalent to symmetric Crossroads with $\epsilon = 0$.
3. Note that the equilibrium is strong, as defined in §1.6.
5. Let $f_1 = Ju_1 + Ku_2 + L$. Then region C in Figure 1.6 (which is drawn for $\gamma > \sigma$) corresponds to $J < 0, K < 0$; region B corresponds to $J < 0, K > 0$ to the right of the line between (α, β) and $(\sigma/\omega, 0)$, to $J = 0, K > 0$ on the line, and to $K > J > 0$ to the left of it; and region A corresponds to $J > K > 0$ below the line between (α, β) and $(0, 1-\theta)$, to $J > 0, K = 0$ on the line, and to $J > 0, K < 0$ above it. Similarly, the seven rows of Table 1.8 correspond, respectively, to $J > 0, K < 0$ or $J > K > 0$; $J < 0, K > 0$ or $K > J > 0$; $J < 0, K < 0$; $J = 0, K < 0$; $J < 0, K = 0$; $J = K > 0$; and $J = K = 0$.
7. $G(s) = \frac{7}{10} + \frac{1}{100}(s - 12)^2$ if $12 \leq s \leq 15$; $\frac{3s}{50} - \frac{11}{100}$ if $15 \leq s \leq 17$; and $1 - \frac{1}{100}(20 - s)^2$ if $17 \leq s \leq 20$. The probability density function is defined by $g(s) = G'(s)$, $0 < s < 20$.
8. To obtain, e.g., (1.50c), note from (1.48) that if $(u, v) \in D_C$ then $\frac{\partial f_1}{\partial u} = \frac{c}{25}\{(u - v + 6)(v - 3u - 6) + 70\}$, $\frac{\partial^2 f_1}{\partial u^2} = \frac{2c}{25}\{2(v - 6) - 3u\}$; whence f_1 has a maximum for $\max(0, v - 6) \leq u \leq v - 1$ at $u = \frac{1}{3}(2v + \sqrt{(v - 6)^2 + 210}) - 4$ provided this number—which clearly exceeds $\max(0, v - 6)$—is less than or equal to $v - 1$, or $v \geq \frac{11}{2}$. Otherwise the maximum is at $u = v - 1$.

9. To obtain, e.g., (1.53c), note that because $(u, v) \in D_C$ implies $\frac{\partial f_2}{\partial v} = \frac{1}{75}c\{(3v - 2u - 12)^2 - (u+6)^2 + 90\} \geq 0$ if $(u+6)^2 \leq 90$, the maximum of f_2 on D_C is at $v = u + 6$ if $u \leq 3(\sqrt{10} - 2)$. If $u > 3(\sqrt{10} - 2)$, however, then for $u + 1 \leq v \leq u + 6$, f_2^C has a local minimum at $v = v_{up}(u)$, where
$$v_{up}(u) = \tfrac{1}{3}(2u + \sqrt{(u+6)^2 - 90}) + 4;$$
and a local maximum at $v = v_{down}(u)$, where
$$v_{down}(u) = \tfrac{1}{3}(2u - \sqrt{(u+6)^2 - 90}) + 4,$$
provided that $v_{down}(u) \geq u + 1$ or $u \leq \frac{9}{2}$. But this local maximum is not the maximum for $3(\sqrt{10} - 2) \leq u \leq \frac{9}{2}$, because $f_2^C(u, u+6) - f_2^C(u, v_{down}(u)) \geq 0$ when $3(\sqrt{10} - 2) \leq u \leq 4$ and $f_2^C(u, 10) - f_2^C(u, v_{down}(u)) \geq 0$ when $4 \leq u \leq \frac{9}{2}$. For $\frac{9}{2} \leq u \leq 9$, the maximum (for $u + 1 \leq v \leq 10$) must be at $v = u + 1$, because $f_2^C(u, u + 1) - f_2^C(u, 10) = \frac{1}{5}c(9 - 2u)(u - 9) \geq 0$.

11. As we move to the left along the interval $3(\sqrt{10} - 2) \leq u \leq \frac{9}{2}$, the points $(u, v_{up}(u))$ and $(u, v_{down(u)})$ defined in the solution to Exercise 1.9 move closer together, until they coalesce at the point $(3\{\sqrt{10} - 2\}, 2\sqrt{10})$. Thus R_2 in Figure 1.11(b) is defined by $v =$
$$
\begin{array}{lcc}
u + 6 & \text{if} & 0 \leq u \leq 2(\sqrt{10} - 3) \\
2\sqrt{10} & \text{if} & 2(\sqrt{10} - 3) \leq u \leq 3(\sqrt{10} - 2) \\
\tfrac{2}{3}u - \tfrac{1}{3}\sqrt{(u+6)^2 - 90} + 4 & \text{if} & 3(\sqrt{10} - 2) \leq u \leq \tfrac{9}{2} \\
\tfrac{1}{2}u + \tfrac{13}{4} & \text{if} & \tfrac{9}{2} \leq u \leq 4\sqrt{10} - \tfrac{13}{2} \\
2\sqrt{10} & \text{if} & 4\sqrt{10} - \tfrac{13}{2} \leq u \leq 2\sqrt{10}.
\end{array}
$$

16. For Player 1 (Nan), we obtain $f_1(u^*, v^*, z^*) - f_1(u, v^*, z^*) = 16ac\pi\left(u - \frac{1}{6}\right)^2 \geq 0$ from (1.63). Similarly for the other players.

19. (a) When $2\delta < \min(\tau_1, \tau_2)$ we have $(\tilde{u}, \tilde{v}) = (1, 1) = (u^*, v^*)$.
(b) When $2\epsilon > \max(\tau_1, \tau_2)$ we have $(\tilde{u}, \tilde{v}) = \frac{1}{\delta + \epsilon}(\epsilon - \frac{1}{2}\tau_2, \epsilon - \frac{1}{2}\tau_1)$, which lies neither in R_1 nor in R_2.
(c) For $2\epsilon > \max(\tau_1, \tau_2)$, we obtain $f_1(\tilde{u}, \tilde{v}) = -(\delta + \frac{1}{2}\tau_2)\theta_2$ and $f_2(\tilde{u}, \tilde{v}) = -(\delta + \frac{1}{2}\tau_1)\theta_1$, where θ_1 and θ_2 are defined by (1.22). Compare with Table 1.5.

20. Because $v, z \geq 0$ we have $v + z$ is minimized with respect to v, z by $v = 0 = z$. Thus $m_1(u) = 8ac\pi u(1/3 - 2u)$, and similarly for m_2, m_3. So the unique max-min strategies for Players 1, 2 and 3 are, respectively, $\tilde{u} = \frac{1}{12}$, $\tilde{v} = \frac{1}{16}$ and $\tilde{z} = \frac{5}{48}$.

21. The unique Nash equilibrium is $(u^*, v^*, z^*) = \left(\frac{17}{120}, \frac{7}{40}, \frac{11}{60}\right)$.

23. Nan's reward in (1.48a) is minimized by taking $v = 0$ if $0 \le u \le 6$ but $v = u$ if $u > 6$. Thus in (1.75) we have $m_1(u) = \frac{1}{5}cu(7 - 2u)$ if $0 \le u \le 1$, $m_1(u) = \frac{1}{25}cu(6 - u)^2$ if $1 < u \le 6$ and $m_1(u) = 0$ if $u > 6$; and the maximum is readily shown to occur where $u = 2$. Similarly, we find $m_2(v) = \frac{1}{5}cv(13 - 2v)$ if $0 \le v \le 1$, $m_2(v) = \frac{1}{25}cv\{30 + (6 - v)^2\}$ if $1 < v \le 6$ and $m_2(v) = \frac{6}{5}cv$ if $v > 6$; and it follows that m_2 is strictly increasing. Thus the unique max-min strategies are $\tilde{u} = 2$ for Player 1 and $\tilde{v} = \alpha$ for Player 2. Note that $(\tilde{u}, \tilde{v}) \notin D$ if $\alpha > 8$.

24. **(a)** I's payoff matrix is now

	H	D	B
H	$\frac{1}{2}(\rho - C)$	ρ	$\frac{1}{4}(3\rho - C)$
D	0	$\frac{1}{2}\rho$	$\frac{1}{4}\rho$
B	$\frac{1}{4}(\rho - C)$	$\frac{3}{4}\rho$	$\frac{1}{2}\rho$

and II's matrix the transpose thereof. Let strategies u and v be defined as in §1.4. Then $B = (0,0)$, and so the payoff to each player when both select B is $f_1(0,0,0,0) = \frac{1}{2}\rho$. If Player 1 now unilaterally adopts $u \ne (0,0)$, then her payoff against B is $f_1(u_1, u_2, 0, 0) = u_1 \cdot \frac{1}{4}(3\rho - C) + u_2 \cdot \frac{1}{4}\rho + (1 - u_1 - u_2) \cdot \frac{1}{2}\rho$; and because either u_1 or u_2 must be positive, we have $f_1(0,0,0,0) - f_1(u_1, u_2, 0, 0) = \frac{1}{4}\{(C - \rho)u_1 + \rho u_2\} > 0$ if $\rho < C$. Similarly for Player 2. Thus B is a strong Nash-equilibrium strategy.

(b) It is perhaps unlikely that both owner and intruder would share the same values of ρ and C, and perhaps even less likely that they would have equal chances of victory if both were to select either H or D (as we have assumed in Tables 1.3-1.4).

25. $f_1(u, v) = Tv + P(1 - v) - \{(P - S)(1 - v) + (T - R)v\}u$, which is maximized by $u = 0$. Similarly for f_2. Thus the only Nash equilibrium is $(0, 0)$.

26. Define dimensionless prices u, v and parameters μ, σ by $u = \frac{p_1}{4cL}$, $v = \frac{p_2}{4cL}$, $\mu = \frac{b-a}{2L}$ and $\sigma = \frac{b+a}{2L}$, where $0 < \mu \le \sigma < 1, \mu + \sigma \le 1$. From $|p_1 - p_2| \le 2c(b - a)$ or $|u - v| \le \mu$, the decision set is $D = \{(u,v) \mid \max(0, v - \mu) \le u \le v + \mu, \max(0, u - \mu) \le v \le u + \mu\}$. Let the residential coordinate of the next customer be the random variable X. Then the next customer will buy from the first store if $X < a$ or $a < X < b$ and $p_1 + 2c(X - a) <$

$p_2 + 2c(b - X)$, i.e., if $0 < X < \frac{p_2 - p_1}{4c} + \frac{1}{2}(a + b)$. Because X is distributed uniformly between $x = 0$ and $x = L$, the probability of this event is $(p_2 - p_1)/4cL + (a + b)/2L$; whence, by reasoning similar to that which yielded (1.42), the first player's reward is $p_1\{(p_2 - p_1)/4cL + (a + b)/2L\}$. Thus $f_1(u, v) = 4cLu(v - u + \sigma)$; and similarly, $f_2(u, v) = 4cLv(u - v - \sigma + 1)$.

In calculating R_1, let us first suppose that $0 \leq v \leq \mu$. Then $\max(0, v - \mu) = 0$, and f_1 has its maximum on $0 \leq u \leq v + \mu$ where $u = (v + \sigma)/2$ if $(v + \sigma)/2 \leq v + \mu$, but where $u = v + \mu$ if $(v + \sigma)/2 > v + \mu$; in other words, where

$$u = \begin{cases} v + \mu & \text{if } v < \sigma - 2\mu \\ \frac{1}{2}(v + \sigma) & \text{if } v \geq \sigma - 2\mu. \end{cases}$$

The boundary segment $u = v + \mu$ is irrelevant if $\mu < \sigma < 2\mu$, whereas the interior segment $u = (v + \sigma)/2$ is irrelevant if $3\mu < \sigma < 1$. Similarly, for $\mu < v < \infty$, we have $\max(0, v - \mu) = v - \mu$; and so f_1 has its maximum on $v - \mu \leq u \leq v + \mu$ where

(i)
$$u = \begin{cases} v + \mu & \text{if } v < \sigma - 2\mu \\ \frac{1}{2}(v + \sigma) & \text{if } \sigma - 2\mu \leq v \leq \sigma + 2\mu \\ v - \mu & \text{if } v > \sigma + 2\mu \end{cases}$$

The lower boundary segment $u = v + \mu$ is irrelevant if $\sigma < 3\mu$. Combining our results for $0 \leq v \leq \mu$ with those for $\mu < v < \infty$, we find that R_1 is defined by

(ii)
$$u = \begin{cases} \frac{1}{2}(v + \sigma) & \text{if } 0 \leq v \leq \sigma + 2\mu \\ v - \mu & \text{if } \sigma + 2\mu < v < \infty \end{cases}$$

if $\sigma < 2\mu$ $(b > 3a)$ and by (i) if $\sigma > 2\mu$ $(b < 3a)$.

Similar arguments can be used to calculate R_2. It is unnecessary to repeat the analysis, because the symmetry between f_1 and f_2 enables us to deduce the results from (i) and (ii) by swapping u with v and σ with $1 - \sigma$. Thus R_2 is defined by

$$v = \begin{cases} \frac{1}{2}(u + 1 - \sigma) & \text{if } 0 \leq u \leq 1 - \sigma + 2\mu \\ u - \mu & \text{if } 1 - \sigma + 2\mu < u < \infty \end{cases}$$

if $1 - \sigma < 2\mu$ $(3b - a > 2L)$, but by

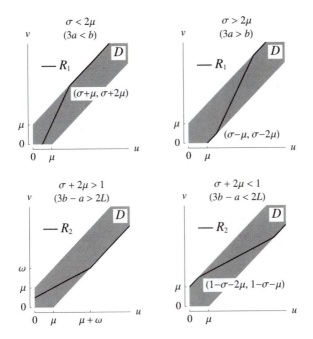

Figure B.1

$$
v = \begin{cases} u + \mu & \text{if } 0 \le u \le 1 - \sigma - 2\mu \\ \frac{1}{2}(u + 1 - \sigma) & \text{if } 1 - \sigma - 2\mu < u \le 1 - \sigma + 2\mu \\ u - \mu & \text{if } 1 - \sigma + 2\mu < u < \infty \end{cases}
$$

if $1 - \sigma > 2\mu$ $(3b - a < 2L)$. R_1 and R_2 are sketched in Figure B.1, where $\omega = 1 - \sigma + \mu$ and the lower extremities of R_1 and R_2 in the left-hand panels are at $\left(\frac{1}{2}\sigma, 0\right)$ and $\left(0, \frac{1-\sigma}{2}\right)$, respectively.

To find all Nash equilibria, we must calculate $R_1 \cap R_2$ in every case. Suppose, to begin with, that $\sigma < 2\mu$. Then, by inspection of Figure B.1, R_1 meets R_2 in a single point if $\sigma + 2\mu > 1$; or if $\sigma + 2\mu < 1$, but $(\sigma + \mu, \sigma + 2\mu)$ lies above $(1 - \sigma - 2\mu, 1 - \sigma - \mu)$ on the line $v = u + \mu$, that is, if $2\sigma + 3\mu > 1$. The point of intersection, the Nash equilibrium, is then

(iii) $$(u^*, v^*) = \tfrac{1}{3}(\sigma + 1, 2 - \sigma).$$

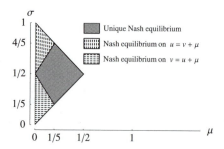

Figure B.2

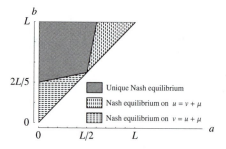

Figure B.3

If $\sigma < 2\mu$ and $\sigma + 2\mu < 1$ but $2\sigma + 3\mu < 1$, however, then $R_1 \cap R_2$ is the line segment joining $(\sigma + \mu, \sigma + 2\mu)$ to $(1 - \sigma - 2\mu, 1 - \sigma - \mu)$.

Now suppose that $\sigma > 2\mu$. If also $2\sigma + 3\mu < 1$, then $R_1 \cap R_2$ is still the line segment joining $(\sigma + \mu, \sigma + 2\mu)$ to $(1 - \sigma - 2\mu, 1 - \sigma - \mu)$. If $2\sigma + 3\mu > 1$ and the point $(\mu + \omega, \omega)$ lies above the point $(\sigma - \mu, \sigma - 2\mu)$ on the line $u = v + \mu$, i.e., if $1 - 3\mu < 2\sigma < 1 + 3\mu$, then (iii) is again the unique Nash equilibrium. If the point $(\mu + \omega, \omega)$ lies below the point $(\sigma - \mu, \sigma - 2\mu)$, however, i.e., if $2\sigma > 1 + 3\mu$, then $R_1 \cap R_2$ is the line segment joining $(\mu + \omega, \omega)$ to $(\sigma - \mu, \sigma - 2\mu)$. Combining these results, we find that the region in the μ-σ plane for which there exists a unique Nash equilibrium is the shaded region in Figure B.2.

Because $a = L(\sigma - \mu), b = L(\sigma + \mu)$, the triangle where $\mu \geq 0, \sigma \geq \mu, \sigma + \mu \leq 1$ in the μ-σ plane is mapped to the triangle where $0 \leq a \leq b \leq L$ in the a-b plane by the matrix $\begin{bmatrix} -L & L \\ L & L \end{bmatrix}$, as shown in Figure B.3. The shaded region, corresponding to store locations for which there is a unique Nash equilibrium, contains points in the a-b plane at which $b + 2L \geq 5a$ and $5b \geq a + 2L$ or, equivalently, $b - a \geq 4 \max \left(a - \frac{1}{2}L, \frac{1}{2}L - b \right)$. In particular, there is a unique Nash equilibrium if the stores are located on opposite sides of $x = \frac{1}{2}L$, but a multiplicity of solutions if both are located near $x = 0$ or $x = L$.

27. **(a)** I's payoff matrix is now

	H	D	R
H	$\frac{1}{2}(\rho - C)$	ρ	$\frac{1}{2}(\rho - C)$
D	0	$\frac{1}{2}\rho$	$\frac{1}{2}\rho(1 - \lambda)$
R	$\frac{1}{2}(\rho - C)$	$\frac{1}{2}\rho(1 + \lambda)$	$\frac{1}{2}\rho$

and II's matrix the transpose thereof. To show that R is a strong Nash-equilibrium strategy, see the solution to Exercise 1.24.
(b) No; see, for example, Exercise 2.8.

29. Because $\mu > \frac{1}{2}\tau$, Figure 1.6 is unchanged, except that $(\sigma/\omega, 0)$ becomes $\left(\frac{\kappa + \sigma}{\kappa + 1}, 0 \right)$, where $\kappa = \mu/\delta$ and α, β, θ are defined by
$$\alpha = \frac{(\gamma + \sigma)(\kappa + \sigma)}{\kappa + (\kappa + 1)\gamma + \sigma^2}, \quad \beta = \frac{(1 - \sigma)(\kappa - \sigma)}{\kappa + (\kappa + 1)\gamma + \sigma^2}, \quad \theta = \frac{\gamma + \sigma}{\gamma + \kappa}$$
instead of by (1.33)-(1.34), but Tables 1.8-1.10 are unchanged. For **(a)**, $\sigma = \frac{1}{3}, \kappa = \frac{4}{3}, \gamma = \frac{2}{3}$ and $(\alpha, \beta) = \left(\frac{5}{9}, \frac{2}{9} \right)$; for **(b)**, $\sigma = \frac{1}{4}, \kappa = \frac{3}{4}, \gamma = \frac{1}{2}$ and $(\alpha, \beta) = \left(\frac{4}{9}, \frac{2}{9} \right)$.

30. To show that $(1, 0)$ is a strong Nash equilibrium, we must show that $f_1(1, 0) > f_1(u, 0)$ for all u such that $0 \leq u < 1$ *and* $f_2(1, 0) > f_2(1, v)$ for all v such that $0 < v \leq 1$. Likewise, to show that $(0, 1)$ is strong, we must show that $f_1(0, 1) > f_1(u, 1)$ for all u such that $0 < u \leq 1$ *and* $f_2(0, 1) > f_2(0, v)$ for all v such that $0 \leq v < 1$. Verifying these inequalites is routine algebra.

31. Figure B.2's speckled region becomes the curvilinear quadrilateral $\frac{1}{9}\max(5 - 5\sigma - \sigma^2, -1 + 7\sigma - \sigma^2) \leq \mu \leq \min(\sigma, 1 - \sigma)$, requiring in particular that $\mu \geq \frac{1}{4}$ or $b - a \geq L/2$. Thus a unique Nash equilibrium exists only if the stores are located sufficiently near to opposite ends of the line segment $0 \leq x \leq L$.

32. **(a)** Let $u = (u_1, u_2)$ be Player 1's strategy, let $v = (v_1, v_2)$ be Player 2's strategy, and let $g_k(u, v)$ be Player k's reward, $k = 1, 2$. Player 1 is equally likely to be either a northbound

Table B.1.

R_1					$R_1 \cap R_2$			
u_1	u_2	v_1	v_2		u_1	u_2	v_1	v_2
1	1	v_1^A	v_2^A		1	1	0	0
1	0	v_1^B	v_2^B		1	0	1	0
0	0	v_1^C	v_2^C		0	0	1	1
0	1	v_1^D	v_2^D		0	1	0	1
1	u_2	v_1^{AB}	v_2^{AB}		1	θ_2	θ_1	0
u_1	0	v_1^{BC}	v_2^{BC}		θ_1	0	1	θ_2
0	u_2	v_1^{CD}	v_2^{CD}		0	θ_2	θ_1	1
u_1	1	v_1^{DA}	v_2^{DA}		θ_1	1	0	θ_2
u_1	u_2	θ_1	θ_2		θ_1	θ_2	θ_1	θ_2

driver who selects G with probability u_1 against a southbound
Player 2 who selects G with probability v_2; or a southbound
driver who selects G with probability u_2 against a northbound
Player 2 who selects G with probability v_1. Thus $g_1(u,v) = \frac{1}{2}f_1(u_1,v_2) + \frac{1}{2}f_2(v_1,u_2)$, where f_1 and f_2 are defined by (1.23);
and $g_2(u,v) = \frac{1}{2}f_1(v_1,u_2) + \frac{1}{2}f_2(u_1,v_2)$, or $g_2(u,v) = g_1(v,u)$.
(b) The rational reaction sets are defined by (1.32) with f replaced by g, and Δ by the unit square. Let A denote the rectangle
in the unit square with vertices $(0,0),(\theta_1,0),(\theta_1,\theta_2)$ and $(0,\theta_2)$,
B the rectangle with vertices $(\theta_1,0),(1,0),(1,\theta_2)$ and (θ_1,θ_2), C
the rectangle with vertices $(\theta_1,\theta_2),(1,\theta_2),(1,1),(\theta_1,1)$ and D the
rectangle with vertices $(0,\theta_2),(\theta_1,\theta_2),(\theta_1,1)$ and $(0,1)$. Then,
using the system of notation introduced in §1.4, we find R_1 as
shown in Table B.1 (with $0 \le u_1, u_2 \le 1$); and R_2 follows by
analogy with Tables 1.8 and 1.9. We see that there are nine
Nash equilibria in all.
33. (a) See [**146**, p. 1023].
(b) From **(a)** and (1.15), we have $f_1(u,v) = (v_2 + v_3 - 2v_4)u_1 + 2v_5 u_2 + (1 - v_1 - 2v_3 - v_4 - 4v_5)u_3 - (v_1 - 3v_2 + 3v_5)u_4 - (v_1 - v_2 + 2v_3 - 3v_4)u_5 - v_2$, implying $f_1(u^*,v^*) = -\frac{6}{35}$. Player
1 has no incentive to depart unilaterally from u^* if the maximum of $f_1(u,v^*) = -\frac{1}{35}(10 - 4\{u_1 + u_2 + u_3 + u_4\} + u_5)$ subject
to $u_1 + u_2 + u_3 + u_4 + u_5 \le 1$ is $-\frac{6}{35}$. The inequality implies
$f_1(u,v^*) \le -\frac{6}{35} - \frac{1}{7}u_5$, which is maximized where $u_5 = 0$. Similarly for Player 2.

Chapter 2

1. Note that a large value of δ is especially likely when there is more than one car in both northbound and southbound lanes, because then another vehicle is likely to have captured the space into which, if GG has been selected, one of the players must reverse before a path can be cleared. Thus $\delta > \frac{1}{2}\tau$ is likely to hold at busy junctions, and we can interpret the inequality as saying that the junction is slow, rather than that the drivers are fast.

2. (a) Because (u^*, u^*) is a Nash-equilibrium strategy combination if $(u^*, u^*) \in R_1 \cap R_2$, we must show that
$$f_1(u^*, u^*) = \max_u f_1(u, u^*), \quad f_2(u^*, u^*) = \max_v f_2(u^*, v),$$
where $f_2(v, u) = f_1(u, v) = f(u, v)$, as in (2.5). These two conditions are equivalent to $f_1(u^*, u^*) \geq f_1(u, u^*)$ for all u and $f_2(u^*, u^*) \geq f_2(u^*, v)$ for all v, both of which are equivalent to $f(u^*, u^*) \geq f(v, u^*)$ for all v. But this is implied by (2.12).

   (b) No (unless $u^* = v^*$). Consider, e.g., strategy combinations $(1, 0)$ and $(0, 1)$ in §2.3's symmetric version of Crossroads.

5. See [**137**, pp. 13-14].

7. The only Nash equilibrium, $(0, 0)$, is also an ESS.

8. We already know from Exercise 1.27 that R is a strong Nash-equilibrium strategy; hence R is a strong ESS. To show that $\left(\frac{1}{2}, \frac{1}{2}\right)$ is an ESS, proceed as for Four Ways at the end of §2.3.

9. From $f(u, v) = (u_1, u_2, 1 - u_1 - u_2)A(v_1, v_2, 1 - v_1 - v_2)^T$ with A defined by Exercise 1.28 and $v^* = \left(\frac{1}{3}, \frac{1}{3}\right)$, we have $f(v^*, v^*) = -\frac{1}{3}\lambda = f(u, v^*)$ and $f(v^*, u) - f(u, u) = \lambda\left(u_1 - \frac{1}{3}\right)^2 + \lambda\left(u_2 - \frac{1}{3}\right)^2 + \lambda\left(u_1 + u_2 - \frac{2}{3}\right)^2$, which is positive for all $u \neq v^*$ if $\lambda > 0$ but negative if $\lambda < 0$. So $\left(\frac{1}{3}, \frac{1}{3}\right)$ is an ESS for $\lambda > 0$; and if v is an ESS for $\lambda < 0$, then $v \neq \left(\frac{1}{3}, \frac{1}{3}\right)$. Define $\hat{f}(u_1, u_2) = f(v, v) - f(u, v)$. Then $\hat{f}(u_1, u_2) \geq 0$ for any mutant strategy u if v is an ESS. But $\hat{f}(0, 0) = \rho(v_2 - v_1) + \lambda\{v_1(1 - v_1) + v_2(1 - v_2) - (v_1 + v_2)^2)\}$ or $\hat{f}(0, 0) = \rho(v_2 - v_1) + O(\lambda)$, where $O(\lambda)$ denotes terms that approach zero as $\lambda \to 0$. Because $|\lambda|$ is very much smaller than ρ, it follows that $\hat{f}(0, 0) \geq 0$ only if $v_1 \leq v_2$. Similarly, $\hat{f}(0, 1) = \rho(2v_1 + v_2 - 1) + O(\lambda)$ implies $2v_1 + v_2 \geq 1$; and $\hat{f}(0, 1) = \rho(1 - v_1 - 2v_2) + O(\lambda)$ implies $v_1 + 2v_2 \leq 1$. So v can

be an ESS only if $v_1 \leq v_2, 2v_1 + v_2 \geq 1$ and $v_1 + 2v_2 \leq 1$, or $v = \left(\frac{1}{3}, \frac{1}{3}\right)$. Hence no ESS exists for $\lambda < 0$.

10. (b) The average delay decreases with η if $\phi = f(v^*, v^*)$ increases with λ, where v^* is defined by (2.30). If $\theta > 1$ and $0 < \lambda \leq \theta - 1$, then (2.27) implies $\phi = f(0,0) = \frac{1}{2}(\delta + \epsilon)\lambda - (1 + \theta)\delta + (1 - \theta)\epsilon$, and so $\partial\phi/\partial\lambda = \frac{1}{2}(\delta + \epsilon) > 0$. If $\theta \leq 1$ (and $\lambda - \theta + 1 \geq 0$), then (2.27) and (2.30) imply $\partial\phi/\partial\lambda = \frac{1}{2}(\delta a + \epsilon b)\{1 + \lambda\}^{-3}$, where $a = b + 4(1 + \lambda)\theta$ and $b = \{3(1 - \theta + \lambda) + \lambda^2 + \theta^2\}\lambda + 3\theta^2 - 3\theta + 1 > 0$ (because $3\theta^2 - 3\theta + 1 \geq \frac{1}{4}$ if $0 \leq \theta \leq 1$). So again $\partial\phi/\partial\lambda > 0$.

(c) The rational reaction set is defined by (2.29) with $G(v) = \int_0^v g(\xi)\,d\xi$ in place of v.

11. Trajectories that begin on the boundary of Δ where $p_1 p_2 = 0$ must remain on that boundary and converge to $(0,0)$ as $t \to \infty$.

14. Define $y(t) = 1 - \sum_{k=1}^m x_k(t)$. Then $y(0) = 1$, by (2.48). On using (2.47), $\frac{dy}{dt} = -\sum_{k=1}^m \frac{dx_k}{dt} = -\kappa \sum_{k=1}^m x_k W_k + \kappa \overline{W} \sum_{k=1}^m x_k = -\kappa \overline{W} y$. The solution subject to $y(0) = 0$ is $y(t) = 0$, or (2.49).

15. Suppose, for example, that $\epsilon = 0.1, \lambda = 0.9, \sigma = 0.7$ and $x(0) = (0,0,1,0) = y(0)$. Then $a_{ij} = \epsilon = 0.1, 1 \leq i, j \leq 4$ for $K = 1$, and the payoff matrices for $K = 2, \dots, 7$ are as shown in Table B.2. DH is the unique ESS for $K = 4, \dots, 7$ and all higher values of K, except in the limit as $K \to \infty$ when $a_{33} \to 1$ and $a_{4j} \to 1$ for $j = 1, \dots, 4$ (because there are more sites than animals, an inveterate Dove is guaranteed to find a site eventually).

16. Let $A(K)$ denote the payoff matrix when the conflict lasts for K units of time. Then the payoff matrix should be replaced by its expected value $\mathrm{E}[A] = \sum_{k=1}^\infty A(k) \cdot \mathrm{Prob}(K = k)$. Suppose, for example, that in every time unit there is constant probability w that the conflict will last for another unit of time; i.e., $\mathrm{Prob}(K \geq k) = w^{k-1}$ for $k \geq 1$. Then $\mathrm{Prob}(K = k) = \mathrm{Prob}(K \geq k) - \mathrm{Prob}(K \geq k+1) = (1 - w)w^{k-1}$, implying $\mathrm{E}[A] = (1 - w)\sum_{k=1}^\infty w^{k-1}A(k)$. For $\epsilon = 0.1, \lambda = 0.9, \sigma = 0.7$ and $x(0) = (0,0,1,0) = y(0)$, it is clear from the previous exercise that DH will be the unique ESS provided only that w is sufficiently large, and similarly for other values of ϵ, λ and σ.

17. See §2.8.

18. With $q(y_3)$ defined by (2.69), we have
$$H_1(x,y) = x_1\big(1 - q(y_3)v_2\{1 - \tfrac{1}{2}(1 - \lambda)u_1\}\big)$$

Table B.2.

	$K = 2$				$K = 3$		
0.1782	0.1840	0.1842	0.1900	0.2403	0.2551	0.2555	0.2710
0.1779	0.1837	0.1779	0.1837	0.2397	0.2548	0.2396	0.2548
0.1779	0.1840	0.1839	0.1900	0.2400	0.2551	0.2553	0.2710
0.1776	0.1837	0.1776	0.1837	0.2395	0.2548	0.2394	0.2548

	$K = 4$				$K = 5$		
0.2902	0.3157	0.3161	0.3439	0.3308	0.3675	0.3676	0.4095
0.2896	0.3161	0.2893	0.3161	0.3306	0.3693	0.3297	0.3693
0.2906	0.3157	0.3168	0.3439	0.3326	0.3675	0.3703	0.4095
0.2901	0.3161	0.2897	0.3161	0.3324	0.3693	0.3314	0.3693

	$K = 6$				$K = 7$		
0.3641	0.4120	0.4116	0.4686	0.3917	0.4505	0.4491	0.5217
0.3648	0.4161	0.3629	0.4161	0.3938	0.4575	0.3905	0.4575
0.3680	0.4120	0.4173	0.4686	0.3984	0.4505	0.4589	0.5217
0.3686	0.4161	0.3667	0.4161	0.4002	0.4575	0.3969	0.4575

Table B.3.

	$\lambda = 0$				$\lambda = 0.2$				$\lambda = 0.9$		
0.620	0.704	0.683	0.794	0.591	0.686	0.663	0.794	0.503	0.623	0.598	0.794
0.593	0.671	0.573	0.671	0.578	0.671	0.556	0.671	0.529	0.671	0.499	0.671
0.602	0.704	0.671	0.794	0.588	0.686	0.671	0.794	0.543	0.623	0.671	0.794
0.570	0.671	0.541	0.671	0.570	0.671	0.541	0.671	0.570	0.671	0.541	0.671

$$+ \epsilon x_3 \big(1 - \sigma(1 - u_2)(y_1 + y_2) - \tfrac{1}{2}\sigma(1 + \lambda)u_2 v_1 y_1\big),$$
$$H_2(x, y) = \tfrac{1}{2}\lambda\{u_1 v_2 x_1 q(y_3) + \epsilon\sigma u_2 v_1 x_3 y_1\} + x_2\{1 - v_2 q(y_3)\},$$
$$H_3(x, y) = \{1 - u_1\}v_2 x_1 q(y_3) + x_3\{1 - \epsilon + \epsilon\sigma(1 - u_2)(y_1 + y_2)\},$$
$$H_4(x, y) = \{x_2 + \tfrac{1}{2}u_1 x_1\}v_2 q(y_3) + \tfrac{1}{2}\epsilon\sigma u_2 v_1 x_3 y_1 + x_4$$
and $H_k(x, y) = H_{k-4}(y, y)$ for $k = 5, \ldots, 8$.

19. Increasing ϵ favors DH; increasing σ favors HH.

20. Suppose, for example, that $\epsilon = 0.1, \lambda = 0.9$ and $\sigma = 0.7$. Then Table 2.3 is unchanged (to four significant figures); whereas Table 2.2 is replaced by Table B.3. The difference is negligible.

21. (a) Let k be an ESS, and let i infiltrate k. Then $W_k - W_i = (a_{ki} - a_{ii})x_i + (a_{kk} - a_{ik})x_k > 0$ by (2.37) and (2.40), and (2.47) or (2.52) implies $x_k(\infty) = 1$.

(b) Let k be a strong ESS, and let i be any infiltrator. Then

$$W_k - W_i = (a_{kk} - a_{ik})x_k + \sum_{\substack{j=1 \\ j \neq k}}^{m} (a_{kj} - a_{ij})x_j.$$

Because x_j is small for every $j \neq k$, $a_{kk} > a_{ik}$ and x_k is approximately 1, the last $m - 1$ terms on the right hand side can be neglected. Thus $W_k > W_i$ for any $i \neq k$, and either (2.47) or (2.52) implies $x_k(\infty) = 1$.

22. **(c)** To 2 significant figures, $\alpha_c = 0.72$. For $\alpha_c - \alpha > 0$ but $\alpha_c \approx \alpha$, x_3 decreases at first but ultimately rises rapidly to 1.

(d) Use of (2.52) with (2.50) and (2.77), though suggestive, is not strictly valid because a_{ij} in (2.77) is the payoff to an i-strategist against a homogeneous population of j-strategists; in a heterogeneous population, the strategy of an opponent would not be fixed. For an analogous problem and its resolution, see §5.5.

23. See [**139**].

24. See [**139**].

25. From the solution to Exercise 1.32, there are three symmetric Nash-equilibrium strategy combinations, namely, (e_1, e_1), (e_2, e_2) and (θ, θ), where $e_1 = (1, 0)$, $e_2 = (0, 1)$, $\theta = (\theta_1, \theta_2)$ and θ_1, θ_2 are defined by (1.22). If $g(u, v)$ is the reward to a u-strategist against a v-strategist, then $g(e_1, e_1) - g(u, e_1) = \frac{1}{2}(\delta + \epsilon)\{(1 - u_1)\theta_2 + u_2(1 - \theta_1)\} > 0$ for all $u \neq e_1$, implying that e_1 is a strong ESS; and similarly for e_2. For a discussion of how either ESS might establish itself (regardless of whether $\tau_1 = \tau_2$), see [**219**]. Note, however, that because $g(\theta, \theta) = g(u, \theta)$ for all u and $g(\theta, u) - g(u, u) = (\delta + \epsilon)(\theta_1 - u_1)(\theta_2 - u_2)$ may be either positive or negative, the mixed strategy θ is not evolutionarily stable. That only pure strategies may be ESSes is a consequence of a theorem due to Selten [**203**].

Chapter 3

2. **(a)** From $\partial f / \partial \xi = -5f_1 + 5f_2 + 20\xi - 8 = 0$, $\xi = \frac{1}{4}(f_1 - f_2) + \frac{2}{5}$. Now substitute into (3.8) and simplify.

(b) The small curvilinear triangles would correspond to $v > 1$ or $v < 0$ under the mapping f defined by (3.4).

3. **(b)** From Figure 3.1, $\left(-\frac{8}{5}, -\frac{12}{5}\right)$ lies on $f(L(3/5);$ and, from (3.7) with $\xi = \frac{3}{5}$, we have $v = \frac{1}{5}$. So $\left(-\frac{8}{5}, -\frac{12}{5}\right)$ yields the rewards from strategy combination $(u, v) = \left(\frac{3}{5}, \frac{1}{5}\right)$. Similarly, $\left(-\frac{12}{5}, -\frac{8}{5}\right)$ yields the rewards from strategy combination $(u, v) = \left(\frac{1}{5}, \frac{3}{5}\right)$.

5. For all δ, ϵ, τ_1 and τ_2 satisfying (3.1), the shape of the reward set is similar to that of Figure 3.1 (except that $f_1 = f_2$ is not an axis of symmetry when $\tau_1 \neq \tau_2$). In fact, (3.12) generalizes to

$$\theta_1(f_1 + \tau_2) - \xi_1 f_2 = 0 \quad \text{if} \quad (\epsilon - \delta)\theta_1 \leq f_2 \leq 0$$
$$\frac{\{f_1 - f_2 - \frac{1}{2}(\tau_1 - \tau_2)\}^2 + \epsilon^2 \nu^2}{\nu\{\tau_1(f_1 + \epsilon) + \tau_2(f_2 + \epsilon) + \tau_1 \tau_2\}} = 1 \quad \text{if} \quad f_1 \leq (\epsilon - \delta)\theta_2, f_2 \leq (\epsilon - \delta)\theta_1$$
$$\xi_2 f_1 - \theta_2(f_2 + \tau_1) = 0 \quad \text{if} \quad (\epsilon - \delta)\theta_2 \leq f_1 \leq 0,$$

where $\nu = (\tau_1 + \tau_2)/(\delta + \epsilon)$ and ξ_1, ξ_2 are defined by (3.45). So, when $\delta = 5$, $\epsilon = 3$, $\tau_1 = 4$ and $\tau_2 = 2$, unimprovable points in $\overline{F}$ are $(-2, 0), (0, -4)$ and the interior of the arc of the parabola $16 (f_1 - f_2 - 1)^2 - 48 f_1 - 24 f_2 - 231 = 0$ that joins $(-3, -\frac{13}{4})$ to $(-\frac{15}{8}, -\frac{35}{8})$.

7. By arguments similar to those in §3.2 we find that, for all δ, ϵ, τ_1 and τ_2 satisfying (3.1), P_{nec} contains $(1, 0), (0, 1)$ and a segment of the straight-line $(\delta + \epsilon)(u + v) = 2\epsilon$ between (θ_1, ξ_2) and (ξ_1, θ_2) inclusive, where ξ_1 and ξ_2 are defined by (3.45). In particular, when $\delta = 5, \epsilon = 3, \tau_1 = 4$ and $\tau_2 = 2$, P_{nec} contains $(1, 0), (0, 1)$ and the straight line from $\left(\frac{5}{8}, \frac{1}{8}\right)$ to $\left(\frac{1}{4}, \frac{1}{2}\right)$. All of these points lie in D^*, because (3.43) yields $f_1(u, v) - \hat{f}_1 = (1 - 4u)(2v - 1) \geq 0$ and $f_2(u, v) - \hat{f}_2 = (1 - 8v)(u - 5/8) \geq 0$ for $(u, v) \in P_{nec}$. Hence $P_{nec}^* = P_{nec}$. But $P_{nec} \neq P$, because $\left(\frac{5}{8}, \frac{1}{8}\right)$ and $\left(\frac{1}{4}, \frac{1}{2}\right)$ are both improvable. At $\left(\frac{1}{4}, \frac{1}{2}\right)$, $\frac{\partial f_1}{\partial v} = 0$, $\frac{\partial f_2}{\partial v} = 3$ and $h = (0, 1)$ is admissible; thus, in the direction of h, f_2 is increasing while f_1 is not decreasing. Similarly for $\left(\frac{5}{8}, \frac{1}{8}\right)$, where $\frac{\partial f_1}{\partial u} = 3$, $\frac{\partial f_2}{\partial u} = 0$ and $h = (1, 0)$ is admissible.

8. (a) This follows readily from (3.43) and $\overline{x}_1 \geq 0, \overline{x}_2 \geq 0$.

(b) From (3.44), we have $(u, v) = g_1(u)g_2(v)$, where g_1 increases from a negative value at $u = 0$ to a positive maximum at $u = \alpha_1$ and then decreases again to a negative value at $u = 1$; and similarly for g_2. Thus the maximum of $\overline{d}$ over the unshaded region in Figure 3.5 must occur at either $(1, 0)$ or $(0, 1)$, or else at (α_1, α_2). By routine algebra, $\overline{d}(1, 0) = \left(\delta - \frac{1}{2}\tau_1\right)\left(\delta + \frac{1}{2}\tau_2\right)\xi_2\theta_2$, $\overline{d}(0, 1) = \left(\delta - \frac{1}{2}\tau_2\right)\left(\delta + \frac{1}{2}\tau_1\right)\xi_1\theta_1$ and $(\delta + \epsilon)^2\overline{d}(\alpha_1, \alpha_2) = \frac{1}{256}(\tau_1 + \tau_2)^4$. The Nash bargaining solution can be identified by comparing these three magnitudes.

(c) No. If $\tau_1 = \tau_2 = \tau$ in Crossroads between fast drivers, then (α, α) is the unique Nash bargaining solution whenever $\delta\epsilon/\sqrt{\delta^2 + \epsilon^2} < \tau/2 < \epsilon$, where $\alpha_1 = \alpha_2 = \alpha$.

9. **(a)** By the method of §3.1, $\overline{F}$ is a quadrilateral or curvilinear triangle whose upper and right-hand boundaries consist of line segments from (S, T) to (R, R) to (T, S). By the method of §1.7, the max-min reward vector is (P, P). Thus P (i.e., the set of all locally unimprovable strategy combinations, not the payoff for mutual defection) and P_G both consist of line segments from $(1, 0)$ to $(1, 1)$ to $(0, 1)$ in D; and P^* is the subset containing points between $(1, a)$ and $(a, 1)$, where $a = \frac{P-S}{R-S}$.

(b) The Nash bargaining solution is at $(1, 1)$ in D, corresponding to (R, R) in $\overline{F}$. By the symmetry of $\overline{F}$ about $f_1 = f_2$, it suffices to show that $\overline{d} = (f_1 - P)(f_2 - P)$ takes its maximum on the line segment from (R, R) to (T, S) at (R, R), which is readily established by using the equation of the line segment to write $\overline{d}$ as a function of either f_1 or f_2 alone.

10. First show that the triangle $\overline{F}$ contains $(\tilde{f}_1, \tilde{f}_1)$. Then the area $(f_1 - \tilde{f}_1)(f_2 - \tilde{f}_2)$ of the rectangle whose diagonal joins (f_1, f_2) to $(\tilde{f}_1, \tilde{f}_2)$ is maximized with one corner on the line segment (3.52), whose equation can be used to write the area as a function of either f_1 or f_2 alone. Thus the Nash bargaining reward vector is
$$f = -\tfrac{1}{2}(\tau_1, \tau_2) + \frac{\{\tau_1 - \tau_2\}\{\tau_1\tau_2 - 4\delta\epsilon\}}{8\tau_1\tau_2\{\delta + \epsilon\}}(\tau_2, -\tau_1),$$
and $\hat{\omega} = (0, -f_2/\tau_1, -f_1/\tau_2, 0)$ is the corresponding correlated strategy. Note that $f_1 = f_2$ when the game is symmetric (but there is no interpersonal comparison of utilities).

11. With the suggested parameterization, setting $c = 1$ in (1.48) implies that $f_1(u, v) = f_1(\xi + t, t) = (t + \xi)(6 - \xi)^2/25$ and $f_2(u, v) = f_2(\xi + t, t) = t(\xi + 4)(16 - \xi)/25$ in D_A. Eliminating t, we find that the image of $L(\xi)$ is a line segment with equation

(i)
$$\frac{25f_1}{(6-\xi)^2} - \frac{25f_2}{(16-\xi)(4+\xi)} = \xi.$$

For $\xi = 6$ we obtain the left-hand rim of $\overline{F}$: $f_1 = 0, 0 \le f_2 \le 16$. For $1 \le \xi < 6$, the upper endpoint of $f(L(\xi))$, i.e., the point with coordinates $(f_1(10, 10 - \xi), f_2(10, 10 - \xi))$, traces out the arc of the upper boundary of $\overline{F}$ between $(0, 16)$ and $(10, 27)$, whose equation is therefore

(ii) $f_1 = \frac{2}{5}(6 - \xi)^2, \quad f_2 = \frac{1}{25}(10 - \xi)(16 - \xi)(4 + \xi).$

For $2 \le \xi \le 6$, these lines do not cross. For $1 \le \xi \le 2$, however, they intersect near $(0, 0)$ in such a way that their envelope forms part of the boundary of $\overline{F}$. Partial differentiation of (i) yields

(iii) $\frac{50 f_1}{(6-\xi)^3} + \frac{50(6-\xi) f_2}{(16-\xi)^2(4+\xi)^2} = 1.$

Solving for f_1 and f_2, we obtain the parametric equations of the associated envelope, which define the boundary of $\overline{F}$ between $\left(\frac{32}{25}, 0\right)$ and $\left(\frac{17}{8}, \frac{27}{8}\right)$:

(iv)
$$f_1 = F_1(\xi) = \frac{1}{5000}(6 - \xi)^3(64 + 24\xi - 3\xi^2)$$
$$f_2 = F_2(\xi) = \frac{3}{5000}(2 - \xi)(64 + 12\xi - \xi^2)^2.$$

The images of the lines covering the lower half of D_B have parametric equations $f_1 = (t + \xi)(7 - 2\xi)/5, f_2 = t(13 + 2\xi)/5$, for which (3.10) yields

(v) $\frac{5 f_1}{7 - 2\xi} - \frac{5 f_2}{13 + 2\xi} = \xi, \quad \frac{10 f_1}{(7 - 2\xi)^2} + \frac{10 f_2}{(13 + 2\xi)^2} = 1$

where $0 \le \xi \le 1$. Thus we obtain parametric equations

(vi)
$$f_1 = F_1(\xi) = \frac{1}{200}(13 + 4\xi)(7 - 2\xi)^2$$
$$f_2 = F_2(\xi) = \frac{1}{200}(7 - 4\xi)(13 + 2\xi)^2$$

for the boundary between $\left(\frac{17}{8}, \frac{27}{8}\right)$ and $\left(\frac{637}{200}, \frac{1183}{200}\right)$. At the same time, the upper ends of the line segments, whose coordinates are $(2\{7 - 2\xi\}, \{13 + 2\xi\}\{10 - \xi\}/5)$, trace out the arc of the upper boundary of $\overline{F}$ between $(10, 27)$ and $(14, 26)$.

Similarly, the images of the lines that cover the upper half of D_B have equations $f_1 = t(7 + 2\xi)/5$ and $f_2 = (t + \xi)(13 - 2\xi)/5$. Thus the boundary of $\overline{F}$ between $\left(\frac{637}{200}, \frac{1183}{200}\right)$ and $\left(\frac{729}{200}, \frac{1331}{200}\right)$ has equations (vi) with plus and minus signs interchanged, and the boundary between $(14, 26)$ and $\left(\frac{81}{5}, 22\right)$ is traced by varying $(\{7 + 2\xi\}\{10 - \xi\}/5, 2\{13 - 2\xi\})$ simultaneously.

The rest of the boundary of $\overline{F}$ is determined by the images of the lines covering D_C. It is a join of two arcs, a lower envelope and an upper endpoint locus, which intersect with a common tangent at the point parameterized by, say, $\xi = \hat{\xi}$. The envelope extends upwards from $\left(\frac{729}{200}, \frac{1331}{200}\right)$ with parametric equations

$$f_1 = F_1(\xi) = \frac{3(22-8\xi+\xi^2)(34+12\xi-\xi^2)^2}{5000(6-\xi)}$$

(vii)

$$f_2 = F_2(\xi) = \frac{(34+24\xi-3\xi^2)(66-12\xi+\xi^2)^2}{5000(6-\xi)},$$

where $1 \leq \xi \leq \hat{\xi}$. The endpoint locus extends downwards from $(81/5, 22)$ with parametric equations

(viii) $\quad f_1 = \frac{1}{25}(10-\xi)(34+12\xi-\xi^2), \quad f_2 = \frac{2}{5}(66-12\xi+\xi^2)$

where again $1 \leq \xi \leq \hat{\xi}$. We obtain $\hat{\xi}$ as the only value of ξ between 1 and 6 for which both $F_1(\hat{\xi}) = f_1(10-\hat{\xi}, 10)$ and $F_2(\hat{\xi}) = f_2(10-\hat{\xi}, 10)$, i.e., as the only root between 1 and 6 of the equation $3\xi^4 - 60\xi^3 + 452\xi^2 - 3176\xi + 9756 = 0$. So $\hat{\xi} \approx 4.729$; i.e., (vii) and (viii) intersect at $(F_1(\hat{\xi}), F_2(\hat{\xi})) \approx (14.42, 12.65)$. Note that the image lines end in the interior of $\overline{F}$ for $\hat{\xi} \leq \xi \leq 6$.

12. (a) The apex of the arc defined by (ii) occurs where its tangent is horizontal: $\partial f_2/\partial \xi = 0$, implying $\xi = 2(11 - \sqrt{79})/3 \approx 1.4079$. The analogous extreme of the arc defined by (viii) occurs where its tangent is vertical: $\partial f_1/\partial \xi = 0$, implying $\xi = (22 - \sqrt{226})/3 \approx 2.3222$. Between (but excluding) these extremes, where $(f_1, f_2) \approx (8.435, 27.12)$ and $(f_1, f_2) \approx (17.34, 17.41)$, respectively, every point on the boundary of $\overline{F}$ is (globally) unimprovable, by Figure 3.3; and clearly these are the only unimprovable points. Moreover, from Exercise 1.23 (with $c = 1$ and $\alpha = 10$) we have $\tilde{f}_1 = m_1(2) = 32/25$ and $\tilde{f}_2 = m_2(10) = 12$, so that every unimprovable point satisfies $f_1 > \tilde{f}_1, f_2 > \tilde{f}_2$. Thus $P^* = P$. It consists of straight-line segments joining $(10, 10)$ in the top right-hand corner of D—the Nash bargaining solution— to $(10, 2\{4 + \sqrt{79}\}/3) \approx (10, 8.592)$ and $(\{8 + \sqrt{226}\}/3, 10) \approx (7.678, 10)$, but excludes these two points.

Chapter 4

2. See (4.56).
3. (b) $\epsilon_1 = -\frac{8}{45}$ and $C^+(\epsilon_1) = \left\{ \left(\frac{19}{45}, \frac{16}{45}, \frac{2}{9} \right) \right\}$.
4. $X^1 = C^+\left(-\frac{1}{9}\right) = \left\{ \left(\frac{1}{3}, \frac{2}{9}, \frac{1}{9}, \frac{1}{3} \right) \right\}$.
5. $X^1 = C^+\left(-\frac{11}{20}\right) = \left\{ \left(\frac{11}{40}, \frac{29}{120}, \frac{5}{24}, \frac{11}{40} \right) \right\}$.

6. $\epsilon_1 = -\frac{7}{80}$ and $C^+(\epsilon_1) = \left\{\left(\frac{21}{80}, \frac{59}{240}, \frac{11}{48}, \frac{21}{80}\right)\right\}$.

7. **(a)** X can be represented as the interval $[0, 1]$ of the x_1-axis with x_1 increasing to the right and $x_2 \ (= 1 - x_1)$ to the left, although it is really that part of the line $x_1 + x_2 = 1$ which extends from the point $(1, 0)$ to the point $(0, 1)$. From $\nu(\{1\}) = 0 = \nu(\{2\})$ and $\nu(\{1, 2\}) = 1$, $C^+(\epsilon)$ is defined by $-x_1 \leq \epsilon$, $-x_2 \leq \epsilon$ and $x_1 + x_2 = 1$; or $-\epsilon \leq x_1 \leq 1 + \epsilon$, which requires $\epsilon \geq -\frac{1}{2}$ but no other restrictions on ϵ. Thus $\epsilon_1 = -\frac{1}{2}$, $X^1 = \left\{\left(\frac{1}{2}, \frac{1}{2}\right)\right\}$.

(b) $x \in C^+(\epsilon)$ requires x_1, x_2 and $x_1 + x_2$ to lie between $-\epsilon$ and $1 + \epsilon$, so that $\epsilon_1 = -\frac{1}{3}$ and $X^1 = \left\{\left(\frac{1}{3}, \frac{1}{3}, \frac{1}{3}\right)\right\}$.

10. Let Jed, Ned and Ted be Players 1, 2 and 3, as usual, and let $\overline{\nu}$ be the time saved in hours. Then $\overline{\nu}(\{1, 2\}) = 3 = \overline{\nu}(\{1, 3\})$, $\overline{\nu}(\{2, 3\}) = 2$ and $\overline{\nu}(\{1, 2, 3\}) = 6$. The core is a pentagon (and coincides with the reasonable set). The least rational core is $X^1 = C^+\left(-\frac{2}{9}\right) = \left\{\left(\frac{4}{9}, \frac{5}{18}, \frac{5}{18}\right)\right\}$. Fair shares of the time saved are $\frac{8}{3}$ hours for Jed and $\frac{5}{3}$ hours each for Ned and Ted. Jed should arrive at 1:20 p.m. and leave at 2:40 p.m., Ned should leave at 1:20 p.m. and Ted should arrive at 2:40 p.m.

12. $\epsilon_1 = 0$; $\phi_1(x) = \frac{1}{4} - 2x_1$ if $\frac{1}{8} \leq x_1 \leq \frac{1}{4}$; $\phi_1(x) = x_1 - \frac{1}{2}$ if $\frac{1}{4} \leq x_1 \leq \frac{1}{2}$; and the nucleolus is $\left\{\left(\frac{1}{4}, \frac{1}{4}, \frac{1}{4}, \frac{1}{4}\right)\right\}$.

13. A 3-player game is improper if and only if $\nu(S) > 1$, where S is one of the 2-player coalitions. Suppose $S = \{a, b\}$, and let $N = \{a, b, c\}$. Then, if the game has a core, there must exist at least one x such that we require $x_a + x_b \geq \nu(S) > 1, x_c \geq 0$; but this is impossible, because $x_a + x_b + x_c = 1$.

14. To calculate X^1, replace $x_1 \geq -\epsilon, x_2 \geq -\epsilon$ and $x_1 + x_2 \leq 1 + \epsilon$ in (4.95) by, respectively, $x_1 \geq 0, x_2 \geq 0$ and $x_1 + x_2 \leq 1$ (equivalent to replacing $x \in \overline{X}$ by $x \in X$), then use the method of §4.3. Because $e(\{2, 3\}, x) = \frac{3}{8}$ on X^1, we have $\Sigma^1 = \{\{1\}, \{2\}, \{3\}, \{1, 2\}, \{1, 3\}\}$. Moreover, because $e(\{2\}, x) = -x_2$ and $e(\{3\}, x) = x_2 - 1$ are never greater than $-\frac{5}{16}$ on X^1, whereas the other three excesses are nonnegative, for $x \in X^1$ we have $\phi_1(x) = \max\left(0, \frac{11}{16} - x_2, x_2 - \frac{5}{16}\right)$, which takes its minimum ϵ_2 where $x_2 = \frac{1}{2}$.

15. If the jumps had all counted, prize money (in dollars) would have been earned as in Table 4.2 with 3 and 1 in place of the first two zeroes. Thus the official proceeds are \$20, with the characteristic function defined by $\nu(\{1,2\}) = \frac{3}{4} = \nu(\{1,3\})$ and $\nu(\{2,3\}) = \frac{11}{10}$; $x \in C(\epsilon)$ if $x \in \overline{X}$, $-\epsilon \le x_1 \le -\frac{1}{10} + \epsilon$, $-\epsilon \le x_2 \le \frac{1}{4} + \epsilon$ and $-\epsilon \le x_1 + x_2 \le 1 + \epsilon$. By the method of §4.3, the least value of ϵ for which the inequalities are consistent is $\overline{\epsilon}_1 = \frac{1}{5}$, which is also ϵ_1. The least core, least rational core and nucleolus are all $\{(\frac{1}{10}, \frac{9}{20}, \frac{9}{20})\}$. Jed gets \$2, Ned and Ted \$9 each.

17. (a) The Shapley value (and the nucleolus) for a 2-player game coincides with (4.101).

(b) The Shapley value for a 3-player CFG is the imputation whose transpose is

$$\left(x^S\right)^T = \frac{1}{6}\begin{bmatrix} \nu(\{1,2\}) \\ \nu(\{2,3\}) \\ \nu(\{3,1\}) \end{bmatrix} + \frac{1}{6}\begin{bmatrix} \nu(\{1,3\}) \\ \nu(\{2,1\}) \\ \nu(\{3,2\}) \end{bmatrix} + \frac{1}{6}\begin{bmatrix} 1 \\ 1 \\ 1 \end{bmatrix} - \frac{1}{3}\begin{bmatrix} \nu(\{2,3\}) \\ \nu(\{3,1\}) \\ \nu(\{1,2\}) \end{bmatrix}.$$

(c) Hence, e.g., the Shapley value for the 3-person car pool is $x^S = \frac{1}{6(3+2d)}(10 + 4d, 7 + 4d, 1 + 4d)$.

18. (a) $12x_2^S = \nu(\{2,3\}) + \nu(\{2,4\}) + \nu(\{2,1\}) + \nu(\{2,3,4\}) - \nu(\{3,4\}) + \nu(\{2,3,1\}) - \nu(\{3,1\}) + \nu(\{2,4,1\}) - \nu(\{4,1\}) + 3\{1 - \nu(\{3,4,1\})\}$, $12x_3^S = \nu(\{3,4\}) + \nu(\{3,1\}) + \nu(\{3,2\}) + \nu(\{3,4,1\}) - \nu(\{4,1\}) + \nu(\{3,4,2\}) - \nu(\{4,2\}) + \nu(\{3,1,2\}) - \nu(\{1,2\}) + 3\{1 - \nu(\{4,1,2\})\}$, $12x_4^S = \nu(\{4,1\}) + \nu(\{4,2\}) + \nu(\{4,3\}) + \nu(\{4,1,2\}) - \nu(\{1,2\}) + \nu(\{4,1,3\}) - \nu(\{1,3\}) + \nu(\{4,2,3\}) - \nu(\{2,3\}) + 3\{1 - \nu(\{1,2,3\})\}$ by cyclic permutation of 1, 2, 3 and 4 in (4.100) where, obviously, $\{2,3,1\}$ is the same as $\{1,2,3\}$, etc.

20. Let x be any imputation in the core (assumed non-empty) and let $C(n,j)$ be the number of j-player coalitions among n players; i.e., $(n-j)!\,j!\,C(n,j) = n!$. For any j-player coalition S we have $\nu(S) = f(j)$, and so (4.16) and (4.19) imply $f(j) = \sum_{i \in S} x_i$. Summing over all j-player coalitions, $C(n,j)f(j) \le \sum\sum_{i \in S} x_i$, where the first summation is over all $S \in \Sigma^0$ such that $\#(S) = j$. Now consider Player k. The number of j-player coalitions containing him is $C(n-1,j-1)$, because this is the number of $(j-1)$-player coalitions among the other $n-1$ players to which he can be added. So x_k appears precisely $C(n-1,j-1)$ times in the double summation for all $k \in N$, implying $\sum\sum_{i \in S} x_i = C(n-1,j-1)\sum_{k=1}^n x_k = C(n-1,j-1)$, by (4.12b). Thus $f(j) \le C(n-1,j-1)/C(n,j)$. But $C(n-1,j-1)/C(n,j) = j/n$. So (4.101) yields $e(S, x^E) = \nu(S) - \#(S)/n = f(j) - j/n \le 0$ for any j-player coalition, and hence for any S. That is, $x^E \in C^+(0)$.

22. Hiring a pure mathematician. If an applied mathematician were hired, then the Shapley-Shubik index would give each professor one sixth of the power—as is obvious intuitively, because there would then be an equal number of pure and applied mathematicians. If a pure mathematician were hired, however, then the Shapley-Shubik index would give each of the two applied mathematicians a fifth of the power (and each of the others only 15%).

23. (a) Suppose, for example, that nation E has committed an outrage, and that nations A, B, C and D are contemplating a military blockade. Suppose that these four nations have, respectively, 50,000, 300,000, 400,000 and 200,000 troops, and that 700,000 troops are needed to enforce the blockade. Then, in terms of enforcing the blockade, a coalition of nations is winning if it has more than 700,000 troops, and otherwise losing; in which case, nations A and D are dummies.
(b) If Player i is a dictator, then $\{i, j\}$ is winning but $\{j\}$ is losing for any $j \neq i$; hence Player j cannot be a dictator. For the illustration, raise one nation's army in (a) to 700,000 troops.

24. Before the change, Glen Cove, Long Beach and Oyster Bay were all dummies.

25. (a) See, e.g., Chapter 5 of [**27**] and [**216**].
(b) From $x^S = \left(\frac{7}{30}, \frac{7}{30}, \frac{7}{30}, \frac{3}{20}, \frac{3}{20}\right)$ and $x^B = \left(\frac{3}{13}, \frac{3}{13}, \frac{3}{13}, \frac{2}{13}, \frac{2}{13}\right)$, slightly more: to two decimal places, the indices agree. Although it is common for x^S and x^B to agree qualitatively, it is unusual for them to agree quantitatively, and it is not unusual for them to assign power in radically different ways; see, e.g., [**27**, p. 193] and [**216**, p. 109].
(c) See [**216**].

26. Suppose that $x \in C^+(0)$ but $x \notin \overline{R}$, where $\overline{R}$ denotes the reasonable set. Then, from (4.15), there exists $i \in N$ such that $x_i > \nu(T) - \nu(T - \{i\})$ for all T to which Player i belongs. In particular, $x_i > \nu(N) - \nu(N - \{i\}) = 1 - \nu(N - \{i\})$. Also $e(N - \{i\}, x) \leq 0$, from (4.19). But (4.12b) and (4.16) together imply that $e(N - \{i\}, x) = \nu(N - \{i\}, x) - (1 - x_i)$, which cannot be positive. So $x_i \leq 1 - \nu(N - \{i\})$, which is a contradiction.

Chapter 5

1. Define $S(n) = \sum_{k=1}^{n} \text{Prob}(k \le M \le n)$. Then, for all $n \ge 2$,

$$
\begin{aligned}
S(n) &= \sum_{k=1}^{n} \sum_{j=k}^{n} \text{Prob}(M = j) \\
&= \sum_{k=1}^{n-1} \sum_{j=k}^{n} \text{Prob}(M = j) + \text{Prob}(M = n) \\
&= \sum_{k=1}^{n-1} \left\{ \sum_{j=k}^{n-1} \text{Prob}(M = j) + \text{Prob}(M = n) \right\} + \text{Prob}(M = n) \\
&= \sum_{k=1}^{n-1} \sum_{j=k}^{n-1} \text{Prob}(M = j) + (n-1)\text{Prob}(M = n) + \text{Prob}(M = n) \\
&= S(n-1) + n \cdot \text{Prob}(M = n).
\end{aligned}
$$

So $\sum_{n=1}^{N} n \cdot \text{Prob}(M = n) = S(1) + \sum_{n=2}^{N} n \cdot \text{Prob}(M = n) = S(1) + \sum_{n=2}^{N} \{S(n) - S(n-1)\} = S(N)$. Now let $N \to \infty$.

2. (a) Let $\text{Prob}(U|V)$ denote the conditional probability of U, given V. Define $\xi_k = \text{Prob}(M \ge k)$. Then, for all $k \ge 1$, $\xi_{k+1} = \text{Prob}(M \ge k + 1 | M \ge k) \cdot \xi_k = w\xi_k$. Solving subject to $\xi_1 = \text{Prob}(M \ge 1) = 1$ produces (5.23).

   (b) Because $\text{Prob}(M \ge k) = w^{k-1}$ and the probability of no further interaction is $1 - w$, $\text{Prob}(M = k) = w^{k-1}(1 - w)$. So $\mu = \text{E}[M] = (1 - w) \sum_{k=1}^{\infty} k w^{k-1} = \frac{1}{1-w}$, on using (5.71).

3. (a) $f(TF2T, TF2T) = R\mu < T + wR\mu = f(STCO, TF2T)$; whereas (5.36) implies $f(TFT, TFT) = R\mu > T + wS + w^2 R\mu = f(STCO, TFT)$.

   (b) If $STCO$ is strategy 1 and $TF2T$ strategy 2, then $W_1 = x_1(P + wR\mu) + x_2(T + wR\mu)$ and $W_2 = x_1(S + wR\mu) + x_2 R\mu$, so that $W_1 - W_2 = (P - S)x_1 + (T - R)x_2 > 0$ for all x_1, x_2. Hence the final composition is all $STCO$. The average expected payoff to the population is thereby reduced from R to $P + wR\mu$.

5. H is a conic section with positive discriminant $(c + d)^2$, hence a hyperbola. The left-hand branch passes through $(0, 0)$ and $\left(0, \frac{b}{d}\right)$; the right-hand branch passes through $\left(-\frac{a}{c}, 0\right)$ and meets the line $x_1 + x_2 = 1$ where $x_1 = \frac{d-b}{d+c-b+a}$ and $x_2 = \frac{a+c}{d+c-b+a}$. The sign of the quantity $(cx_1 - dx_2)(x_1 + x_2) + ax_1 + bx_2$ is either negative in the shaded region and positive in the unshaded region, or vice versa. To determine which, simply evaluate the quantity at a single point (not on the hyperbola), e.g., $(1, 0)$, where the quantity takes the positive value $a + c = 1 - w^2$.

9. (a) Use $(P-S)(T-P) - (R-S)(T-R) \equiv (R-P)(P+R-S-T)$.

10. The payoff to $ALLD$ against $RNDM$ is T or P with equal probability. The payoff to $RNDM$ is equally likely to be S or P against $ALLD$, and equally likely to be any of R, S, T or P against itself. Thus, for $1 \leq k < \infty$, $\phi_{45}(k) = \frac{1}{2}(T + P)$, $\phi_{54}(k) = \frac{1}{2}(S + P)$ and $\phi_{55}(k) = Q$, where Q is defined by $Q = \frac{1}{4}(R + S + T + P)$; and, from arguments similar to those employed in §5.4, $\phi_{15}(k) = Q\sigma_k + \frac{1}{2}(R + S)(1 - \sigma_k)$, $\phi_{51}(k) = Q\sigma_k + \frac{1}{2}(R + T)(1 - \sigma_k)$, $\phi_{25}(k) = Q\gamma_k + \frac{1}{2}(R + S)(1 - \gamma_k)$, $\phi_{52}(k) = Q\gamma_k + \frac{1}{2}(R+T)(1-\gamma_k)$, $\phi_{35}(k) = Q\sigma_k + \frac{1}{2}(T+P)(1-\sigma_k)$ and $\phi_{53}(k) = Q\sigma_k + \frac{1}{2}(S + P)(1 - \sigma_k)$.

11. **(b)** Observe that the recurrence equation has an equilibrium (independent of k) solution $\epsilon_k = 1/2$; but it fails to satisfy the starting condition $\epsilon_1 = 1$. So define the perturbation $\eta_k = \epsilon_k - 1/2$, and obtain for it both a recurrence equation and a starting condition. Solve for η_k, and hence obtain $\epsilon_k = \eta_k + 1/2$.

13. The frequency of third encounters depends on N. From (5.75) with $x_4 = 0$, $W_1 - W_2 = \frac{1}{2}\alpha\lambda_1 w x_3\{\lambda_2(T - S) - \mu(2R - S - T)\}$, which is negative whenever $N < \frac{(2R-S-T)w}{(1-w)(T-R)}$. In the case of Axelrod's prototype, this condition requires $N + 1 \leq 145\ TFT$, which would have to be satisfied if TFT were stable against $STFT$—see the remarks following (5.97).

14. From (5.66) with $m = 5$ and $x_2 = x_3 = x_4 = 0$, (5.67), (5.70) and Exercise 5.10, $W_1 = \mu(x_1 - \alpha)R + \frac{1}{2}x_5(R + S + 2\mu w Q)$ and $W_5 = \frac{1}{2}x_1(R + T + 2\mu w Q) + \mu(x_5 - \alpha)Q$, where $x_5 = N_5/N$ and $Q = \frac{1}{4}(R + S + T + P)$. We need $W_1 > W_5$ for $N_1 = N$, $N_5 = 1$ or $x_1 = 1$, $x_5 = \alpha$; then $4(1-w)(W_1 - W_5) = (T + R - P - S)w - 2(T - R) - \alpha\{(T + R - P - S)(1 - w) + 3R - S - T - P\}$. This expression is positive when both $(T+R-P-S)w > 2(T-R)$ and $\{(T+R-P-S)w - 2(T-R)\}N > (T+R-P-S)(1-w)+3R - S-T-P$. Now, if $S+T \geq P+R$, then $\frac{T-R}{R-S} \geq \frac{2T-2R}{T+R-P-S} \geq \frac{T-R}{T-P}$; whereas if $S + T \leq P + R$, then the inequalities are reversed. So (5.34) implies that TFT is stable for large sufficiently N.

15. In a population of $TF2T$ and $ALLD$, we have $N_1 = 0 = N_3$; and (5.72) implies $W_2 = \mu(x_2 - \alpha)R + x_4\{S(1 + w) + \mu w^2 P\}$, $W_4 = x_2\{T(1 + w) + \mu w^2 P\} + \mu(x_4 - \alpha)P$. For $TF2T$ to be stable, we require $W_2 > W_4$ when $N_2 = N$ and $N_4 = 1$. Set $x_2 = 1$ and $x_4 = \alpha$. Then $W_2 - W_4 > 0$ implies (5.86).

16. In a population of $TF2T$ and $RNDM$, we have $N_1 = N_3 = N_4 = 0$, $W_2 = \mu(x_2 - \alpha)R + \frac{1}{2}x_5\{(1 + w)(R + S) + 2\mu w^2 Q\}$ and $W_5 = \frac{1}{2}x_2\{(1 + w)(T + R) + 2\mu w^2 Q\} + \mu(x_5 - \alpha)Q$. Now $W_2 > W_5$ when $x_2 = 1$ and $x_5 = \alpha$ if $w > \sqrt{2(T - R)}/\sqrt{T + R - P - S}$ and $\{w^2(T + R - P - S) - 2(T - R)\}N > (1 - w^2)(T - S) + 2R - S - T + w^2(R - P)$. Because $TF2T$ is nicer than TFT, a greater chance of further interaction is required than in Exercise 5.14.

19. We have $\xi_1 - \frac{T-R}{T-P} = \frac{(R-P)\{K + 2(T-P)\sqrt{(T-R)(T-S)} + 2(T-R)(2R-P-T)\}}{(T-P)\{(2R-P-S)^2 + 4(T-R)(R-P)\}}$, where $K = (2R - P - S)(2R - S - T)$. But (5.2) implies $T - S > T - R$, hence $\sqrt{(T - R)(T - S)} > T - R$. So $\xi_1 - \frac{T-R}{T-P} > \frac{(R-P)\{(2R-P-S)(2R-S-T) + 4(T-R)(R-P)\}}{(T-P)\{(2R-P-S)^2 + 4(T-R)(R-P)\}} > 0$; and similarly for ξ_2.

20. With $RNDM$ as strategy 5 (and $N_2 = N_3 = N_4 = 0$), $W_1 - W_5 = \mu(R - Q) - \frac{1}{2}\lambda_1(1 + \alpha)(R + T - 2Q) + \frac{1}{2}x_5\{2\mu(Q - R) + \lambda_1(R - P)\}$, where $x_5 = N_5/N$ and $Q = \frac{1}{4}(R + S + T + P)$. Proceeding in the usual way, we require for stability that $W_1 - W_5 > 0$ when $x_1 = 1$ and $x_5 = \alpha$; and this condition readily reduces to $\zeta_3(N) < 0$, where ζ_3 is defined by (5.87) and (5.89) with $j = 3$. Thus TFT is stable against infiltration by $RNDM$ if $w > \xi_3$ and $w(3R - S - T - P)/\{2(1 - w)(T - R)U_3(w)\} < N < U_3(w)$, where ξ_3 is defined by (5.89) and (5.91), and U_3 by (5.89) and (5.93). For example, in Axelrod's prototype we have $\xi_3 = 0.899$ and $U_3(w) = 213.5$; and the lower bound on N in (5.92b) is then $N > 1.012$. Thus the population is stable against $RNDM$ if $3 \leq N + 1 \leq 214$.

21. The rewards are $W_2 = \mu(x_2 - \alpha)R + x_5\{\mu Q - \lambda_1^2(2Q - R - S)/2\lambda_2\}$ and $W_5 = x_2\{\mu Q - \lambda_1^2(2Q - R - T)/2\lambda_2\} + \mu(x_5 - \alpha)Q$. We require $W_2 > W_5$ when $x_2 = 1$ and $x_5 = \alpha$ or $2(1 - w)^2(T - R)N^3 + 2(1 - w)\{2w(T - R) + (1 - w)(R - S)\}N^2 + 4w\{(1 - w)(R - S) - w(R - Q)\}N + 4w^2(R - Q) < 0$, which requires in particular that $(7R - 5S - T - P)w > 4(R - S)$ for a negative coefficient of N. For Axelrod's prototype, we require $3 \leq N + 1 \leq 91$.

22. Consider a population of $N + 1$ indistinguishable individuals, of whom N_1 play TFT and $N_4 = N - N_1 + 1$ play $ALLD$ in an IPD with opponents drawn at random. Then (5.66) still yields correct expressions, but ϕ_{11}, ϕ_{14} and ϕ_{41} in (5.67)-(5.68) must be modified for absence of recognition. Let θ_k denote the probability that a TFT-strategist cooperates on its k-th move. Then $\theta_1 = 1$, because TFT is a nice strategy; and $\theta_k = (\{N_1 - 1\}/N)^{k-1} = (x_1 - \alpha)^{k-1}$ for $k \geq 2$ because interaction is random, the probability of meeting another TFT-strategist is still $x_1 - \alpha$ and absence of recognition implies that a TFT-strategist will cooperate on move k only if her previous partner was a TFT-strategist who cooperated—in other words, because $\theta_k = (x_1 - \alpha)\theta_{k-1}$. Thus, because $1 - \theta_k$ is the probability that a TFT-strategist will defect on move k, we have $\phi_{11}(k) = (1 - \theta_k)\{T\theta_k + P(1 - \theta_k)\} + \theta_k\{R\theta_k + S(1 - \theta_k)\} = (P + R - S - T)\theta_k^2 + (S + T - 2P)\theta_k + P$. Similarly, $\phi_{14}(k) = S\theta_k + P(1 - \theta_k)$, $\phi_{41}(k) = T\theta_k + P(1 - \theta_k)$ and $\phi_{44}(k) = P$. On substituting into (5.66), with $m = 4$ and $x_2 = 0 = x_3$, and on using (5.31) and (5.64), we readily obtain $W_4 = \mu P + (x_1 - \alpha)(T - P)/(1 - w\{x_1 - \alpha\})$ and $W_1 = W_4 - (x_1 - \alpha)(S + T - P - R)/(1 - w\{x_1 - \alpha\}^2) - (P - S)/(1 - w\{x_1 - \alpha\})$. It follows that $W_4 > W_1$ (immediately for $S + T \geq P + R$, and after a little algebra for $P + R > S + T$).

23. Note that, if $q < \epsilon Q$, then $2R - S - T > q/\epsilon + \epsilon q - 2q = q(1 - \epsilon)^2/\epsilon > 0$.

24. With $S = 0$ and $T = P + R$, (5.101) reduces to $w > \xi(q)$ and $N < \overline{U}(q, w)$ where $q = P/R \ (< 1)$, $(4 - 3q^2)\xi(q) = q + 2(1 - q)\{1 + q + \sqrt{q(1 + q)}\}$ and $2(1 - w)q\overline{U}(q, w) = (2 - q)w + \sqrt{(4 - 3q^2)w^2 - 2(2 - 2q^2 + q)w + 1} - 1$. As q increases from 0 to 1, ξ increases monotonically from $\xi(0) = \frac{1}{2}$ to $\xi(1) = 1$. In a laboratory, N could not be too large; and it would be impractical to vary w, because each new value of w would require the animals to be retrained. On the other hand, it should be possible to alter q quite readily. So the sign of $w - \xi(q)$ should be controlled as a function of q, not w. Suppose, for example, that we choose $w = \frac{4}{5}$. Then $w - \xi(q) = 0.8 - \xi(q)$ is positive if $q < 0.294$ (to three significant figures) and negative if $q > 0.294$. Thus we could test the model of §5.7 by choosing two values of q, one larger than 0.294 but the other smaller than 0.294, and imbedding the experiment of §5.1 in an iterated prisoner's dilemma among $N + 1$ animals who interact at random. Suppose, for example, that we choose $q = \frac{1}{2}$ and $q = \frac{1}{4}$. Then, with w held fixed at $\frac{4}{5}$, (5.92b) and (5.101) predict that TFT is unstable against $ALLD$ when $q = \frac{1}{2}$ for any value of N, but stable against $ALLD$ when $q = \frac{1}{4}$ if

$2 < N < 6$. Because this inequality is satisfied if $4 \leq N+1 \leq 6$, we could test the theory with four to six animals.

25. (b) $P + R = S + T$.

(c) No. The equations are satisfied by infinitely many values of $\sigma_c, \sigma_d, \omega_c$ and ω_d.

26. (a) The payoff matrix is (5.1) with $R = b - c, S = -c, T = b$ and $P = 0$; $b > c$ ensures that (5.2) is satisfied.

(b) Cheat $= ALLD$ and Sucker $= ALLC$ by definition. Grudger is not by definition the same as TFT because, though nice, it is completely unforgiving; however, Grudger $= TFT$ in this particular IPD because $ALLC$ and $ALLD$ are the only other strategies.

(c) From (5.95), for Grudger to be uninvadable we require

$$w > \frac{2(b-c)\{c+\sqrt{c(b+c)}\}+b(2b-c)}{4b^2-3c^2},$$

$$N < \frac{\sqrt{\{(2b-c)w-b\}^2-4cw(1-w)(b-c)}+(2b-c)w-b}{2c(1-w)}.$$

(d) $\phi_1(x_1, x_2) = x_1 x_2 \{(\mu - \lambda_1)(b-c)x_1 - \mu\alpha(b-c) - \lambda_1(1+\alpha)c\}$ in (5.44), where μ, λ_1, x_1, x_2 and α are defined by (5.24), (5.76) and p. 201. The coefficient of x_2 in the squiggly bracket is in the first instance $(S + T - P - R)\mu$, but vanishes because the prisoner's dilemma is decomposable. Thus a sufficient condition for Grudger to eliminate both Cheat and Sucker is that $\phi_1 > 0$ when $x_2 > 0$, or $(\mu - \lambda_1)(b-c)x_1 > \mu\alpha(b-c) + \lambda_1(1+\alpha)c$, which after some rearrangement is the desired condition.

28. (b) Use mathematical induction.

(c) See, e.g., [**144**, p. 196].

(d) The first two rows of U for $ALLC$ versus TFT are identical to the first row of U for TFT versus itself, and the last two rows to the third row. Every row of U^k is $(1 - 3\epsilon, 2\epsilon, \epsilon, 0) + O(\epsilon^2)$ for $k \geq 2$. Hence, to first order in ϵ, $x(1) = (1 - 2\epsilon, \epsilon, \epsilon, 0)$ and $x(k) = (1 - 3\epsilon, 2\epsilon, \epsilon, 0)$. Similarly, the first two rows of U for $ALLD$ versus TFT are identical to the second row of U for TFT versus itself, and the last two rows to the fourth row. Every row of U^k is $(0, \epsilon, 2\epsilon, 1 - 3\epsilon) + O(\epsilon^2)$ for $k \geq 2$. Hence, to first order in ϵ, $x(1) = (\epsilon, 0, 1 - 2\epsilon, \epsilon)$ and $x(k) = (0, \epsilon, 2\epsilon, 1 - 3\epsilon)$.

Chapter 6

1. Note that with $v = \frac{1}{2}$ we have $f(u,v) = f(v,v) = C^2\sigma_f$ but $f(v,u) - f(u,u) = 2C^2\sigma_f(u-v)^2/u > 0$ for all $u \neq v$.

3. (a) See [**145**, p. 260].
   (c) $v^* = \frac{1}{1+\theta}$ for all $\theta < 1$.

5. (a) Note that $f(v^*,v^*) = \frac{2\alpha\mu(108-169\theta)}{35(24+13\theta)} \geq 0$ if (6.20) is satisfied.
   (b) For all $u \neq \frac{26}{35}$ such that $0 \leq u \leq 1$, $f(26/35,u) - f(u,u)$ $= \frac{\alpha\mu}{207924080}(114244 + 63700u - 35525u^2)(26 - 35u)^2 > 0$.

7. See [**145**, pp. 256-257].

9. (a) $v^* = \frac{3}{3+2\theta}$ for all $\theta < 1$
   (b) $v^* = \frac{2}{2+\theta}$ for $\theta \leq \frac{6}{5-2^{4/3}} - 2 \approx 0.4192$. See [**156**, pp. 68-69].

10. (b) $\phi(v,v) - \phi(u,v) = f(v,v) - (1-r)f(u,v) - rf(u,u)$. If this expression equals zero, then $f(v,v) = (1-r)f(u,v) + rf(u,u)$. It follows that $\phi(v,u) - \phi(u,u) = (1-r)f(v,u) + rf(v,v) - f(u,u) = (1-r)\{f(v,u) + rf(u,v) - (1+r)f(u,u)\}$.

13. (a) $v^* = \frac{1-r}{1+(1+r)\theta}$.
    (b) See [**145**, p. 261].

14. See [**145**, pp. 263-264].

15. From Exercise 1.25, $f(u,v) = -\{(P-S)(1-v) + (T-R)v\}u + Tv + P(1-v)$, where u is the probability of cooperation.
    (a) Here $\partial^2\phi/\partial u^2 = -2r(S+T-P-R) < 0$. With $r_1 = \frac{P-S}{T-P}$, $r_2 = \frac{T-R}{R-S}$ the unique strong ESS among kin is $v^* = 0$ if $0 < r \leq r_1$, $v^* = \frac{(1+r_2)(r-r_1)}{(r_2-r_1)(1+r)}$ if $r_1 < r < r_2$ and $v^* = 1$ if $r_2 \leq r < 1$.
    (b) The unique Nash equilibrium is (v^*,v^*); it is strong for $r \leq r_1$ or $r \geq r_2$ but weak for $r_1 < r < r_2$. In either case, the probability of cooperation increases with r for $r_1 < r < r_2$; there is no cooperation for $r \leq r_1$, and full cooperation for $r \geq r_2$.

16. (a) $\phi_1(u,v) = \frac{u+rbv}{u+bv}(1-u-v)K$, $\phi_2(u,v) = \frac{ru+bv}{u+bv}(1-u-v)K$.
    (c) See [**191**, p. 269].

17. (6.48c) is non-intuitive [**174**, p. 125], but see [**148**, pp. 104-105].

18. Consider $f(\bar{u},\bar{v})$ defined by (6.49) as a function of $\bar{u}$ on $(-\infty,\infty)$; set $a = -\frac{\epsilon+\bar{v}}{\alpha}$, $b = -\epsilon - \alpha\bar{v}$ and assume that $\alpha < 1$, so that $a < b$. Then $\lim_{\bar{u}\to\pm\infty} f(\bar{u},\bar{v}) = \mp\infty = \lim_{\bar{u}\to a\pm} f(\bar{u},\bar{v}) = \lim_{\bar{u}\to b\pm} f(\bar{u},\bar{v}) = \mp\infty$; but elsewhere f is continuous. So the

non-zero root of $\frac{\partial f}{\partial u} = 0$ on (a, ∞), say $\hat{u}$, maximizes f on that interval; and if $\overline{u}^*$ maximizes f on $[0, \infty)$, then $\overline{u}^* = \hat{u}$ if $\frac{\partial f}{\partial u}\big|_{\overline{u}=0} > 0$ but otherwise $\overline{u}^* = 0$ (even if $\alpha = 1$, although f then has no zero between a and b). For $\delta^2 \geq \alpha + (1-\alpha)p$, $\overline{u}^* = 0$, implying $\overline{v}^* = 0$. For $\delta^2 < \alpha + (1-\alpha)p$, $\frac{\partial f}{\partial u}\big|_{\overline{u}=0} = 0$ has a positive root, say $\hat{v}$, such that $\overline{u}^* = 0$ for $v \geq \hat{v}$ but $\overline{u}^* = \hat{u}$ for $v < \hat{v}$; and so $\overline{v}^*$ is the only positive root of $\frac{\partial f}{\partial u}\big|_{\overline{u}=\overline{v}} = 0$. The result is (6.50).

20. (a) On using (6.55a) and (6.55d) with $p = b$, (6.56) reduces to
$b \sum_{k=0}^{N} \left\{1 + \frac{\alpha(N-k)}{1+k}\right\} \text{Prob}(D = k) + (1-b)\{1 - \text{Prob}(D = 0)\} =$
$b \sum_{k=0}^{N} \left\{1 - \alpha + \frac{\alpha(N+1)}{k+1}\right\} \binom{N}{k} b^k (1-b)^{N-k} + (1-b)\{1 - (1-b)^N\} =$
$b(1-\alpha) \sum_{k=0}^{N} \binom{N}{k} b^k (1-b)^{N-k} + \alpha b(N+1) \sum_{k=0}^{N} \binom{N}{k} \frac{b^k(1-b)^{N-k}}{k+1} +$
$(1 - b)\{1 - (1 - b)^N\} = b(1 + \alpha)\left(1 - (1-b)^{N+1}\right) - \alpha b$. On multiplying by $\gamma(N + 1) = 1 - \delta(1 - \sigma)$, we obtain (6.57).

(b) On using (6.55a)-(6.55c) with $p = 1 - G(Z)$, we readily find that $\phi_T(Z) = \sum_{k=0}^{N} \left(1 - \delta\left(1 - \frac{\sigma r k}{N}\right)\right)(1 + \alpha r k)\text{Prob}(F(Z) = k)$
$= \sum_{k=0}^{N} \left(1 - \delta + r\{\alpha(1 - \delta) + \frac{\delta\sigma}{N}\}k + \frac{\alpha\delta\sigma r^2 k^2}{N}\right)\text{Prob}(F(Z) = k) =$
$(1 - \delta) \sum_{k=0}^{N} \binom{N}{k}\{1 - G(Z)\}^k\{G(Z)\}^{N-k}$
$\quad + r\{\alpha(1 - \delta) + \frac{\delta\sigma}{N}\} \sum_{k=0}^{N} k\binom{N}{k}\{1 - G(Z)\}^k\{G(Z)\}^{N-k}$
$\quad + \frac{\alpha\delta\sigma r^2}{N} \sum_{k=0}^{N} k^2\binom{N}{k}\{1 - G(Z)\}^k\{G(Z)\}^{N-k}$,
which reduces to (6.60). Also, because $G(L) = b$ and $G'(z) = g(z)$, (6.61) follows easily from the substitution $x = 1 - G(z)$.

(c) See [**155**, p. 388].

21. See [**155**, pp. 383-389].

22. (a) Note that $\psi'(v_0) = -4 - b(1 + 6v_0) - 6r(19 - 8v_0) - 6(1-b) < 0$ if $\psi(v_0)$ is the difference between the left- and right-hand sides of (6.106a) when $v_1 = 1$; and $\psi(0) > 0 > \psi(1)$ reduces to (6.107).

(b) Now let $\psi(v_2)$ be the difference between the right- and left-hand sides of (6.114a). Then $\psi(0) = 1 + 13b > 0$, and $\psi(1) < 0$ reduces to $12r > 4b + 1$.

(c) See [**149**, p. 1165].

24. (a) $J - I$ decreases with a but increases with b and t. It is least as $a \to 1$, $b \to 0$ and $t \to 0$; then $J - I \to \frac{1}{2}$, by (A.42) of [**2**].

(b) $\frac{I}{I-J+1}$, which approaches 1 in the limit as $a \to 1$.

25. (b) $\eta(s)$ decreases in a piecewise-linear fashion between $s = 0$ and $s = L$, then increases linearly; $\eta(I) = 0$ and $\eta(J) = 0$ are merely rearrangements of $I = \theta_1$ and $J = \theta_2$, respectively.

26. (a) See [2, p. 419].

(b) There are two possibilities. First, suppose that a resident who threatens pays no threat cost if the intruder flees. Then $f(u, v)$ is modified because $\tau = -T$ if $X > Y$ in (6.125b). This modification has no effect on f_3 or f_4; but (for a uniform distribution) the second integral in (6.130a) becomes $-p_r V t u_1 (1 - v_4)$, with (6.130b) changing accordingly. R is affected if $v_3 < v_4 < 1$. Then $f_1(u_1, v) = \frac{1}{2} V (v_4 - v_3)(v_4 + v_3 + u_1\{2(a + b) - bu_1\}) - V t u_1 (1 - v_4)$ when $v_4 \leq u_1 \leq 1$, so that the concavity of f_1 is always downward for $u_1 \geq v_4$ (it no longer depends on Δ). Now f_1 decreases monotonically between $u_1 = 0$ and $u_1 = 1$ if $\Delta \geq 1 + a + b$; f_1 increases monotonically between $u_1 = 0$ and $u_1 = 1$ if $\Delta \leq a$; and, if $a < \Delta < 1 + a + b$, then f_1 increases to a maximum and then decreases again, the maximum occurring where $u_1 = \theta_1$ if $1 + a + b > \Delta \geq 1 + a + b(1 - v_3)$; where $u_1 = \omega_1$ if $1 + a + b(1 - v_3) > \Delta > a + b(1 - v_4)$; and where $u_1 = 1 - (\Delta - a)/b$ if $a + b(1 - v_4) \geq \Delta > a$. From (6.130b) and the corresponding variation of f_2, we deduce that the maximum of f_r for $0 \leq u_1 \leq u_2 \leq 1$ must occur where $u_2 = 1$ if $\Delta > a$. Setting $u = v$ in Tables 6.14 and 6.15 as before, we find that a strong ESS must correspond to the first three rows of Table 6.14. But $v_2 = 1$ implies $\delta = a + b - 1 < b$, which eliminates the third row; $v_4 > v_3$ eliminates the first; and $v_4 > v_3$ also eliminates the second, because $v_2 = 1$ implies $(1 + b\{1 - v_1\})\theta_3 = v_1 + (a + b)(1 - v_1)$ and $\theta_4 = \omega_3$, so that $\theta_3 < \theta_4$. There is therefore no strong ESS; and the proof that there is no weak ESS is unchanged.

(c) See [2, pp. 420-421].

27. (a) The $(1 - u)C$ females in a mutant brood produce C children apiece, regardless. Replacing $N(1 - v)$ by $N(1 - v)(1 - \beta)$ and Nv by $Nv(1 - \alpha)$ in §6.1's expressions for d and s, respectively, we find that $\frac{d_O}{s_O} = \frac{(1 - \beta)\sigma_f(1 - v)}{(1 - \alpha)\sigma_m v}$ is the ratio of daughters to sons among outbreeders, in lieu of (6.3). So expected number of grandchildren per mutant brood becomes $\sigma_f(1 - u)C \cdot C + \sigma_m u(1 - \alpha)C \cdot Cd_O/s_O = \sigma_f\{1 + (1 - \beta - \{2 - \beta\}v)u/v\}C^2$. But some of these grandchildren are doubly mutant by virtue of having a u-strategist for each parent, and so we should count them twice. Because $\alpha u C$ inbreeding males and $\beta(1 - u)C$ inbreeding females mate, the ratio of daughters to sons among inbreeders is $\frac{d_I}{s_I} = \frac{\beta\sigma_f(1-u)}{\alpha\sigma_m u}$, implying that $\sigma_m u\alpha C\cdot C\frac{d_I}{s_I} = \sigma_f\beta(1-u)C^2$ must be added. Now expected number of genes transmitted to the second generation is proportional to $\sigma_f\{1+\beta+(1-\beta-2v)u/v\}C^2$; and it is readily shown that the ESS becomes $v^* = \frac{1}{2}(1-\beta)$.

(b) Use (6.5) in (6.27) with $r = \beta$.

(c) $\partial v^*/\partial\beta < 0$ but $\partial v^*/\partial\alpha = 0$. See [**130**, pp. 160-161].

28. (a) $P(v_2) = 0$, where $P(\xi) = 199\xi^{12}(5+\xi)Q(\xi) - 50(13\xi+11)$ and $Q(\xi) = 1352078 - 13728792\xi + 63740820\xi^2 - 178474296\xi^3 + 334639305\xi^4 - 440936496\xi^5 + 416440024\xi^6 - 281801520\xi^7 + 133855722\xi^8 - 42493880\xi^9 + 8112468\xi^{10} - 705432\xi^{11}$.

(b) No. There are at least two solutions satisfying $0 < v_2 < 1$ because $P(0) = -550$, $P(0.7) \approx 105$ and $P(1) = -6$; and the graph of P shows that there are precisely two solutions, which are readily found by numerical means to be $v_2 = 0.6036$ and $v_2 = 0.9867$. Both correspond to an ESS; see [**149**, p. 1166].

Bibliography

1. E. S. Adams and R. L. Caldwell, *Deceptive communication in asymmetric fights of the stomatopod crustacean Gonodactylus bredini*, Animal Behaviour **39** (1990), 706–716.

2. E. S. Adams and M. Mesterton-Gibbons, *The cost of threat displays and the stability of deceptive communication*, Journal of Theoretical Biology **175** (1995), 405–421.

3. R. D. Alexander, *The Biology of Moral Systems*, Aldine De Gruyter, New York, 1987.

4. J. Apaloo, *Revisiting strategic models of evolution: The concept of neighborhood invader strategies*, Theoretical Population Biology **52** (1997), 71–77.

5. P. Auger and D. Pontier, *Fast game theory coupled to slow population dynamics: the case of domestic cat populations*, Mathematical Biosciences **148** (1998), 65–82.

6. R. Axelrod, *The Evolution of Cooperation*, Basic Books, New York, 1984.

7. ———, *The Complexity of Cooperation*, Princeton University Press, Princeton, New Jersey, 1997.

8. R. Axelrod and D. Dion, *The further evolution of cooperation*, Science **242** (1988), 1385–1390.

9. R. Axelrod and W. D. Hamilton, *The evolution of cooperation*, Science **211** (1981), 1390–1396.

10. P. R. Y. Backwell, J. H. Christy, S. R. Telford, M. D. Jennions, and N. I. Passmore, *Dishonest signalling in a fiddler crab*, Proceedings of the Royal Society of London B **267** (2000), 719–724.

11. M. A. Ball and G. A. Parker, *Sperm competition games: a general approach to risk assessment*, Journal of Theoretical Biology **194** (1998), 251–262.

12. Z. Barta and L.-A. Giraldeau, *The effect of dominance hierarchy on the use of alternative foraging tactics: a phenotype-limited producing-scrounging game*, Behavioral Ecology and Sociobiology **42** (1998), 217–223.

13. P.A. Bednekoff, *Mutualism among safe, selfish sentinels: a dynamic model*, American Naturalist **150** (1997), 373–392.

14. J. Bendor and P. Swistak, *Types of evolutionary stability and the problem of cooperation*, Proceedings of the National Academy of Sciences USA **92** (1995), 3596–3600.

15. _____ , *The evolutionary stability of cooperation*, American Political Science Review **91** (1997), 290–307.

16. K. Binmore, *Essays on the Foundations of Game Theory*, Basil Blackwell, Oxford, 1990.

17. D. T. Bishop and C. Cannings, *Ordinal conflicts with random rewards*, Journal of Theoretical Biology **122** (1986), 225–230.

18. P. G. Blackwell, *The n-player war of attrition and territorial groups*, Journal of Theoretical Biology **189** (1997), 175–181.

19. M. C. Boerlijst, M. A. Nowak, and K. Sigmund, *The logic of contrition*, Journal of Theoretical Biology **185** (1997), 281–293.

20. E. von Böhm-Bawerk, *Capital and Interest, Volume II*, Libertarian Press, South Holland, Illinois, New York, 1959, translation of 1889 French original.

21. E. Borel, *On games that involve chance and the skill of the players*, Econometrica **21** (1953), 101–115, translation of 1924 French original.

22. R. Boyd, *Mistakes allow evolutionary stability in the repeated prisoner's dilemma game*, Journal of Theoretical Biology **136** (1989), 47–56.

23. R. Boyd and P. J. Richerson, *No pure strategy is evolutionarily stable in the repeated prisoner's dilemma game*, Nature **327** (1987), 58–59.

24. _____ , *The evolution of reciprocity in sizable groups*, Journal of Theoretical Biology **132** (1988), 337–356.

25. _____ , *The evolution of indirect reciprocity*, Social Networks **11** (1989), 213–236.

26. J. W. Bradbury and S. L. Vehrencamp, *Economic models of animal communication*, Animal Behaviour **59** (2000), 259–268.

27. S. J. Brams, *Game Theory and Politics*, Macmillan, New York, 1975.

28. K. Brauchli, T. Killingback, and M. Doebeli, *Evolution of cooperation in spatially structured populations*, Journal of Theoretical Biology **200** (1999), 405–417.

29. B. Brembs, *Chaos, cheating and cooperation: potential solutions to the prisoner's dilemma*, Oikos **76** (1996), 14–24.

30. J. L. Brown, *Helping and Communal Breeding in Birds*, Princeton University Press, Princeton, New Jersey, 1987.

31. J. S. Brown, *Game theory and habitat selection*, in [**55**], pp. 188–220.

32. M. Bulmer, *Theoretical Evolutionary Ecology*, Sinauer, Sunderland, Massachusetts, 1994.

33. J. W. Burgess, *Social spiders*, Scientific American **234** (1976), no. 3, 100–106.

34. T. Caraco and J. L. Brown, *A game between communal breeders: when is food-sharing stable?*, Journal of Theoretical Biology **118** (1986), 379–393.

35. E. Castillo, A. Cobo, F. Jubete, and R. E. Pruneda, *Orthogonal Sets and Polar Methods in Linear Algebra*, John Wiley, New York, 1999.

36. E. L. Charnov, *The Theory of Sex Allocation*, Princeton University Press, Princeton, New Jersey, 1982.

37. B. Child, *The practice and principles of community-based wildlife management in Zimbabwe: the CAMPFIRE programme*, Biodiversity and Conservation **5** (1996), 369–398.

38. C. W. Clark and M. Mangel, *Dynamic State Variable Models in Ecology*, Oxford University Press, New York, 2000.

39. K. C. Clements and D. W. Stephens, *Testing models of non-kin cooperation: mutualism and the prisoner's dilemma*, Animal Behaviour **50** (1995), 527–549.

40. A. M. Colman, *Game Theory and Experimental Games*, Pergamon Press, Oxford, 1982.

41. A. A. Cournot, *Researches into the Mathematical Principles of the Theory of Wealth*, Macmillan, New York, 1897, translation of 1838 French original.

42. V. P. Crawford, *Nash equilibrium and evolutionary stability in large- and finite-population "playing the field" models*, Journal of Theoretical Biology **145** (1990), 83–94.

43. R. Cressman, *The stability concept of evolutionary game theory*, Lecture Notes in Biomathematics, vol. 94, Springer-Verlag, Berlin, 1992.

44. _____ , *Evolutionary stability for two-stage Hawk-Dove games*, Rocky Mountain Journal of Mathematics **25** (1995), 145–155.

45. R. Cressman and G. T. Vickers, *Spatial and density effects in evolutionary game theory*, Journal of Theoretical Biology **184** (1997), 359–369.

46. P. H. Crowley, T. Cottrell, T. Garcia, M. Hatch, R. C. Sargent, B. J. Stokes, and J. M. White, *Solving the complementarity dilemma: evolving strategies for simultaneous hermaphroditism*, Journal of Theoretical Biology **195** (1998), 13–26.

47. P. H. Crowley, L. Provencher, S. Sloane, L. A. Dugatkin, B. Spohn, L. Rogers, and M. Alfieri, *Evolving cooperation: the role of individual recognition*, Biosystems **37** (1996), 49–66.

48. P. H. Crowley and R. C. Sargent, *Whence tit-for-tat?*, Evolutionary Ecology **10** (1996), 499–516.

49. R. Dawkins, *The Selfish Gene*, second ed., Oxford University Press, Oxford, 1989.

50. R. L. Devaney, *An Introduction to Chaotic Dynamical Systems*, second ed., Addison-Wesley, Menlo Park, California, 1989.

51. _____, *Chaos, Fractals and Dynamics*, Addison-Wesley, Menlo Park, California, 1990.

52. R. L. Devaney and L. Keen (eds.), *Chaos and Fractals: The Mathematics Behind the Computer Graphics*, American Mathematical Society, Providence, Rhode Island, 1989.

53. L. A. Dugatkin, *Cooperation Among Animals: An Evolutionary Perspective*, Oxford University Press, New York, 1997.

54. L. A. Dugatkin, M. Mesterton-Gibbons, and A. I. Houston, *Beyond the prisoner's dilemma: toward models to discriminate among mechanisms of cooperation in nature*, Trends in Ecology and Evolution **7** (1992), 202–205.

55. L. A. Dugatkin and H. K. Reeve (eds.), *Game Theory and Animal Behavior*, Oxford University Press, New York, 1998.

56. F. Y. Edgeworth, *Mathematical Psychics*, Kegan Paul, London, 1881.

57. M. Enquist and O. Leimar, *Evolution of fighting behaviour: the effect of variation in resource value*, Journal of Theoretical Biology **127** (1987), 187–205.

58. I. Eshel, M. W. Feldman, and A. Bergman, *Long-term evolution, short-term evolution, and population-genetic theory*, Journal of Theoretical Biology **191** (1998), 391–396.

59. G. W. Evans and C. M. Crumbaugh, *Effects of prisoner's dilemma format on cooperative behavior*, Journal of Personality and Social Psychology **3** (1966), 486–488.

60. R. M. Fagen, *When Doves conspire: evolution of nondamaging fighting tactics in a nonrandom-encounter animal conflict model*, American Naturalist **115** (1980), 858–869.

61. M. W. Feldman and E. A. C. Thomas, *Behavior-dependent contests for repeated plays of the prisoner's dilemma II: dynamical aspects of the evolution of cooperation*, Journal of Theoretical Biology **128** (1987), 297–315.

62. R. Ferriere and R. E. Michod, *The evolution of cooperation in spatially heterogeneous populations*, American Naturalist **147** (1996), 692–717.

63. R. A. Fisher, *The Genetical Theory of Natural Selection*, Oxford University Press, Oxford, 1930.

64. M. Flood, K. Lendenmann, and A. Rapoport, *2 × 2 games played by rats: different delays of reinforcement as payoffs*, Behavioral Science **28** (1983), 65–78.

65. S. A. Frank, *Foundations of Social Evolution*, Princeton University Press, Princeton, New Jersey, 1998.

66. D. Fudenberg and D. K. Levine, *The Theory of Learning in Games*, The MIT Press, Cambridge, Massachusetts, 1998.

67. R. Gadagkar, *Survival Strategies: Cooperation and Conflict in Animal Societies*, Harvard University Press, Cambridge, Massachusetts, 1997.

68. J. Garay and Z. Varga, *When will a sexual population evolve to an ESS?*, Proceedings of the Royal Society of London B **265** (1998), 1007–1010.

69. C. C. Gibson and S. A. Marks, *Transforming rural hunters into conservationists: an assessment of community-based wildlife management programs in africa*, World Development **23** (1995), 941–957.

70. L.-A. Giraldeau and T. Caraco, *Social Foraging Theory*, Princeton University Press, Princeton, New Jersey, 2000.

71. R. Gomulkiewicz, *Game theory, optimization, and quantitative genetics*, in [**55**], pp. 283–303.

72. L. M. Gosling and W. J. Sutherland (eds.), *Behaviour and Conservation*, Cambridge University Press, Cambridge, 2000.

73. J. D. Goss-Custard and W. J. Sutherland, *Individual behaviour, populations and conservation*, in [**115**], pp. 373–395.

74. A. Grafen, *The hawk-dove game played between relatives*, Animal Behaviour **27** (1979), 905–907.

75. ———, *Biological signals as handicaps*, Journal of Theoretical Biology **144** (1990), 517–546.

76. ———, *Modelling in behavioural ecology*, Behavioural Ecology: An Evolutionary Approach (J. R. Krebs and N. B. Davies, eds.), Blackwell Science, Oxford, third ed., 1991, pp. 5–31.

77. J. Greenberg, *The Theory of Social Situations: An Alternative Game Theoretic Approach*, Cambridge University Press, Cambridge, 1990.

78. P. Grim, *Spatialization and greater generosity in the stochastic prisoner's dilemma*, Biosystems **37** (1996), 3–17.

79. B. Grofman and H. Scarrow, *Iannucci and its aftermath: the application of the Banzhaf index to weighted voting in the state of New York*, Applied Game Theory (S. J. Brams, A. Schotter, and G. Schwödiauer, eds.), Physica-Verlag, Vienna, 1979, pp. 168–183.

80. M. R. Gross and J. Repka, *Game theory and inheritance in the conditional strategy*, in [**55**], pp. 168–187.

81. W. D. Hamilton, *The genetical theory of social behaviour, I & II*, Journal of Theoretical Biology **7** (1964), 1–52.

82. _____, *Extraordinary sex ratios*, Science **156** (1967), 477–488.

83. P. Hammerstein, *What is evolutionary game theory?*, in [**55**], pp. 3–15.

84. _____, *Darwinian adaptation, population genetics and the streetcar theory of evolution*, Journal of Mathematical Biology **34** (1996), 511–532.

85. P. Hammerstein and G. A. Parker, *The asymmetric war of attrition*, Journal of Theoretical Biology **96** (1982), 647–682.

86. P. Hammerstein and R. Selten, *Game theory and evolutionary biology*, Handbook of Game Theory (R. J. Aumann and S. Hart, eds.), vol. 2, Elsevier Science, Amsterdam, 1994, pp. 929–993.

87. R. Hannesson, *Fishing as a supergame*, Journal of Environmental Economics and Management **32** (1997), 309–322.

88. C. B. Harley, *Learning rules, optimal behaviour, and evolutionary stability*, Journal of Theoretical Biology **127** (1987), 377–379.

89. J. C. Harsanyi, *Games with incomplete information played by "Bayesian" players. Part I: the basic model*, Management Science **14** (1967), 159–182.

90. _____, *Games with incomplete information played by "Bayesian" players. Part II: Bayesian equilibrium points*, Management Science **14** (1968), 320–334.

91. _____, *Games with incomplete information played by "Bayesian" players. Part III: the basic probability distribution of the game*, Management Science **14** (1968), 486–502.

92. _____, *The tracing procedure: a Bayesian approach to defining a solution for n-person games*, International Journal of Game Theory **4** (1975), 61–94.

93. J. C. Harsanyi and R. Selten, *A General Theory of Equilibrium Selection in Games*, The MIT Press, Cambridge, Massachusetts, 1988.

94. W. N. Hazel and R. Smock, *Modeling selection on conditional strategies in stochastic environments*, Adaptation in Stochastic Environments (J. Yoshimura and C. W. Clark, eds.), Lecture Notes in Biomathematics, vol. 98, Springer-Verlag, Berlin, 1993, pp. 147–154.

95. B. Heinrich, *Ravens in Winter*, Simon and Schuster, New York, 1989.

96. W. G. S. Hines, *Evolutionarily stable strategies: a review of basic theory*, Theoretical Population Biology **31** (1987), 195–272.

97. W. G. S. Hines and J. Maynard Smith, *Games between relatives*, Journal of Theoretical Biology **79** (1979), 19–30.

98. W. G. S. Hines and M. Turelli, *Multilocus evolutionarily stable strategy effects: additive effects*, Journal of Theoretical Biology **187** (1997), 379–388.

99. J. Hofbauer and K. Sigmund, *Evolutionary Games and Population Dynamics*, Cambridge University Press, Cambridge, UK, 1998.

100. A. I. Houston and J. M. McNamara, *Models of Adaptive Behaviour*, Cambridge University Press, Cambridge, 1999.

101. A. I. Houston and B. H. Sumida, *Learning rules, matching and frequency dependence*, Journal of Theoretical Biology **126** (1987), 289–308.

102. D. M. Hugie and L. M. Dill, *Fish and game: a game theoretic approach to habitat selection by predators and prey*, Journal of Fish Biology (Supplement A) **45** (1994), 151–169.

103. P. L. Hurd and M. Enquist, *Conventional signalling in aggressive interactions: the importance of temporal structure*, Journal of Theoretical Biology **192** (1998), 197–211.

104. P. L. Hurd and R. C. Ydenberg, *Calculating the ESS level of information transfer in aggressive communication*, Evolutionary Ecology **10** (1996), 221–232.

105. V. C. L. Hutson and G. T. Vickers, *The spatial struggle of tit-for-tat and defect*, Philosophical Transactions of the Royal Society of London B **348** (1995), 393–404.

106. M. Jeter, *Mathematical Programming*, Marcel Dekker, New York, 1986.

107. R. A. Johnstone, *The evolution of animal signals*, in [**115**], pp. 155–178.

108. ———, *Game theory and communication*, in [**55**], pp. 94–117.

109. E. Kalai and D. Samet, *Persistent equilibria in strategic games*, International Journal of Game Theory **13** (1984), 129–144.

110. T. Killingback, M. Doebeli, and N. Knowlton, *Variable investment, the continuous prisoner's dilemma, and the origin of cooperation*, Proceedings of the Royal Society of London B **266** (1999), 1723–1728.

111. Y.-G. Kim, *Status signaling games in animal contests*, Journal of Theoretical Biology **176** (1995), 221–231.

112. J. H. Koeslag, *Sex, the prisoner's dilemma game, and the evolutionary inevitability of cooperation*, Journal of Theoretical Biology **189** (1997), 53–61.

113. D. P. Kraines and V. Y. Kraines, *Natural selection of memory-one strategies for the iterated prisoner's dilemma*, Journal of Theoretical Biology **203** (2000), 335–355.

114. J. R. Krebs and N. B. Davies (eds.), *An Introduction to Behavioural Ecology*, third ed., Blackwell Science, Oxford, 1993.

115. J. R. Krebs and N. B. Davies (eds.), *Behavioural Ecology: An Evolutionary Approach*, fourth ed., Blackwell Science, Oxford, 1997.

116. D. M. Kreps, *Game Theory and Economic Modelling*, Clarendon Press, Oxford, 1990.

117. T. Kura and K. Kura, *War of attrition with individual differences on RHP*, Journal of Theoretical Biology **193** (1998), 335–344.

118. O. Leimar, *Repeated games: a state space approach*, Journal of Theoretical Biology **184** (1997), 471–498.

119. S. A. Levin and H. C. Muller-Landau, *The evolution of dispersal and seed size in plant communities*, Evolutionary Ecology Research **2** (2000), 409–435.

120. R. Levins, *The strategy of model building in population biology*, American Scientist **54** (1966), 421–431.

121. B. Lomborg, *Nucleus and shield: the evolution of social structure in the iterated prisoner's dilemma*, American Sociological Review **61** (1996), 278–307.

122. R. D. Luce and H. Raiffa, *Games and Decisions*, John Wiley, New York, 1957.

123. D. Luenberger, *Linear and Nonlinear Programming*, second ed., Addison-Wesley, Reading, Massachusetts, 1984.

124. O. L. Mangasarian, *Nonlinear Programming*, McGraw-Hill, New York, 1969.

125. J. H. Marden and R. A. Rollins, *Assessment of energy reserves by damselflies engaged in aerial contests for mating territories*, Animal Behaviour **48** (1994), 1023–1030.

126. J. H. Marden and J. K. Waage, *Escalated damselfy territorial contests are energetic wars of attrition*, Animal Behaviour **39** (1990), 954–959.

127. P. Marrow, R. A. Johnstone, and L. D. Hurst, *Riding the evolutionary streetcar: where population genetics and game theory meet*, Trends in Ecology and Evolution **11** (1996), 445–446.

128. M. Maschler, B. Peleg, and L. S. Shapley, *Geometric properties of the kernel, nucleolus and related solution concepts*, Mathematics of Operations Research **4** (1979), 303–338.

129. J. Maynard Smith, *The theory of games and the evolution of animal conflicts*, Journal of Theoretical Biology **47** (1974), 209–219.

130. _____, *The Evolution of Sex*, Cambridge University Press, Cambridge, 1978.

131. _____, *Game theory and the evolution of behavior*, Proceedings of the Royal Society of London B **205** (1979), 475–488.

132. _____, *Evolution and the Theory of Games*, Cambridge University Press, Cambridge, 1982.

133. _____, *Can a mixed strategy be stable in a finite population?*, Journal of Theoretical Biology **130** (1988), 247–251.

134. J. Maynard Smith and G.R. Price, *The logic of animal conflict*, Nature **246** (1973), 15–18.

135. J. Maynard Smith and E. Szathmáry, *The Origins of Life*, Oxford University Press, Oxford, 1999.

136. A. Mehlmann, *The Game's Afoot! Game Theory in Myth and Paradox*, Student Mathematical Library, vol. 5, American Mathematical Society, Providence, Rhode Island, 2000.

137. M. Mesterton-Gibbons, *A game-theoretic analysis of a motorist's dilemma*, Mathematical and Computer Modelling **13** (1990), no. 2, 9–14.

138. _____, *An escape from the prisoner's dilemma*, Journal of Mathematical Biology **29** (1991), 251–269.

139. _____, *Ecotypic variation in the asymmetric Hawk-Dove game: when is Bourgeois an ESS?*, Evolutionary Ecology **6** (1992), 198–222 and 448.

140. _____, *An Introduction to Game-Theoretic Modelling*, Addison-Wesley, Redwood City, California, 1992.

141. _____, *On the iterated prisoner's dilemma in a finite population*, Bulletin of Mathematical Biology **54** (1992), 423–443.

142. _____, *Game-theoretic resource modeling*, Natural Resource Modeling **7** (1993), 93–147.

143. _____, *The Hawk-Dove game revisited: effects of continuous variation in resource-holding potential on the frequency of escalation*, Evolutionary Ecology **8** (1994), 230–247.

144. _____, *A Concrete Approach to Mathematical Modelling*, John Wiley, New York, 1995, reprint of 1989 original.

145. _____, *On the war of attrition and other games among kin*, Journal of Mathematical Biology **34** (1996), 253–270.

146. _____, *Game theory*, Handbook of Discrete and Combinatorial Mathematics (K. H. Rosen, J. G. Michaels, J. L. Gross, J. W. Grossman, and D. R. Shier, eds.), CRC Press, Boca Raton, Florida, 1999, pp. 1016–1028.

147. _____, *On sperm competition games: incomplete fertilization risk and the equity paradox*, Proceedings of the Royal Society of London B **266** (1999), 269–274.

148. _____, *On sperm competition games: raffles and roles revisited*, Journal of Mathematical Biology **39** (1999), 91–108.

149. _____, *On the evolution of pure winner and loser effects: a game-theoretic model*, Bulletin of Mathematical Biology **61** (1999), 1151–1186.

150. M. Mesterton-Gibbons and E. S. Adams, *Animal contests as evolutionary games*, American Scientist **86** (1998), 334–341.

151. M. Mesterton-Gibbons and M. J. Childress, *Constraints on reciprocity for non-sessile organisms*, Bulletin of Mathematical Biology **181** (1996), 65–83.

152. M. Mesterton-Gibbons and L. A. Dugatkin, *Cooperation among unrelated individuals: evolutionary factors*, Quarterly Review of Biology **67** (1992), 267–281.

153. _____, *Toward a theory of dominance hierarchies: effects of assessment, group size, and variation in fighting ability*, Behavioral Ecology **6** (1995), 416–423.

154. _____, *Cooperation and the prisoner's dilemma: toward testable models of mutualism versus reciprocity*, Animal Behaviour **54** (1997), 551–557.

155. _____, *On the evolution of delayed recruitment to food bonanzas*, Behavioral Ecology **10** (1999), 377–390.

156. M. Mesterton-Gibbons, J. H. Marden, and L. A. Dugatkin, *On wars of attrition without assessment*, Journal of Theoretical Biology **58** (1996), 861–875.

157. M. Mesterton-Gibbons and E. J. Milner-Gulland, *On the strategic stability of monitoring: implications for cooperative wildlife management programs in Africa*, Proceedings of the Royal Society of London B **265** (1998), 1237–1244.

158. R. E. Michod, *Darwinian Dynamics*, Princeton University Press, Princeton, New Jersey, 1999.

159. E. J. Milner-Gulland and R. Mace, *Conservation of Biological Resources*, Blackwell Science, Oxford, 1998.

160. D. W. Mock, G. A. Parker, and P. L. Schwagmeyer, *Game theory, sibling rivalry, and parent-offspring conflict*, in [55], pp. 146–167.

161. A. Muthoo, *Bargaining Theory with Applications*, Cambridge University Press, Cambridge, 1999.

162. R. B. Myerson, *Game Theory: Analysis of Conflict*, Harvard University Press, Cambridge, Massachusetts, 1991.

163. J. F. Nash, *The bargaining problem*, Econometrica **18** (1950), 155–162.

164. ———, *Non-cooperative games*, Annals of Mathematics **54** (1951), 286–295.

165. J. A. Newman and T. Caraco, *Cooperative and non-cooperative bases of food-calling*, Journal of Theoretical Biology **141** (1989), 197–209.

166. R. Noë and P. Hammerstein, *Biological markets: supply and demand determine the effect of partner choice in cooperation, mutualism and mating*, Behavioural Ecology and Sociobiology **35** (1994), 1–11.

167. G. Nöldeke and L. Samuelson, *How costly is the honest signaling of need?*, Journal of Theoretical Biology **197** (1999), 527–539.

168. M. A. Nowak, *Stochastic strategies in the prisoner's dilemma*, Theoretical Population Biology **38** (1990), 93–112.

169. M. A. Nowak, S. Bonhoeffer, and R. M. May, *More spatial games*, International Journal of Bifurcation and Chaos **4** (1994), 33–56.

170. M. A. Nowak and K. Sigmund, *The evolution of stochastic strategies in the prisoner's dilemma*, Acta Applicandæ Mathematicæ **20** (1990), 247–265.

171. ———, *A strategy of win-stay, lose-shift that outperforms tit-for-tat in the prisoner's dilemma game*, Nature **364** (1993), 56–58.

172. ———, *Evolution of indirect reciprocity by image scoring*, Nature **393** (1998), 573–577.

173. G. Owen, *Game Theory*, second ed., Academic Press, New York, 1982.

174. G. A. Parker, *Sperm competition games*, Proceedings of the Royal Society of London B **242** (1990), 120–133.

175. ———, *Sperm competition and the evolution of ejaculates: toward a theory base*, Sperm Competition and Sexual Selection (T. R. Birkhead and A. P. Møller, eds.), Academic Press, San Diego, 1998, pp. 3–54.

176. ———, *Sperm competition games between related males*, Proceedings of the Royal Society of London B **267** (2000), 1027–1032.

177. G. A. Parker and J. Maynard Smith, *Optimality theory in evolutionary biology*, Nature **348** (1990), 27–33.

178. G. A. Parker and D. I. Rubinstein, *Role assessment, reserve strategy, and acquisition of information in asymmetric animal conflicts*, Animal Behaviour **29** (1981), 221–240.

179. R. J. H. Payne, *Gradually escalating fights and displays: the cumulative assessment model*, Animal Behaviour **56** (1998), 651–662.

180. I. Pen and F. J. Weissing, *Sperm competition and sex allocation in simultaneous hermaphrodites: a new look at Charnov's invariance principle*, Evolutionary Ecology Research **1** (1999), 517–525.

181. L. Phlips, *The Economics of Imperfect Information*, Cambridge University Press, Cambridge, 1988.

182. R. Pool, *Putting game theory to the test*, Science **267** (1995), 1591–1593.

183. D. G. Pruitt, *Reward structure and cooperation: the decomposed prisoner's dilemma game*, Journal of Personality and Social Psychology **7** (1967), 21–27.

184. A. E. Pusey and C. Packer, *The ecology of relationships*, in [**115**], pp. 254–283.

185. D. A. Rand, *Correlation equations and pair approximations for spatial ecologies*, Advanced Ecological Theory (J. M. McGlade, ed.), Blackwell Science, Oxford, 1999, pp. 100–142.

186. A. Rapoport and A. M. Chammah, *Prisoner's Dilemma: A Study in Conflict and Cooperation*, University of Michigan Press, Ann Arbor, Michigan, 1965.

187. E. Rasmusen, *Games and Information*, second ed., Basil Blackwell, Oxford, 1994.

188. J. C. Reboreda and A. Kacelnik, *The role of autoshaping in cooperative two-player games between starlings*, Journal of the Experimental Analysis of Behavior **60** (1993), 67–83.

189. H. K. Reeve, *Game theory, reproductive skew, and nepotism*, in [**55**], pp. 118–145.

190. H. K. Reeve and L. A. Dugatkin, *Why we need evolutionary game theory*, in [**55**], pp. 304–311.

191. H. K. Reeve, S. T. Emlen, and L. Keller, *Reproductive sharing in animal societies: reproductive incentives or incomplete control by dominant breeders?*, Behavioral Ecology **9** (1998), 267–278.

192. S. E. Riechert, *Game theory and animal contests*, in [**55**], pp. 64–93.

193. G. Roberts and T. N. Sherratt, *Development of cooperative relationships through increasing investment*, Nature **394** (1998), 175–179.

194. A. J. Robson, *Evolutionary game theory*, Recent Developments in Game Theory (J. Creedy, J. Borland, and J. Eichberger, eds.), Edward Elgar, Aldershot, U. K., 1992, pp. 165–178.

195. G. Romp, *Game Theory*, Oxford University Press, Oxford, 1997.

196. A. Rubinstein, *Modeling Bounded Rationality*, MIT Press, Cambridge, Massachusetts, 1998.

197. L. Samuelson, *Evolutionary Games and Equilibrium Selection*, MIT Press, Cambridge, Massachusetts, 1997.

198. M. E. Schaffer, *Evolutionarily stable strategies for a finite population and a variable contest size*, Journal of Theoretical Biology **132** (1988), 469–478.

199. T. C. Schelling, *Micromotives and Macrobehavior*, Norton, New York, 1978.

200. H.R. Schiffman, *Sensation and Perception*, second ed., John Wiley, New York, 1982.

201. D. Schmeidler, *The nucleolus of a characteristic function game*, SIAM Journal of Applied Mathematics **17** (1969), 1163–1170.

202. P. L. Schwagmeyer and G. A. Parker, *Male mate choice as predicted by sperm competition in thirteen-lined ground squirrels*, Nature **348** (1990), 62–64.

203. R. Selten, *A note on evolutionarily stable strategies in asymmetric animal conflicts*, Journal of Theoretical Biology **84** (1980), 93–101, reprinted in [**204**, pp. 67-75].

204. _____, *Models of Strategic Rationality*, Kluwer, Dordrecht, 1988.

205. L. S. Shapley, *Cores of convex games*, International Journal of Game Theory **1** (1971), 11–26.

206. M. Shubik, *Game Theory in the Social Sciences: Concepts and Solutions*, The MIT Press, Cambridge, Massachusetts, 1982.

207. _____, *A Game-Theoretic Approach to Political Economy*, The MIT Press, Cambridge, Massachusetts, 1984, Volume 2 of Game Theory in the Social Sciences.

208. _____, *Cooperative game solutions: Australian, Indian, and U.S. opinions*, Journal of Conflict Resolution **30** (1986), no. 1, 63–76.

209. K. Sigmund, *Games of Life: Explorations in Ecology, Evolution and Behaviour*, Oxford University Press, Oxford, 1993.

210. J. B. Silk, E. Kaldor, and R. Boyd, *Cheap talk when interests conflict*, Animal Behaviour **59** (2000), 423–432.

211. M. Sjerps and P. Haccou, *A war of attrition between larvae on the same host plant: stay and starve or leave and be eaten?*, Evolutionary Ecology **8** (1994), 269–287.

212. W.C. Stebbins, *Principles of animal psychophysics*, Animal Psychophysics (W.C. Stebbins, ed.), Appleton-Century-Crofts, New York, 1970, pp. 1–19.

213. D. W. Stephens, *Cumulative benefit games: achieving cooperation when players discount the future*, Journal of Theoretical Biology **205** (2000), 1–16.

214. D. W. Stephens and K. C. Clements, *Game theory and learning*, in [**55**], pp. 239–260.

215. D. W. Stephens, K. Nishimura, and K. B. Toyer, *Error and discounting in the iterated prisoner's dilemma*, Journal of Theoretical Biology **176** (1995), 457–469.

216. P. D. Straffin, *Homogeneity, independence, and power indices*, Public choice **30** (1977), 107–118.

217. ———, *The prisoner's dilemma*, UMAP Journal **1** (1980), no. 1, 101–103.

218. P. D. Straffin and J. P. Heaney, *Game theory and theTennessee Valley Authority*, International Journal of Game Theory **10** (1981), 35–43.

219. R. Sugden, *The Economics of Rights, Co-operation and Welfare*, Basil Blackwell, Oxford, 1986.

220. S. Számadó, *Cheating as a mixed strategy in a simple model of aggressive communication*, Animal Behaviour **59** (2000), 221–230.

221. A. D. Taylor and W. S. Zwicker, *Simple Games*, Princeton University Press, Princeton, New Jersey, 1999.

222. P. D. Taylor and L. Jonker, *Evolutionarily stable strategies and game dynamics*, Mathematical Biosciences **40** (1978), 145–156.

223. R. Trivers, *The evolution of reciprocal altruism*, Quarterly Review of Biology **46** (1971), 35–57, (reprinted in Clutton-Brock & Harvey, pp. 189–226).

224. ———, *Social evolution*, Benjamin/Cummings, Menlo Park, California, 1985.

225. E. E. C. van Damme, *Stability and Perfection of Nash Equilibria*, Springer-Verlag, Berlin, 1987.

226. F. Vega-Redondo, *Evolution, Games, and Economic Behaviour*, Oxford University Press, Oxford, 1996.

227. W. L. Vickery, *How to cheat against a simple mixed strategy ESS*, Journal of Theoretical Biology **127** (1987), 133–139.

228. ———, *Reply to Maynard Smith*, Journal of Theoretical Biology **132** (1988), 375–378.

229. T. L. Vincent and J. S. Brown, *The evolution of ESS theory*, Annual Review of Ecology and Systematics **19** (1988), 423–443.

230. T. L. Vincent and W. J. Grantham, *Optimality in Parametric Systems*, John Wiley, New York, 1981.

231. J. von Neumann and O. Morgenstern, *Theory of Games and Economic Behavior*, third ed., Princeton University Press, Princeton, New Jersey, 1953.

232. L. M. Wahl and M. A. Nowak, *The continuous prisoner's dilemma, I & II*, Journal of Theoretical Biology **200** (1999), 307–338.

233. J. Wang, *The Theory of Games*, Oxford University Press, Oxford, 1988.

234. J. W. Weibull, *Evolutionary Game Theory*, MIT Press, Cambridge, Massachusetts, 1995.

235. M. J. West Eberhard, *The evolution of social behavior by kin selection*, Quarterly Review of Biology **50** (1975), 1–35.

236. S. Wiggins, *Introduction to Applied Nonlinear Dynamical Systems and Chaos*, Springer-Verlag, New York, 1990.

237. D. S. Wilson, *The Natural Selection of Populations and Communities*, Benjamin/Cummings, Menlo Park, California, 1980.

238. S. Wolfram, *Cellular Automata and Complexity*, Addison-Wesley, Reading, Massachusetts, 1994.

239. _____, *The Mathematica Book*, third ed., Wolfram Media/ Cambridge University Press, Champaign, Illinois/Cambridge, 1996.

240. J. Wu and R. Axelrod, *How to cope with noise in the iterated prisoner's dilemma*, Journal of Conflict Resolution **39** (1995), 183–189.

241. H. P. Young, *Individual Strategy and Social Structure*, Princeton University Press, Princeton, New Jersey, 1998.

242. A. Zahavi and A. Zahavi, *The Handicap Principle*, Oxford University Press, New York, 1997.

243. F. Zeuthen, *Problems of Monopoly and Economic Welfare*, G. Routledge, London, 1930.

244. V. I. Zhukovskiy and M. E. Salukvadze, *The Vector-Valued Maximin*, Mathematics in Science and Engineering, vol. 193, Academic Press, Boston, 1994.

Index